Industrial and Laboratory
Alkylations

Industrial and Laboratory Alkylations

Lyle F. Albright, EDITOR
Purdue University

Arthur R. Goldsby, EDITOR

Based on a symposium jointly
sponsored by the Division of
Petroleum Chemistry, Inc. and
the Division of Organic Chemistry
at the 173rd Meeting of the
American Chemical Society,
New Orleans, La.,
Mar. 21–22, 1977.

ACS SYMPOSIUM SERIES 55

AMERICAN CHEMICAL SOCIETY
WASHINGTON, D. C.   1977

Library of Congress CIP Data

Industrial and laboratory alkylations.

(ACS symposium series; 55 ISSN 0097-6156)
"Based on a symposium sponsored by the Division of
Petroleum Chemistry at the 173rd meeting of the Ameri-
can Chemical Society, New Orleans, La."

Includes bibliographical references and index.

1. Alkylation—Congresses.
I. Albright, Lyle Frederick, 1921–   . II. Goldsby,
Arthur R., 1904–   . III. American Chemical Society.
Division of Petroleum Chemistry. IV. Series: American
Chemical Society. ACS symposium series; 55.

TP690.45.I5   661'.8   77-23973
ISBN 0-8412-0385-7   ACSMC8 55 1–462

ACS Symposium Series

Robert F. Gould, *Editor*

FOREWORD

The ACS Symposium Series was founded in 1974 to provide
a medium for publishing symposia quickly in book form. The
format of the Series parallels that of the continuing Advances
in Chemistry Series except that in order to save time the
papers are not typeset but are reproduced as they are submitted by the authors in camera-ready form. As a further
means of saving time, the papers are not edited or reviewed
except by the symposium chairman, who becomes editor of
the book. Papers published in the ACS Symposium Series
are original contributions not published elsewhere in whole or
major part and include reports of research as well as reviews
since symposia may embrace both types of presentation.

CONTENTS

PREFACE

This book consists of 27 papers—all relating in some way to alkylation. Slightly over half of the papers deal with the alkylation of isobutane with light olefins to produce high quality gasoline blending hydrocarbons. New information is presented for isobutane alkylation relative to the chemistry and mechanism, process improvements, recovery of acid catalyst, and status of commercial units. Papers are also presented for the alkylation of aromatics, heterocyclics, coal, and other hydrocarbons. Alkylations using transition metal catalysts, strong acids, free radicals, and bases are also reported.

All of the papers in this book were presented at meetings of the American Chemical Society. Twenty-two were presented at the alkylation symposium sponsored by the Division of Petroleum Chemistry in New Orleans on March 21 and 22, 1977. This symposium was chaired by the editors of this book. Four of the papers were presented earlier at the Symposum on New Hydrocarbon Chemistry; this symposium, also sponsored by the Division of Petroleum Chemistry, was held in San Francisco, August 29 to September 6, 1976. G. A. Olah and L. Schmerling chaired the symposium. One paper was presented at the First Chemical Congress of the North American Continent in Mexico City in December 1975.

When the editors of this book began planning for the alkylation symposium that was held in New Orleans, there was considerable interest in alkylation but many uncertainties relative to petroleum and natural gas supplies, costs, future needs for alkylation products, and phasing out of lead from gasoline. Although serious attempts at developing an energy policy and program (and one for petroleum and natural gas) are now in progress, many uncertainties still exist. However, there now seems to be a realization that improvements can be realized that will result in savings of energy and raw material and in improved quality of alkylates (and hence better utilization of reactants); a healthy optimism now exists for the future of alkylation. The editors were delighted to secure authors from a wide variety of backgrounds including chemistry, engineering, and administration; academia, research institutes, and industry; and from this country and three European countries. A healthy interchange of ideas occurred at the symposium. Interest in alkylation in the past was to some extent at least restricted to petroleum and natural gas type hydrocarbons. Presently, however, there is an increased interest in alkylating coals and hetrocyclic hydrocarbons. Each paper of this book was carefully reviewed by the editors and sometimes by others; several papers were revised to a rather appreciable extent as a result of the reviews.

The first group of papers (chapters 1–12) covers the more theoretical and fundamental aspects of alkylation, including the chemistry, mechanism, and various techniques and catalysts that can be used besides sulfuric acid and hydrogen fluoride. Most papers of this group deal with isobutane alkylation for production of high quality fuels.

The second group of papers (chapters 13–20) discusses the more practical aspects of isobutane alkylation including mixing, reaction variables, computer modeling, recovery of catalyst, and an alternate fuel to alkylate.

The third group of papers (chapters 21–24) covers the alkylation of aromatics and heterocyclics. A paper describing what is probably the world's largest ethylbenzene plant is included; this plant is based on a unique new alkylation process.

The fourth group of papers (chapters 25–27) covers the liquefaction of coal as an alternate source of liquid fuels. The chemistry of coal liquefaction is highly complicated, involving hydrogenation, pyrolysis or cracking, depolymerization, and polymerization. In addition, alkylation and dealkylation steps are occurring. In some cases at least, alkylation is a significant aid in obtaining liquid products from coal.

In conclusion, the editors would like to thank all those who contributed to the success of the symposia and this subsequent book. First, our thanks and appreciation go to the Division of Petroleum Chemistry, Inc. of the American Chemical Society for its sponsorship and encouragement, and, equally important, to the authors without whom it could never have existed, and finally to Mrs. Phyllis Beck (secretary at West Lafayette) who, in addition to her normal secretarial duties, contributed substantially with the secretarial aspects involved in the preparations for the symposium and this book.

LYLE F. ALBRIGHT   ARTHUR R. GOLDSBY
West Lafayette, Indiana   Chappaqua, New York
June 1977

Alkylation Studies

G. M. KRAMER

Exxon Research and Engineering Co., P.O. Box 45, Linden, NJ 07036

The alkylation of C_3, C_4 and C_5 olefins with isobutane in sulfuric acid is a large scale commercial process. The reaction has been extensively studied and its general features are well understood and documented in several lucid reviews (1-3).

In addition to being of commercial interest, the reaction has stimulated much fundamental research that is of enormous value in carbonium ion theory. The overall reaction is usually viewed as proceeding through an ionic chain mechanism in which a t-butyl ion adds to an olefin forming a larger cation that subsequently abstracts a hydride ion from isobutane forming product and a new t-butyl ion. Through the use of isotopic labeling, much has been learned about the behavior of the ions while in the acid (4-11). Nevertheless, there are a number of important questions about the alkylation reaction which have not been dealt with as extensively as might be desired and some of these will be discussed in this paper.

One of these is the question of where does the reaction occur? It is often assumed that alkylation proceeds in the bulk acid phase (12a), but one of the aims of this report is to show that alkylation must proceed in at least two phases and that the reaction occurring at the hydrocarbon-acid interface is by far the most important in controlling the quality or selectivity of alkylate; i.e. the formation of a C_8 fraction from isobutane plus butenes while minimizing the production of side products. The fact that alkylation does not occur uniformly throughout the acid has been recently suggested by Doshi and Albright (12b).

A second question is what might be done to improve selectivity beyond the usual practice of refiners to maximize mixing, maximize the isobutane to olefin ratio, lower the temperature and reduce the olefin space velocity. One approach is to decide what's rate determining and then to develop a chemical solution. This paper will be concerned with developing evidence that hydride transfer from a tertiary paraffin is generally slow and may be considered to be the rate determining step. The fact that a cation abstracts $H^{\ominus}$ from isobutane relatively slowly compared to other reactions it may undergo (deprotonation, racemization and certain isomerizations), can also be deduced from the exchange studies (4-11).

Doshi and Albright, and earlier Hoffmann, Schriesheim and this author (13) have recognized that alkylation performance is related to the presence of oil soluble hydrocarbons, commonly called red oil or conjunct polymers. These species are usually considered to be saturated and unsaturated cations which can function as intermediates in the transfer of hydride ions from isobutane to other alkyl cations. Assuming that hydride transfer is a limiting factor, the discovery of means to augment the rate should result in improved alkylation. This report deals with research which has led to the successful application of cationic surfactants for this purpose in commercial plants.

The report first summarizes hydride transfer information determined during:

a. the reaction of t-butyl chloride with methylcyclopentane,
b. the reaction of 2,3,4-trimethylpentane in tritiated sulfuric acid, and
c. the isomerization of 3-methylpentane in tritiated sulfuric acid.

Some of the reactions have been conducted with "acid modifiers" present and their role is believed to be primarily to change the stability and hence reactivity of the cationic intermediates. Some additives have a marked ability to increase the "steady state" hydride transfer rate and these have proved useful in improving the commercial alkylation reaction.

The routine use of the additives has followed pilot plant and commercial studies which indicate improved alkylation selectivity under well mixed or poorly mixed conditions and a simultaneous reduction in acid consuming side reactions. Typical data from these tests will also be presented.

The Reaction of t-Butyl Chloride with Methylcyclopentane

The major products of the commercial alkylation of isobutane with butenes are trimethylpentanes. This indicates that the products of alkylation are kinetically controlled because thermodynamics would predict a minor proportion of trimethylpentanes if the octanes were to isomerize to equilibrium.

The probability that hydride transfer from isobutane to carbonium ion intermediates is the kinetically slow step affecting product quality was raised by an experiment in which a stream of 96 percent H_2SO_4 was rapidly mixed with a stream of isobutane plus isobutylene in a mixing tee, after which the products were immediately quenched in a large vessel containing cold caustic (13). The C_8 fraction of the product contained trimethylpentenes but no trimethylpentanes.

In view of these results a study of the reaction of t-butyl chloride,which should act as an ion source,with methylcyclopentane which should be an effective hydride donor was undertaken. The reaction was followed manometrically by observing the pressure generated after the halide was added to a stirring emulsion of methylcyclopentane and 96 percent H_2SO_4 (14).

By using high stirring rates it is possible to operate under conditions in which the kinetics of the reaction can be studied in a regime not controlled by mass transfer limitations. Figure 1 shows a typical curve of the generation of pressure with time in a well mixed system. In this experiment t-butyl chloride was added to an 0.1 M emulsion of methylcyclopentane and the acid being stirred at 1000 rpm.

Three distinct regions of pressure generation were detected. First there is an immediate rise when t-butyl chloride contacts the emulsion. This is followed by a transition period leading to the third region in which there is a nearly linear pressure rise. The transition lasts about 0.5 minutes and is probably strongly associated with HCl evolution.

By connecting the apparatus directly to a time of flight mass spectrometer, it was found that the initial hydrocarbon product was isobutylene, not isobutane. After about 10 seconds, however, the gas was mainly isobutane, and olefin was no longer detected. The other major gaseous product is HCl but adsorption on the spectrometer walls made it impossible to quantitatively determine the HCl concentration.

The slope of region III extending from about 0.75 to 2.0 minutes in this experiment is taken as the rate of the hydride transfer reaction. A conventional kinetic analysis showed the rate to be first order in t-butyl chloride and first order in the concentration of methylcyclopentane in the emulsion, Figures 2 and 3.

It was also found that the methylcyclopentane concentration in the acid phase was about 60 ppm. An order of magnitude calculation indicates that the diffusion of methylcyclopentane into the bulk acid phase occurs much faster than the rate of formation of isobutane. Thus the acid phase should be considered to be saturated with methylcyclopentane throughout the reaction in all the kinetic experiments.

The fact that hydride transfer shows a dependence on the methylcyclopentane concentration in the emulsion is consistent with a reaction occurring at the acid-hydrocarbon interface. It would be inconsistent with reaction occurring only in the bulk acid phase since there the methylcyclopentane concentration is constant.

The hydride transfer reaction was found to have an activation energy of 9.9 Kcal/mole. Transfer from methylcyclopentane occurred much more easily than from methylcyclohexane where an Ea of 21.3 Kcal/mole was obtained. The difference between the two hydride donors can be rationalized by noting that solvolysis of the corresponding 1-chloro-1-methylcycloalkanes favors the cyclopentyl

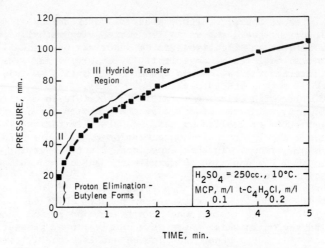

Figure 1. Hydride transfer appears in a distinct region of a pressure–time plot. The concentrations of MCP and t-C_4H_9Cl in the H_2SO_4 (96%) emulsion are 0.1M and 0.2M.

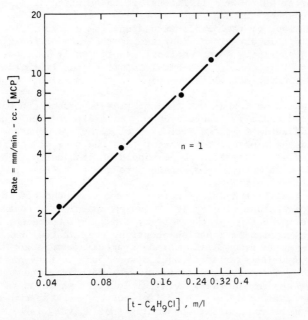

Figure 2.   Hydride transfer is first order in [t-C_4H_9Cl]

system by about 4 Kcal/mole and that in a bimolecular transition state between $t\text{-}C_4H_9^+$ and methylcyclohexane strong steric interactions between the 3-5 hydrogens and the t-butyl ion develop that should raise the transition state energy considerably. The latter interactions are not present in the methylcyclopentane reaction.

The preceding data are consistent with hydride transfer being predominantly an interfacial reaction in sulfuric acid. This view is supported by studies of the effect of acid modifiers on the reaction in 96% H_2SO_4. Table I shows the effect of halogenated acetic acids, methanesulfonic acid and water upon the hydride transfer rate. The rates were estimated in two ways and all rates are relative to that in 96 percent H_2SO_4.

Table I

Medium Effect on Relative Hydride Transfer Rates

Additive	Vol. %	A	B	Additive	Vol. %	A	B
CH_3COOH	0	1.00	1.0	CF_3COOH[(a)]	0	1.00	
	10	0.74	0.49		2	2.31	
	20	0.34	0.31		6	2.47	
	30	0	0.13		20	2.39	
$CH_2ClCOOH$	0	1.00	1.00	CH_3SO_3H	0	1.00	1.00
	4	1.23	1.28		2	1.15	1.04
	12	1.18	1.18		10	1.44	1.09
	20	1.31	1.05		20	0.73	0.52
$CHCl_2COOH$	0	1.00	1.00	H_2SO_4 wt %			
	4	1.06	1.26		96.6	1.31	1.28
	12		1.87		96	1.00	1.00
	20		1.90		90*	0.62	0.41
					85*	0.37	0.26
CCl_3COOH	0		1.00				
	4		1.68		80*	0.15	0.19
	12		1.63				
	20		1.85				

A   Estimated from dp/dt slopes.
B   Estimated from total gas evolution in 1 minute.

(a) $14°C$.
(*) Product is olefinic.

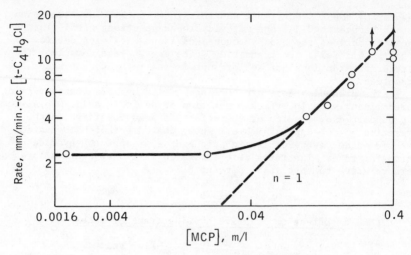

Figure 3. Hydride transfer is first order in MCP

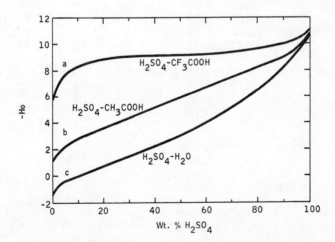

Figure 4. Typical acidity functions for modified sulfuric acid.
(a) H. H. Hyman and R. A. Garber, J. Am. Chem. Soc., 81, 1847
(1959); (b) N. F. Hall and W. F. Spengeman, J. Am. Chem. Soc.,
62, 2487 (1940); (c) L. P. Hammett and A. J. Deyrup, J. Am. Chem.
Soc., 54, 2721 (1932); L. P. Hammett and M. A. Paul, J. Am.
Chem Soc., 56, 827 (1934).

The relative rates were estimated both from dp/dt slopes and
from the total gas evolved after 1 minute. The data indicate that
as the acid is weakened by H_2O addition, the hydride transfer
reaction slows down. On the other hand, the addition of many of
the halogenated acetic acids and methanesulfonic acid results, at
least initially, in a slight rate increase.

The rate changes should reflect a change in the concentration
and stability of the cationic intermediates in the acid. In all
cases the acidity of modified H_2SO_4 decreases as the additive
concentration is increased, typical relationships being shown in
Figure 4. The change in acidity must be reflected in the steady
state R^+ concentration which therefore always decreases in these
systems.

The concentration of methylcyclopentane in the acid depends
strongly on the modifier, Figure 5. The data were obtained by
infrared analyses of CCl_4 extracts of saturated solutions of
methylcyclopentane in H_2SO_4 after standing 24 hours. There is
clearly no simple relationship between the solubility of methylcyclopentane and the relative hydride transfer rates. 2% Methanesulfonic acid for example, increases the solubility about 10-fold
but yields only a slight rate increase.

The fact that some increased rates are observed suggests
that we have increased cationic reactivity by weakening the acid
and destabilizing carbonium ion intermediates. The lack of
correlation with solubility of the hydride donor again indicates
that significant hydride transfer does not occur in the bulk acid
phase. Rather, the data are more easily understood if the reaction
occurs primarily at the acid-hydrocarbon interface under well
mixed conditions.

2,3,4 Trimethylpentane in
Tritiated Sulfuric Acid

The reaction of 2,3,4 trimethylpentane in tritiated sulfuric
acid provides additional evidence indicating the importance of an
interfacial hydride transfer in the formation of products during
alkylation.

This compound reacts readily in concentrated acid. Its
behavior was studied in 88.92 to 98.53 percent acid (15). The
reaction is preceded by an induction period whose duration is
inversely related to the acid strength. In this period one can
detect the formation of small amounts of SO_2 due to the reduction
of the acid as the hydrocarbon is being oxidized to a carbonium
ion. It is possible to relate the conversion of 2,3,4-trimethyl-
pentane to the production of ions and a chain length of 860 has
been determined for this reaction (16).

Once a steady state ion concentration is obtained, the
reaction exhibits kinetics which are first order in the 2,3,4-
trimethylpentane concentration in the emulsion and second order in
proton activity measured on the H_0 scale, Table II and Figure 7.

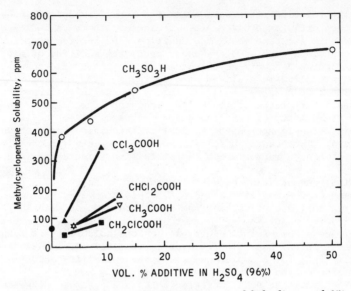

Figure 5. Solubility of methylcyclopentane in modified sulfuric acid, 25°

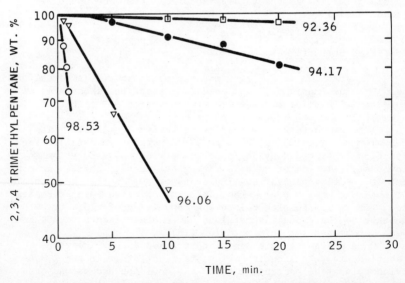

Figure 6. Reaction of 2,3,4-trimethylpentane in HTSO₄ exhibited on induction period

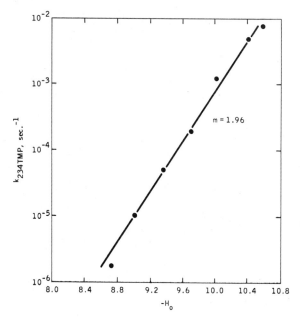

Figure 7. *First-order rate constants for the reaction 2,3,4-trimethylpentane in concentrated sulfuric acid*

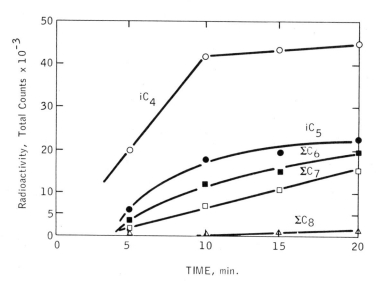

Figure 8. *Reaction products obtained from 2,3,4-trimethylpentane in HTSO₄ at 94%*

Table II

Reaction Rate Constants As A Function Of Acidity

Acidity wt %	$-H_o$ [a]	k_{234} TMP sec^{-1}
88.92	8.72	1.77×10^{-6}
90.64	9.05	1.05×10^{-5}
92.36	9.35	5.00×10^{-5}
94.17	9.70	1.95×10^{-4}
96.06	10.05	1.26×10^{-3}
97.78	10.40	5.00×10^{-3}
98.53	10.58	7.81×10^{-3}

(a) Jorgenson and Hartter, JACS 85, 878 (1963).

Of special interest with respect to alkylation are the product distributions shown as a function of time in Figures 8, 9, 10 and 11. The earliest product is isobutane in both 94 and 98.5% H_2SO_4. All other products including the C_5, C_6 and C_7 paraffins and the C_8 fraction are secondary.

The data can be rationalized by assuming that 2,3,4-trimethyl-pentane is initially converted to an ion by hydride transfer to either a carbonium ion or an oxidizing agent in the acid. The TMP^+ ion then rearranges and cleaves to a $t\text{-}C_4H_9^+$ ion and iC_4H_8. Rapid proton exchange with the acid equilibrates the ion with isobutylene before the cation extracts a hydride ion from another 2,3,4-trimethylpentane molecule or an olefin in the acid.

As the ion concentration in the acid grows toward a steady state value, alkylation and polymerization-cracking reactions occur which generate a distribution of C_5^+ to C_8^+ ions in the acid. An estimate of this "homogeneous" alkylate distribution in the bulk acid can be made from the product distribution at the earliest times shown in Figures 8 to 11.

The "homogeneous" alkylate is distinctly different from typical alkylate made in a well stirred pilot unit or commercial reactor. In particular it exhibits a very high ratio of dimethyl-hexanes to trimethylpentanes, and this ratio is higher in weaker acid than in stronger, Table III.

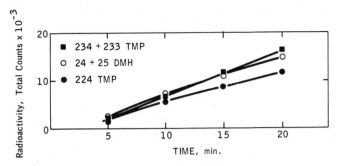

Figure 9.   *Reaction products obtained from 2,3,4-trimethylpentane in HTSO₄ at 94%*

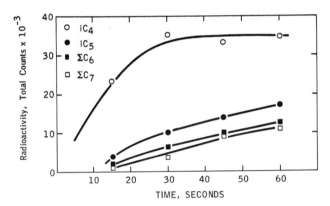

Figure 10.   *Reaction products obtained from 2,3,4-trimethylpentane in HTSO₄ at 98.5%*

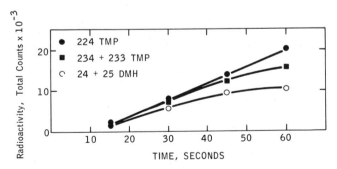

Figure 11.   *Reaction products obtained from 2,3,4-trimethylpentane in HTSO₄ at 98.5%*

Table III

Octane Distribution Vs
2,3,4-Trimethylpentane Conversion[a]

2,3,4 TMP Conv. wt %	% H_2SO_4	DMH/TMP
3.2	94.17	0.62
9.5		0.58
12.0		0.55
20.3		0.53
6.0	98.53	0.33
11.7		0.37
14.2		0.36
27.0		0.31

(a) Typical alkylate, DMH/TMP = 0.15/1.

The high DMH/TMP ratio is probably mainly due to reactions 1 to 3.

$$(1)$$

$$(2)$$

$$(3)$$

Production of dimethylhexanes in this way seems reasonable since chain branching rearrangements are not facile in H_2SO_4, i.e. methylpentanes will equilibrate without forming dimethylbutane. On the other hand 2,5-dimethyl-1,5-hexadiene can be converted to dimethylhexanes, (∿50% selectivity), and cyclic alkanes when reacted with a good hydride donor in concentrated H_2SO_4 indicating that multiple protonation and hydride transfer occurs faster than the rearrangement to trimethylpentane.

Other contributing routes to the dimethylhexanes and the cracking products may include a polymerization-rearrangement-cracking sequence involving $C_{12}H_{25}^+$ ions (17), a rather unlikely rearrangement of $t-C_4H_9^+$ to $s-C_4H_9^+$ followed by its addition to isobutylene, or by the addition of $t-C_4H_9^+$ to 1-butene. The important point is that if the dimethylhexanes are produced as a result of a sequence of reactions in the bulk acid, then this reaction should become more important as the ion and olefin levels in the acid increase. This should lead to

increasing DMH/TMP ratios as reaction proceeds, but in fact the
DMH/TMP ratio decreases as the 2,3,4-trimethylpentane conversion
increases.

The drop in the DMH/TMP ratio suggests that a second
reaction site which preferentially yields trimethylpentanes
develops with time. This should be the interfacial reaction. A
rough measure of the relative importance of the second site can
be made by assuming that bulk acid yields a DMH/TMP ratio of 0.33
and that the interface yields pure TMP. In order to obtain a
typical product with DMH/TMP = 0.15/1 the ratio of product formed
at the interface to that formed in the bulk acid is about 1:1.
This ratio will be much higher if the interfacial reaction
produces some of the dimethylhexanes as seems likely.

The behavior of 2,3,4-trimethylpentane reinforces the conclusions of the study of the t-butylchloride reaction with methylcyclopentane in suggesting that a means of increasing the hydride
transfer rate in H_2SO_4 could lead to improved selectivity in
alkylation. The next reaction was chosen to investigate this
possibility.

3-Methylpentane Isomerization in
Tritiated Sulfuric Acid. Controlling
The Rate with The Triphenylmethyl Cation.

The isomerization of 3-methylpentane occurs slowly in
concentrated H_2SO_4. When using tritiated acid at tracer levels,
it is readily observed that the tertiary 2 and 3-methylpentyl ions
equilibrate quickly once an ion forms. Each ion is assumed to
undergo fast reversible exchange of all protons adjacent to the
cationic center so that to a good approximation the radioactive
methylpentanes will contain 13 exchanged protons.

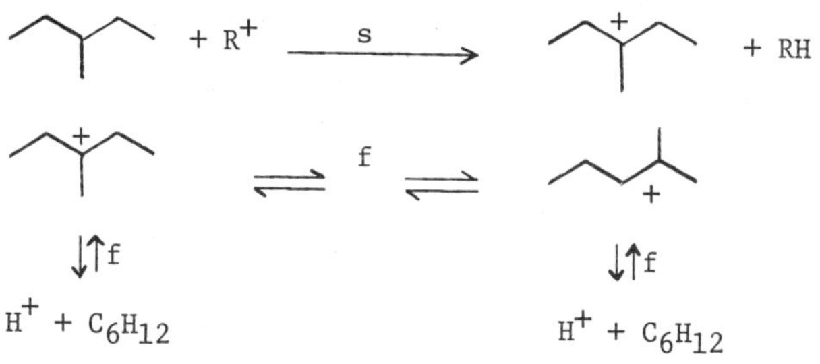

$$+ R^+ \xrightarrow{\ s\ } \qquad + RH \qquad (4)$$

$$\rightleftharpoons \xrightarrow{\ f\ } \rightleftharpoons \qquad (5)$$

$$\downarrow\uparrow f \qquad\qquad\qquad \downarrow\uparrow f$$

$$H^+ + C_6H_{12} \qquad\qquad H^+ + C_6H_{12} \qquad (6)$$

The relative rates indicated in equations 4, 5 and 6 show the
chain carrying hydride transfer reaction as the rate determining
step. Hence determination of means to control the rate of
incorporation of radioactivity in the methylpentanes is synonomous
with being able to control hydride transfer.

If the hydride transfer reaction involves a bimolecular
encounter between a cation and a donor, one way of increasing its

rate would be to destabilize the cation. In the case of an interfacial reaction, this might in principle be accomplished by adding cationic surfactants. In the bulk acid the addition of large cations that might cause a reorganization of solvent structure might also change the stability of the cationic intermediates. Ideally, one might add a large cation that in addition to the preceding might be a bonafide intermediate in accepting hydride ions from paraffins and passing them along to other cations in the acid system, and such a role has occasionally been postulated for "red oil" in alkylation (12b,13). The utility of the triphenylmethyl cation in controlling hydride transfer which might be due to several of these reasons is the topic of this section.

Before discussing the "trityl" ion it is important to note that the methylpentane isomerization is slow in fresh acid (there's an induction period as with 2,3,4-trimethylpentane), but may be accelerated by the addition of small amounts of an olefin like 2-methyl-1-butene. However, the reaction initiated by the olefin is rapidly quenched as the acid reduces a momentarily high intermediate ion concentration to a steady state level. The reduced rate is now a measure of the hydride transfer rate in the acid. The data to be discussed were obtained at 23°C with emulsions containing equal volumes of methylpentane and 95 percent H_2SO_4.

Control of the 3-methylpentane reaction was found to be a complex function of the amount of olefin used to initiate reaction and the amount of triphenylmethyl ion present. Figures 12 and 13 show typical data over a range of compositions and the relative hydride transfer rates as a function of the reagent concentration is shown in Figure 14.

One mole percent triphenylmethyl$^{\oplus}$ in the acid quenches the "background" exchange or isomerization of 3-methylpentane in the acid. Taking the steady state rate of the isomerization initiated with 0.05 mole percent olefin as unity one finds that it may be doubled with the large cation in the acid. The accelerated rate has been found to be stable throughout an 8 hour day in a properly conditioned acid. Still higher controlled rates are attainable as shown in Figure 14.

Phenomenologically, as the olefin concentration is increased at a given triphenylmethyl$^{\oplus}$ concentration, the rates maximize and then descend slowly. At high olefin concentrations the rates decrease toward the lined out value with olefin alone. The deleterious effect of too high olefin concentrations may be partially offset by increasing the triphenylmethyl$^{\oplus}$ level.

These results unambiguously show that it is possible to significantly increase hydride transfer rates in H_2SO_4. Why does this cation work and can increased hydride transfer be realized in steady state alkylation in a commercial unit?

The triphenylmethyl ion exerts its effect at a relatively high concentration in the acid. With a ring balance it has not

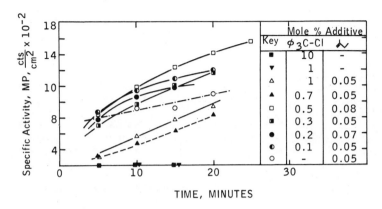

Figure 12. *Effect of 0.1–1.0 mol % triphenylmethylcation on controlling olefinic initiation of 3-methylpentane exchange*

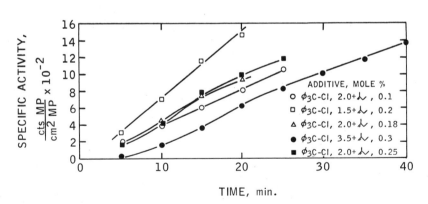

Figure 13. *Using ϕ_3C^+ to control olefin initiation of 3-MC$_5$ exchange*

been possible to find a surface excess of the ion at either the
acid-air or acid-hydrocarbon interface so that the ion is
uniformly distributed throughout the acid. It may be causing
structural changes in the acid and may be an intermediate in
hydride transfer, but it is not an effective surfactant and should
not be very efficient at influencing the surface reaction.

Cationic surfactants which should affect this reaction are
however, readily available. Figure 15 shows typical graphs of
surface tensions of the hexadecyltrimethylammonium ion and an
isotridecyldimethyl benzylammonium ion at air-acid and hexane-
acid interfaces. The surfactants behave much as in water and form
films containing about one ion/130 A^2. The cationic surfactants
would be expected to control carbonium ion behavior at the acid-
hydrocarbon interface at much lower concentrations than the triphenylmethyl ion. A complete comparison of exchange in the 3-
methylpentane system is not available but the surface active
cations have been found to markedly improve the alkylation reaction in continuous pilot plant and commercial apparatus.

Cationic Surfactants in Alkylation

The utility of cationic surfactants in increasing hydride
transfer would be expected to be shown by an increased yield of
octanes during butene alkylation. This follows if alkylate
selectivity is decided by the ratio of the rate at which intermediate ions are captured by hydride transfer to the rate at which
they add to olefins and polymerize, and if the effect of the
additives is to selectively raise the specific rate constant for
hydride transfer, k_H^-.

$$\text{Selectivity} \quad = \quad \frac{k_H^-(iC_4H_{10})(R^+)}{k_p\,(R^+)(R^=)}$$

To study this effect a small microalkylation pilot unit
(MAPU) was constructed. The unit contained a positive displacement pump for accurately passing feed into the bottom of a ½"
diameter glass reactor containing 5 ml of acid, 98% H_2SO_4 fresh.
The reactor was mixed with a high speed rotary "Magna Drive" unit
that reached within 1/16" of the feed inlet. Temperature was
controlled with a circulating water bath. The acid-hydrocarbon
emulsion would visually separate while the system was stirred and
the products left the reactor through a back pressure regulator.
All the products were vaporized and passed through a gas sampling
valve that automatically drew a sample into a gas chromatograph
about once an hour throughout the run. The analyses were used to
calculate a motor octane number assuming the MON was a linear
function of its components. All runs were of the "dying acid"
type in that a single initial acid charge was made. The runs
lasted about a week and the acid was then measured for its carbon

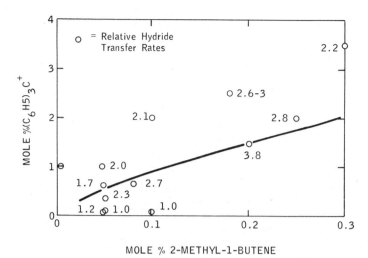

Figure 14. *The trityl ion controls olefin-initiated isomerization of 3-methylpentane*

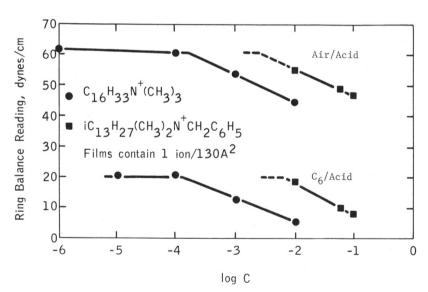

Figure 15. *Unsymmetrical quaternary ammonium ions—active at air–acid and oil–acid surfaces*

content and titratable acidity. In this study the effects of
stirring, olefin space velocity, olefin type and modifier type
were examined.

In preliminary experiments it was found that alkylate quality
was essentially independent of stirring speeds above 1000 rpm.
Accordingly all runs were made at 1200 rpm except when stirring
was being investigated.

One series was made with an isobutane – 2-butene feed, 6.6%
olefin, at 10°C and olefin space velocities of 0.077 to 0.7 V/Hr/V.
The additives studied are listed in Table IV.

Table IV

Alkylation Additives	Concentration, wt %
Tetramethylammonium Chloride (TMAC)	1.1
Dimethyltridecylbenzylammonium chloride (DMTDB)	0.7
FC-95	0.01
FX-161	0.01
Cetyltrimethylammonium bromide (CTMAB)	0.01
Cetylamine	0.01
Octylamine	0.01

The long chain ammonium salt and amine surfactants showed
better selectivity and calculated MON's. The improvement in MON
ranged from about 0.4 at an OSV of 0.08 V/Hr/V to more than 2 at
0.7 V/Hr/V. The space velocity was normally 0.077 and
was cycled during the course of a run between this value and
higher rates which were held until the product quality was steady.
A comparison of selectivity to C_8's in the C_5^+ fraction is shown
in Figure 16. Octylamine was intermediate in its ability to
improve selectivity while two of the surfactants which were
anionic (FC-95 and FX-161), showed no improvement under well mixed
conditions. In addition the tetramethylammonium ion which is not
surface active was also an ineffective additive.

Under poorly mixed conditions the surface active cations
show much more noticeable improvements, Figure 17. Typical data
obtained at 400 to 600 rpm are illustrated in Figure 17 and
suggest that the surfactants might be useful in overcoming mixing
limitations, if they exist, in commercial units.

The surface active cations also improve product quality when
alkylating with a typical refinery feed. Table V contains a list
of additives studied with a feed containing 94 percent isobutane
and 6 percent butenes. The butenes contained 42 percent isobutylene and an equilibrium mixture of 1- and 2-butene.

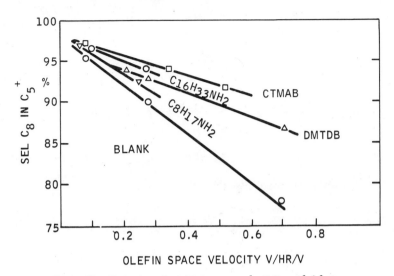

Figure 16. *Cationic surfactants improve selectivity with 2-butene*

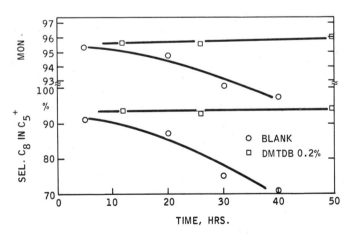

Figure 17. *Cationic surfactant improves alkylate in poorly mixed reactor*

Table V

Test Conditions With Refinery Olefins

$10^{\circ}C$, 0.077 to 0.28 V/Hr/V, OSV

Additive	Concentration, wt %
Cetylamine	0.01
CTMAB	0.005, 0.01, 0.05
Aromax 18/12, $C_{18}H_{37}(C_2H_4OH)_2NO$	0.01
Deriphat 151C, N-Coco β-aminopropionic acid	0.01
Armeen HT, RNH_2 60% C_{18}, 30% C_{16}	0.01
Armeen 12D, RNH_2, 97% C_{12}	0.01

In all cases some improvement in selectivity and octane quality was obtained. Cetylamine and CTMAB showed the best results with typical improvements of about 0.5 MON and 1.5% in selectivity throughout the range of olefin space velocities. Figure 18 shows the improvements obtained with the C_{16} amine.

In commercial alkylation one generally finds that improved reaction conditions lead to more selective and higher octane product, less acid consumption and higher yields based on olefin fed to the unit. Analyses of the spent acid obtained in the pilot plant also indicate that acid consumption is strongly reduced when the cationic surfactants are used.

Table VI

Acid Consumption Reduced With Cationic Surfactants

$10^{\circ}C$, 80 psig, 87 Vol % iC_4 in HC Phase, 98% H_2SO_4

	2-Butene			Mixed Olefins		
	V/V[a]	Acidity wt %	% C	V/V[a]	Acidity wt %	% C
Blank	18.0	89.0	3.6	14.4	91.5	3.3
0.01% $C_{16}H_{33}NH_2$	19.2	93.3	2.8	10.6	94.9	2.2
0.01% CTMAB[b]	19.1	92.7	2.2	13.3	93.3	2.9

(a) Cumulative volume of olefin fed per volume of acid.
(b) In the run with mixed olefins, 0.05% was used.

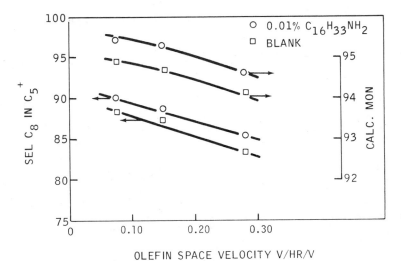

Figure 18.   *Alkylation selectivity improved with refinery olefins*

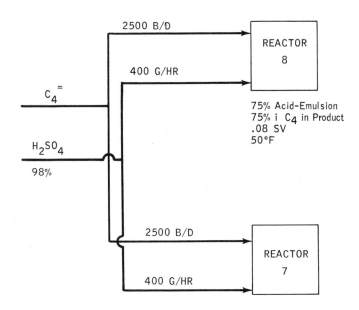

Figure 19.   *Plant test conditions*

Table VI shows that the titratable acidity of used acid is higher
and the carbon content lower in the runs containing additives with
a 2-butene feed. The difference is somewhat less but still
substantial in the experiments with the mixed olefin feed. For
comparison one can estimate the acidity of the blank after
feeding 10.6 volumes of olefin/volume acid as 93.2 percent.

The reactions involved in consuming acid are not well understood at this time and the simplest rationalization of the data is
to note that as alkylation conditions improve, less of the feed
has the opportunity of entering side reactions that deplete the
acid.

Commercial Test

One of the cationic surfactants was evaluated in a commercial
alkylation unit at the Baytown refinery. A parallel test was
conducted in which two reactors received the same feed and fresh
acid, Figure 19. The modifier was injected into an acid recycle
line on reactor 8, to rapidly bring its concentration to working
strength and then the rate was lowered to maintain the concentration. After 11 days the additive concentration was doubled and
after 19 days its addition was switched from reactor 8 to reactor
7. The additive concentration in reactors 7 and 8 is shown in
Figure 20. Note that after the switch the concentration in
reactor 8 depletes in accord with the acid replacement rate of the
unit and so this reactor will continue to receive the benefit of
the additive until the concentration drops to a level estimated
as about 0.005 wt percent.

After this time all improvements due to the additive should
be seen as changes developing in reactor 7.

The difference in titratable acidity of the recycle acid
from the reactors was taken as the best parameter for measuring
the effectiveness of the additive in the test. It can be shown
that the titratable acidity, C, (wt % H_2SO_4) will respond
according to equation 7.

$$C = C_A - (C_A - C_o)e^{-kt} \tag{7}$$

Here k is a constant, t is the time, C_A is the steady state
acidity in the presence of the additive and C_o is the steady state
acidity in the absence of additive. The rate constant k is
determined by the acid make up rate and the acid inventory in the
system and has a value of 0.12 per day in this study.

The difference in titratable acidity, Δ, between reactors 7
and 8 should follow equations 8, 9 and 10.

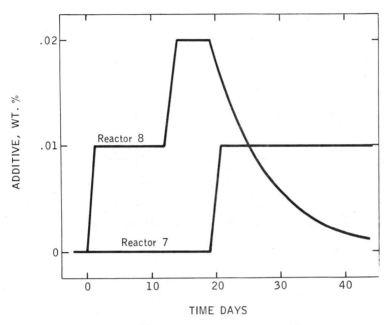

Figure 20.   *Additive concentration in commercial test*

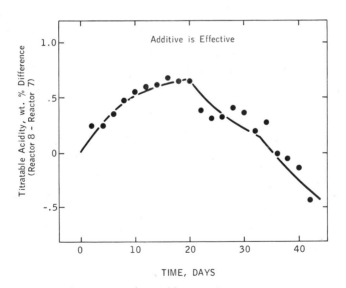

Figure 21.   *Additive is effective*

$$\Delta = \left[C_A - C_o\right]\left[1 - e^{-kt}\right], \ 0\text{-}20 \text{ days} \tag{8}$$

$$\Delta = \left[C_A - C_o\right]\left[e^{-k(t-20)} - e^{-kt}\right], \ 20\text{-}31 \text{ days} \tag{9}$$

$$\Delta = \left[C_A - C_o\right]\left[e^{-k(t-31)} - e^{-k(t-20)} - 1\right], \ 31\text{-}43 \text{ days} \tag{10}$$

During the first twenty days the additive was in reactor 8 and
not reactor 7 and the difference in titratable acidity should
follow equation 8. During the next eleven days the additive is
in both reactors and the Δ decreases according to equation 9.
On the 31st day the additive is below its effective concentration
level, (~ 0.005 wt %), and the difference in acidity decreases at
a faster rate. The solid curve in Figure 21 was calculated from
these equations using a value of $C_A - C_o$ of 0.7.

The circles in Figure 21 are experimental points, each
representing a minimum of 36 titrations. The titratable acidity
of each reactor was measured in triplicate every eight hours and
the points average two-day periods. These data show an
uncorrected decrease of 14 percent in acid consumption (derived
from Δ developed after 18 to 20 days). However the average
temperature in reactor 8 was 4 to 5°F higher than in reactor 7
during the entire test and we estimate that the difference in acid
consumption is closer to 20 percent at constant temperature. The
plant test was unfortunately terminated before the full Δ in the
opposite direction could be obtained but the data clearly shows
the additive to be strongly beneficial in alkylation.

In addition to saving acid the additive appeared to improve
the octane number by more than 0.1 MON as was indicated by a few
spot checks of alkylate during the run. The improvements
generally arose from a slight increase in the C_8 fraction, a rise
in the trimethylpentane concentration and changes of the trimethylpentane distribution. The octane analyses are not nearly
as extensive as the titratable acidity determinations and the
improvements are noted as being consistent with what would be
estimated from plant correlations and the observed reduction in
acid composition.

The additive used in the commercial test is being used in
nearly all of Exxon's alkylation units. No operating problems
have been encountered and it generally has been found to reduce
acid consumption by 15 to 20 percent and to generate slightly
higher octane number product. The cost of the additive is small
relative to the acid savings alone and it is available for
license.

The alkylation model developed in this work is one in which
the reaction is viewed as occurring in the acid phase and at the
acid-hydrocarbon interface. The formation of C_8's and trimethylpentanes occurs preferentially at the interface. Adding cationic
surfactants reduces the stability of the carbonium ion

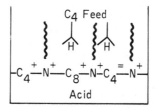

Figure 22. An alkylation model. Cationic surfactants should block surface reactions and destabilize reaction intermediates.

intermediates and causes them to abstract hydride ions more rapidly from isobutane or any other potential donor. Increased hydride transfer converts more of the carbonium ions at the acid interface to saturates faster, yielding product while minimizing polymerization and side reactions. It is also likely that the surfactants physically block alkyl ions from one another in the surface film and thus impede ion + olefin polymerization. In such a film the carbonium ion concentration must also be lower than in the absence of surfactant and mass law effects will therefore also lead to less polymerization and cracking. The fact that steady state hydride transfer rates in H_2SO_4 are subject to control through the use of acid modifiers which act in the bulk acid and at the acid-hydrocarbon interface is the key to the control of sulfuric acid alkylation.

Literature Cited

1. Kennedy, R. M. in "Catalysis," VI, ed. P. H. Emmett, Chapter 1, 1, Reinhold (1958).
2. Condon, F. E. in "Catalysis," VI, ed. P. H. Emmett, Chapter 2, 43, Reinhold (1958).
3. Schmerling, L. in "The Chemistry of Petroleum Hydrocarbons," III, ed. B. T. Brooks, S. S. Kurtz, Jr., C. E. Board, L. Schmerling, Chapter 34, 363, Reinhold (1955).
4. Burwell, R. L. and Gordon, G. S., III, J. Am. Chem. Soc. (1948) 70, 3128.
5. Burwell, R. L. Jr., Maury, L. G. and Scott, R. B., J. Am. Chem. Soc. (1954) 76, 5828.
6. Burwell, R. L., Scott, R. B., Maury, L. G. and Hussey, A. S., J. Am. Chem. Soc. (1954) 76, 5822.
7. Gordon, G. S., III and Burwell, R. L., J. Am. Chem. Soc. (1949) 71, 2355.
8. Ingold, C. K., Raisin, C. G. and Wilson, C. L., J. Am. Chem. Soc. (1936) 58, 1643.
9. Otvos, J. W., Stevenson, D. P., Wagner, C. D. and Beeck, O., J. Am. Chem. Soc. (1951) 73, 5741.

10. Stekina, V. N., Jursanov, D. N., Sterligov, O. D. and
     Liberman, A. L., Doklady Akad. Nauk. S.S.S.R. (1952) 85,
     1045.
11. Stephenson, D. P., Wagner, C. D., Beeck, O. and Otvos, J. W.,
     J. Am. Chem. Soc. (1952) 74, 3269.
12a. Thomas, C. L., "Catalytic Processes and Proven Catalysts,"
     Chapter 9, 87 Acad. Press (1970).
12b. Doshi, B. and Albright, L. F., Ind. Eng. Chem., Proc. Des.
     Dev. (1976) 15, 53.
13. Unpublished results of G. M. Kramer. Mentioned in Hoffmann,
     J. E. and Schriesheim, A., J. Am. Chem. Soc. (1962) 84, 953.
14. Kramer, G. M., J. Org. Chem. (1965) 30, 2671.
15. Kramer, G. M., J. Org. Chem. (1967) 32, 920.
16. Kramer, G. M., J. Org. Chem. (1967) 32, 1916.
17. Hoffmann, J. E. and Schriesheim, A., J. Am. Chem. Soc.,
     (1962) 84, 957.

Reaction Mechanisms for Hydrofluoric Acid Alkylation

T. HUTSON, JR. and G. E. HAYS

Phillips Petroleum Company, Bartlesville, OK 74004

1. Conclusions

1. All four butene isomers are believed to be capable of undergoing isomerization and polymerization under alkylation conditions. These ionic reactions are extremely rapid and precede or accompany isobutane alkylation.

2. Since alkylate compositions from the four butene isomers are basically similar, the butenes are thought to isomerize considerably, approaching equilibrium composition prior to isobutane alkylation. Such a postulation is at variance with previously published alkylation mechanisms. The isomerization step yields predominantly isobutene which then polymerizes and forms a 2,2,4-trimethylpentyl carbonium ion, a precursor of 2,2,4-trimethylpentane, the principal end product. The 2,2,4-trimethylpentyl ion is also capable of isomerization to other trimethylpentyl ions and thus yields other trimethylpentanes, principally 2,3,4-trimethylpentane and 2,3,3-trimethylpentane.

3. Dimethylhexanes are believed to result mainly from codimerization of butene-1 and isobutene, followed by abstraction of a hydride ion from isobutane. This mechanism differs significantly from previously published theory. Thus, high initial concentrations of butene-1 favor dimethylhexane formation. Some isomerization of dimethylhexyl carbonium ions occurs.

4. 2,2,3-Trimethylpentane and its ionic form are not primary reaction products or intermediates of isobutane and 2-butene alkylation. Found in alkylation product in very small concentrations, 2,2,3-trimethylpentane probably results mainly from isomerization of 2,2,4-trimethylpentyl carbonium ions.

5. Excessive olefin polymerization followed by hydride ion abstraction from isobutane probably accounts for the formation of saturated residue. 1-Butene, propylene, the catalysts's low water content, and low isobutane-to-olefin ratio favor the production of residue.

6. In the case of propylene-isobutane alkylation, primary alkylation,

27

hydrogen transfer (chain initiation), and polymerization reactions predominate -- in that relative order of importance. Product analysis indicates
concentrations of about 52, 28, and 12 weight percent of C_7, C_8, and
residue, respectively.

7. Disproportionation of primary C_7 alkylate (from propylene) with
isobutane under alkylation conditions probably accounts for the near-equal
amounts of isopentane and isohexanes found in propylene alkylate.

8. The effect of water in HF alkylation catalyst is an important one;
it appears to be that of slowing polymerization. The production of low-
octane-number residue can be reduced by 50 percent when catalyst water
content is optimized.

9. In studies with propylene, increasing isobutane-to-olefin ratio
suppressed the formation of high-molecular-weight residue, indicating a
substantial reduction in the role of olefin polymerization to large ions.
At the same time, the concentration of the C_7 fraction (primary product)
decreased, and the C_8 fraction (from chain initiation and subsequent hydrogen transfer) increased markedly.

10. HF alkylate shows a significant composition advantage over that
made with conventional H_2SO_4 catalyst. This is believed to be due to a
lesser amount of polymerization occurring with HF catalyst. Also, higher
octane-fraction content, lower dimethylhexane content, and lower heptane
content in HF alkylate indicate that HF catalyst favors a greater amount of
hydrogen transfer from propylene.

11. Good hydrocarbon dispersion in hydrofluoric acid is an important
factor in producing alkylate rich in trimethylpentanes and thus favors
olefin isomerization (to isobutene), isobutene dimerization, and maximizes
hydrogen transfer and primary alkylation reactions. Excess olefin polymerization to form residue is suppressed by good dispersion.

II. Introduction

Our purposes in studying HF alkylation reaction mechanisms are
several fold. The works of Ciapetta (1945) and Schmerling (1946,1955),
although classic in the field of alkylation in their time, were limited by
the then available analytical tools. Today, by using greatly improved
analytical procedures, the product makeup can be much better defined.

Elucidation and better understanding of these mechanisms will make
current experimental work more effective, facilitating the development of
improved HF catalysts, process conditions, mass transfer, etc. Fundamen-
tally, the proposed mechanisms for HF-catalyzed alkylation explain the
production of alkylates of better quality than those obtained from H_2SO_4-
catalyzed alkylation.

A large number of investigators have offered theories on reaction
mechanisms in HF- or H_2SO_4- catalyzed isobutane-olefin alkylation.

All of these theories were based on a carbonium-ion, chain mechanism. Previous investigators have ignored the possibility of olefin isomerization and polymerization as substantial reaction steps in the overall alkylation mechanism. In separate articles Schmerling (1955) and Cupit et al. (1961) summarized the most generally accepted theories similarly. Reliance was heavy on "methyl shift" and "hydride ion shift" to explain the many isomers typically present in various alkylates. High-boiling by-products were postulated to result from polymerization reactions requiring more than one olefin molecule per molecule of isoparaffin. Ciapetta (1945) and Schmerling (1955) attributed dimethylhexanes to isobutane alkylation with 1-butene, involving formation of the dimethylhexyl ion which underwent rearrangement before being converted to octane molecules. Hofmann and Schriesheim (1962) using tagged molecules in isobutane-butene alkylation, attributed the formation of all products other than octanes to the formation of a C_{12} carbonium ion intermediate.

III. Types of Reactions

Chain Initiation. The theory postulated by a number of investigators (Cupit et al., 1961, Schmerling, 1955) is that carbonium ions are generated by addition of a proton (H+) to an olefin molecule in the presence of HF. Albright and Li, 1970, and Hofmann and Schriesheim, 1962, indicate that initiation steps with H_2SO_4 catalyst may involve red oil hydrocarbons. However, only the tertiary butyl carbonium ion performs the chain carrying function in isobutane alkylation. Reactions follow:

(1)
$$\begin{array}{c} C \\ | \\ C=C \\ | \\ C \end{array} \quad + \quad H^+ \quad \longrightarrow \quad \begin{array}{c} C \\ | \\ C-C^+ \\ | \\ C \end{array}$$

(2)
$$C=C-C-C \quad + \quad H^+ \quad \longrightarrow \quad \underset{+}{C-C-C-C}$$

(2-A)
$$\underset{+}{C-C-C-C} \quad + \quad \begin{array}{c} C \\ | \\ C-C-H \\ | \\ C \end{array} \quad \longrightarrow \quad C-C-C-C \quad + \quad \begin{array}{c} C \\ | \\ C-C^+ \\ | \\ C \end{array}$$

(3)
$$C=C-C \quad + \quad H^+ \quad \longrightarrow \quad \underset{+}{C-C-C}$$

$$
(3\text{-}A) \quad
\begin{array}{c} C\text{-}C\text{-}C \\ + \end{array}
\ + \
\begin{array}{c} C \\ | \\ C\text{-}C\text{-}H \\ | \\ C \end{array}
\ \longrightarrow \
\begin{array}{c} C \\ | \\ C\text{-}C\text{-}C \\ | \\ H \end{array}
\ + \
\begin{array}{c} C \\ | \\ C\text{-}C^+ \\ | \\ C \end{array}
$$

Thus, the most direct route to chain-carrying, tertiary butyl carbonium ions is offered in isobutene-isobutane alkylation (Equation I). When initiating with either a linear butene or propylene, a second step is necessary to form the tertiary butyl carbonium ion, i.e., abstraction of a hydride ion from an isobutane molecule while forming a molecule of normal alkane. (Equation 2, 2-A, 3, 3-A). Reaction sequences in these equations are often referred to as hydrogen- or hydride transfer reactions and will be discussed subsequently.

Chain Propagation. In the chain propagation step, an olefin molecule reacts with a tertiary butyl carbonium ion as postulated by Whitmore (1934). This addition reaction produces a larger carbonium ion which then either undergoes isomerization or abstracts a hydride from an isobutane molecule. (Under some circumstances, the larger carbonium ion may add a second molecule of olefin; this reaction will be discussed under "Polymerization".) Hydride abstraction regenerates a chain-carrying, tertiary butyl carbonium ion and also forms a molecule of isoparaffin. Reactions follow:

$$
(4) \quad
\begin{array}{c} C \\ | \\ C\text{-}C^+ \\ | \\ C \end{array}
\ + \
\begin{array}{c} C \\ | \\ C\text{=}C \\ | \\ C \end{array}
\ \longrightarrow \
\begin{array}{c} C \quad\ C \\ | \quad\ | \\ C\text{-}C\text{-}C\text{-}C\text{-}C \\ | \quad\ + \\ C \end{array}
$$

$$
(4\text{-}A) \quad
\begin{array}{c} C \quad\ C \\ | \quad\ | \\ C\text{-}C\text{-}C\text{-}C\text{-}C \\ | \quad\ + \\ C \end{array}
\ + \
\begin{array}{c} C \\ | \\ C\text{-}C\text{-}H \\ | \\ C \end{array}
\ \longrightarrow \
\begin{array}{c} C \quad\ C \\ | \quad\ | \\ C\text{-}C\text{-}C\text{-}C\text{-}C \\ | \quad\ H \\ C \end{array}
\ + \
\begin{array}{c} C \\ | \\ C\text{-}C^+ \\ | \\ C \end{array}
$$

Chain Termination. Chain termination in isobutane alkylation is any reaction sequence which results in the elimination of a tertiary butyl carbonium ion. Specifically, two tertiary butyl carbonium ions are consumed and only one is regenerated. Reactions follow:

$$
(5) \quad
\begin{array}{c} C \\ | \\ C\text{-}C^+ \\ | \\ C \end{array}
\ \longrightarrow \
\begin{array}{c} C \\ | \\ C\text{=}C \\ | \\ C \end{array}
\ + \ H^+
$$

(5-A)

$$
\underset{\overset{\displaystyle C}{|}}{\overset{\displaystyle C}{|}}C\text{-}C^+ \quad + \quad \underset{\overset{\displaystyle C}{|}}{\overset{\displaystyle C}{}}C\text{=}C \quad \longrightarrow \quad C\text{-}\underset{\overset{\displaystyle C}{|}}{\overset{\displaystyle C}{|}}C\text{-}C\text{-}\underset{\overset{\displaystyle +}{}}{\overset{\displaystyle C}{|}}C\text{-}C
$$

(5-B)

$$
C\text{-}\underset{\overset{\displaystyle C}{|}}{\overset{\displaystyle C}{|}}C\text{-}C\text{-}\underset{\overset{\displaystyle +}{}}{\overset{\displaystyle C}{|}}C\text{-}C \quad + \quad \underset{\overset{\displaystyle C}{|}}{}C\text{-}\overset{\displaystyle C}{}C \quad \longrightarrow \quad C\text{-}\underset{\overset{\displaystyle C}{|}}{\overset{\displaystyle C}{|}}C\text{-}C\text{-}\underset{\overset{\displaystyle C}{|}}{\overset{\displaystyle C}{|}}C\text{-}C \quad + \quad C\text{-}\underset{\overset{\displaystyle C}{|}}{}C^+
$$

When alkylating isobutane, chain termination forms primarily, but not entirely, 2,2,4-trimethylpentane; the alkylate from chain termination very closely resembles isobutene alkylate. The similarity of alkylate compositions, particularly their C_8 fractions, originating from various olefins and the distance from thermodynamic equilibrium composition indicates that alkylate molecules, once formed, are relatively stable under alkylation conditions and undergo little isomerization. Undesirable side products, e.g., dimethylhexanes and residue, are probably formed by butene isomerization and polymerization (rather than by isomerization of alkylate or by isomerization of the C_8 carbonium ion which subsequently becomes alkylate).

   Olefin Isomerization.   Olefin isomerization plays an important role in butene–isobutane alkylation reaction mechanisms. Normal butenes are largely isomerized to isobutene before alkylation. This is believed to take place in ionic form, i.e., immediately following olefin protonation, since a number of olefins have been found to add HF across their double bonds quite readily at room temperature (Grosse and Linn, 1938). Thus, the likelihood of olefin molecules being present for very long under alkylation conditions is not great.
   The following facts are the basis for butene isomerization: (1) There is a basic similarity in the composition of alkylates produced from all four butene isomers. (2) Alkylate molecules, once formed, are relatively stable under alkylation conditions and do not isomerize to any appreciable extent; alkylate fractions having the same carbon number are not equilibrated (see Table I). (3) Thermodynamic equilibrium between the butene olefins highly favors isobutene formation at alkylation temperatures. (4) Normal butenes produce only small and variable amounts of normal butane, thus indicating only a small and variable amount of chain initiation from normal butenes. Yet the alkylate composition shows a high concentration of trimethylpentanes and a low concentration of dimethylhexanes. (5) A few of the octane isomers can be explained only by isomerization of the eight-carbon skeletal structure; this isomerization occurs while isobutene dimer is in ionic form. For example, 2,3,3- and 2,3,4-trimethylpentanes

are postulated to be formed by isomerization of the 2, 2, 4-trimethylpentyl carbonium ion. Similarly, 2, 3-dimethylhexane is probably formed by isomerization of isobutene-normal butene codimer, the 2, 4-dimethylhexane structure, while in the ionic form; Scharfe (1973) reports finding these dimethylhexenes, along with trimethylpentenes, in a codimer not made in an alkylation unit.

It is obvious that any isobutane formed by isomerization of normal butenes to isobutene, followed by a hydride transfer to the isobutene, cannot be distinguished from the isobutane charge unless a tracer technique is used. Hofmann and Schriesheim (1962) found, in alkylation studies with C^{14}- labeled 1-butene and isobutene, that C^{14} did indeed appear in the isobutane fraction as well as in the pentane and heavier fractions at what they called "steady-state". (They did not distinguish whether normal butane with C^{14} was present in the isobutane fraction.) The percentage of total radioactivity appearing in the isobutane fraction was 23 percent for 1-butene and 38 percent for isobutene. Neither this laboratory nor others (Albright and Li, 1970) has found normal butane production during normal butenes alkylation to be more than a tenth of that 23 percent value. Clearly this order of magnitude discrepancy suggests olefin isomerization of normal butenes to isobutene. Schmerling (1946) suggested such isomerization of sec-butyl cation to tert-butyl cation before hydride transfer and cited other authors who had made the same assumption; until the C^{14} tracer work was done, this isomerization remained an unverified assumption. Hofmann and Schriesheim's reasoning and conclusion about the importance of the C_{12} carbonium ion are not supported by their data, unless one assumes that labeled normal butenes do not isomerize and that labeled isobutane does not react. The 38 percent value cited in isobutene akylation confirms hydride transfer back and forth between labeled isobutene and isobutane.

Olefin Dimerization. Other investigators (Schaad, 1955, Sparks et al., 1939) have reported that the catalytic polymerization of isobutene produced a liquid polymer consisting mainly of isooctenes. Hydrogenation of these isooctenes gave isooctanes consisting of 70 to 90 percent 2, 2, 4-trimethylpentane; the remainder was reported as mainly 2, 3, 4-trimethylpentane. That earlier work used analytical techniques inferior to those now available. Typically, it is now found that the C_8 fraction of isobutene-isobutane alkylate catalyzed by HF contains about 62.6 percent 2, 2, 4-trimethylpentane, 13.4 percent 2, 3, 4-trimethylpentane, 11.7 percent 2, 3, 3-trimethylpentane, 1.3 percent 2, 2, 3-trimethylpentane, and 11 percent dimethylhexanes (see normalized data column for isobutene feed in Table VII). Thus it appears that, under alkylation conditions, isobutene dimerization followed by hydride abstraction from isobutane is another reaction route to the large amount of 2, 2, 4-trimethylpentane.

The lesser amounts of other trimethylpentanes are formed mainly by isomerization of trimethylpentyl carbonium ions followed by hydride abstraction.
These reactions are illustrated as follows:

Isobutene Dimerization

$$
(6) \qquad 2 \;
\begin{array}{c}
\text{C} \\
\| \\
\text{C-C} \\
|\\
\text{C}
\end{array}
+ \; \text{H}^+ \;\longrightarrow\;
\begin{array}{cc}
\text{C} & \text{C} \\
| & | \\
\text{C-C-C-C-C} \\
| & + \\
\text{C}
\end{array}
$$

Hydride Abstraction

$$
(6\text{-A}) \quad
\begin{array}{cc}
\text{C} & \text{C} \\
| & | \\
\text{C-C-C-C-C} \\
| & + \\
\text{C}
\end{array}
\; + \;
\begin{array}{c}
\text{C} \\
| \\
\text{C-C} \\
| \\
\text{C}
\end{array}
\;\longrightarrow\;
\begin{array}{cc}
\text{C} & \text{C} \\
| & | \\
\text{C-C-C-C-C} \\
| \\
\text{C}
\end{array}
\; + \;
\begin{array}{c}
\text{C} \\
| \\
\text{C-C}^+ \\
| \\
\text{C}
\end{array}
$$

Carbonium Ion Isomerization

$$
(7) \qquad
\begin{array}{cc}
\text{C} & \text{C} \\
| & | \\
\text{C-C-C-C-C} \\
| & + \\
\text{C}
\end{array}
\;\longrightarrow\;
\begin{array}{ccc}
\text{C} & \text{C} & \text{C} \\
| & | & | \\
\text{C-C-C-C-C} \\
& & +
\end{array}
$$

$$
(7\text{-A}) \quad
\begin{array}{ccc}
\text{C} & \text{C} & \text{C} \\
| & | & | \\
\text{C-C-C-C-C} \\
& + &
\end{array}
\; + \;
\begin{array}{c}
\text{C} \\
| \\
\text{C-C} \\
| \\
\text{C}
\end{array}
\;\longrightarrow\;
\begin{array}{ccc}
\text{C} & \text{C} & \text{C} \\
| & | & | \\
\text{C-C-C-C-C} \\
\end{array}
\; + \;
\begin{array}{c}
\text{C} \\
| \\
\text{C-C}^+ \\
| \\
\text{C}
\end{array}
$$

Reaction sequence 1, 4, and 4-A, and sequence 6 and 6-A shown
above are believed to represent the predominant ones involved in isobutane-butene alkylation, since the greater portion (80 to 90 percent) of
C_8 fractions from all four butenes is made up of trimethylpentanes which
are predominantly 2,2,4-trimethylpentane. Hofmann and Schriesheim
(1962) in alkylation studies with radio-labeled butenes, concluded that
a majority of trimethylpentanes does not arise as a result of expected
isobutane-isobutene alkylation, but that rapid isobutene polymerization
followed by hydride transfer predominates. These investigators also found
about the same levels of radioactivity in all of the trimethylpentanes,
which is consistent with the isobutene dimerization and C_8 carbonium ion
isomerization route rather than the postulation that some trimethylpentanes
are formed by other routes. Both of the above reaction sequence groups

(6 and 7) give the same overall net reaction, i.e., one mole of isobutene combined with one mole of isobutane to yield one mole of alkylate. Thus, one mole of isobutane reacts to produce each mole of product; such yields are substantiated in pilot plant and commercial plant operation (Hutson and Logan, 1975).

The following reasons do not support the early postulation (Ciapetta, 1945, Schmerling, 1955) that the 2,2,3-trimethylpentane configuration is a primary reaction product of isobutane and butene-2 and that it isomerizes readily: (1) In degradation reactions using sulfuric acid catalyst, 2,2,3-trimethylpentane showed very great stability, i.e., considerably less reactivity than the other three trimethylpentanes (Doshi and Albright, 1976). (2) Based on thermodynamic equilibrium composition, the 2,2,3-isomer would be substantially favored over the 2,3,4- and 2,3,3-isomers (Table I). (3) The 2,2,3-trimethylpentane isomer appears in very small amounts in the alkylation product from isobutane and 2-butene; these concentrations are much lower than those for 2,3,4- and 2,3,3-trimethylpentane.

Schmerling (1955) observed that normal butene polymers when alkylated gave products similar to those obtained by alkylation of the monomers with H_2SO_4. Based on the yields of trimethylpentanes, he concluded that with trimers depolyalkylation predominated, rather than hydrogen transfer (and subsequent chain terminations). He concluded that trimers depolymerized to butylenes prior to alkylation with isobutane. He rationalized that, since the yield of trimethylpentanes was about 50 percent greater than the theoretical (stoichiometric) based on hydrogen transfer alone, the reason had to be depolymerization of the trimers. With trimers, he observed that 75–83 percent of the trimer depolyalkylated. He rationalized further that dimers could behave similarly; this is a moot point, because yields of alkylate from butene monomers and dimers are identical when the alkylate formed during chain termination is considered. It was observed (Phillips Petroleum Company, 1946) that isobutene-butene copolymer yielded alkylate identical to that from the monomers.

Hydrogen Transfer.   Hydrogen transfer (sometimes called self alkylation of isobutane) occurs with propylene-isobutane mixtures using HF catalyst. This is a chain initiative reaction in that tertiary butyl carbonium ions are formed. End products are (1) propane and (2) 2,2,4-trimethylpentane. Reactions follow:

(8) $$C_3H_6 \ + \ H^+ \ \rightleftharpoons \ C_3H_7^+$$

$$(8\text{-}A) \qquad C_3H_7^+ \; + \; iC_4H_{10} \qquad\qquad \longrightarrow \qquad C_3H_8 \; + \; iC_4H_9^+$$

$$(8\text{-}B) \qquad\qquad\qquad iC_4H_9^+ \qquad\qquad \rightleftharpoons \qquad iC_4H_8 \; + \; H^+$$

$$(8\text{-}C) \qquad iC_4H_8 \; + \; iC_4H_9^+ \qquad\qquad \rightleftharpoons \qquad iC_8H_{17}^+$$

$$(8\text{-}D) \qquad iC_8H_{17}^+ \; + \; iC_4H_{10} \qquad\qquad \longrightarrow \qquad IC_8H_{18} \; + \; iC_4H_9^+$$

Overall Reaction:

$$(9) \qquad C_3H_6 \; + \; 2iC_4H_{10} \qquad\qquad \longrightarrow \qquad C_3H_8 \; + \; iC_8H_{18}$$

Similarly, hydrogen transfer reactions occur when isobutane is alkylated with n-butenes or with amylenes. There is no chain termination taking place in hydrogen transfer; hydrogen transfer represents either chain initiation or chain transfer.

Excess Polymerization. A small amount of high-boiling heavy "tail" or residue is formed in isobutane alkylation, even under the most favorable reaction conditions. The polymer molecule is in reality an isoparaffin formed from two or more molecules of olefin plus one molecule of isobutane. Polymer is formed because of the inherent tendency of larger carbonium ions, e.g., C_7 or C_8 ions, to complete with tertiary butyl carbonium ions for addition of olefin molecules before abstracting hydride ions and becoming isoparaffin molecules. Reactions follow:

$$(10) \qquad 2\,C_4H_8 \; + \; iC_4H_9^+ \qquad \longrightarrow \qquad iC_{12}H_{25}^+$$

$$(10\text{-}A) \qquad iC_{12}H_{25}^+ \; + \; iC_4H_{10} \qquad \longrightarrow \qquad iC_{12}H_{26} \; + \; iC_4H_9^+$$

Polymer or residue formation is minimized by maintaining proper reaction conditions, i.e., good mass transfer, high isobutane-to-olefin ratio, proper catalyst activity, and minimum concentration of alkylate in the reaction zone.

The other important reaction of the polymer ion is that of cracking or scission to form a lower-molecular-weight carbonium ion plus an

olefin molecule. For example: $iC_{12}H_{25}^+ = iC_7H_{15}^+ + iC_5H_{10}$.
Both of these products are subject to further reaction under alkylation
conditions and can account for many isoparaffins commonly found in
alkylates in minor amounts. Similar equations may be written for
$iC_{16}H_{33}^+$ ions.

<u>Disproportionation</u>. Disproportionation is believed to play only a
minor role in the formation of alkylate components. What does occur
is probably via a carbonium ion mechanism, i.e., when the precursor
is in ionic form. Disproportionation reactions could account for the formation of the small concentrations of isopentane, isohexanes, and isoheptanes which are usually found in butene-isobutane alkylates. An
example follows:

$$
(11) \quad
\begin{array}{c} C \\ | \\ C\text{-}C \\ | \\ C \end{array}
\; + \;
\begin{array}{c} C \quad\;\; C \\ | \quad\;\; | \\ C\text{-}C\text{-}C\text{-}C \\ | \\ C \end{array}
\;\longrightarrow\;
\begin{array}{c} C \\ | \\ C\text{-}C\text{-}C\text{-}C \end{array}
\; + \;
\begin{array}{c} C \quad\;\; C \\ | \quad\;\; | \\ C\text{-}C\text{-}C\text{-}C\text{-}C \end{array}
$$

<u>Conjunct Polymers</u>. Conjunct polymers (frequently called acid-
soluble oils in HF alkylation, red oils in sulfuric acid alkylation) are an
exceedingly complex mixture of highly unsaturated, cyclic hydrocarbons.
These polymers are by-products of tertiary butyl carbonium ions, and
their formation undoubtedly involves a complexity of reactions. Miron
and Lee (1963) found the bulk of an HF conjunct polymer to be made up
of molecules containing 2-4 rings with an average ring size of 5-6 carbon
atoms. They estimated the number of double bonds per molecule of polymer at about 2.5 to 3. Thus, these polymers are hydrogen-deficient.
The hydrogen lost during their formation apparently goes into chain termination, i.e., the formation of isobutane most probably and possibly some
propane when propylene is present in alkylation feed. HF alkylation has
found no benefit from having acid-soluble oils present in the catalyst.
When they are present in amounts greater than about one weight percent,
they have a detrimental effect on alkylate quality and yield.

IV. <u>Experimental</u>

<u>Normal Butene Reactions</u>. Under alkylation conditions, all four
butene isomers are believed to undergo isomerization, dimerization, and
co-dimerization when first coming in contact with HF catalyst, i.e.,
immediately following protonation. These are very rapid, ionic reactions
and take place competitively along with isobutane alkylation. Alkylate
compositions from the four butenes are basically similar (see Table VII).
However, 1-butene produces a C_8 fraction containing nearly two times

the amount of dimethylhexanes produced from the other three butenes (21.77 vs about 10.5 percent; see normalized data columns in Table VII). The remaining C_8 material, largely trimethylpentanes, is of a composition similar to that produced from the other three butenes. Since thermodynamic equilibrium at alkylation temperature (about 77F) highly favors isobutene formation (the following thermodynamic equilibrium composition was calculated from API Project 44 data at 77F: Isobutene = 84.5 percent, trans-2-butene = 11.6 percent, cis-2-butene = 3.6 percent, and 1-butene = 0.3 percent), it is postulated that the similarity of C_8 compositions results from either (1) the rapid rate of olefin isomerization and then isobutane alkylation of isobutene -- both of these reactions take place in ionic form -- or (2) olefin isomerization followed by olefin dimerization and hydride abstraction from isobutane.

Isoparaffin and Carbonium Ion Isomerization. It is well known that 2,2,4-trimethylpentane is relatively stable when contacted with HF; reaction conditions must be quite severe (high temperatures and long contact times) to obtain any appreciable conversion of this isoparaffin. Therefore, isoparaffin isomerization is believed to be of small significance in the overall reaction scheme. Isomerization of trimethylpentyl carbonium ions is believed to account for the production of minor amounts of the other three trimethylpentanes. However, the approach to equilibrium composition within the C_8 fraction is very poor as shown in Table I.

TABLE I. OCTANES FRACTION FROM ISOBUTENE ALKYLATE

	Isobutene (1)	Alkylate (3)	Equilibrium Composition (20°C) (2)	(3)
2,2,4-Trimethylpentane	62.63	70.37	8.0	69.77
2,3,4-Trimethylpentane	13.35	15.00	1.1	9.58
2,3,3-Trimethylpentane	11.67	13.12	0.77	6.71
2,2,3-Trimethylpentane	1.34	1.51	1.6	13.94
Dimethylhexanes + Methylheptanes	11.01	–	84.06	–
Other	–	–	4.47	–

(1) Data for isobutene, Table VII.
(2) Prosen, E. J., Pitzer, K. S., Rossini, F. D., J. Res. Nat. Bur. Stand., 34, 255 (1945).
(3) Trimethylpentanes only, normalized to 100 percent.

As shown in Table I, the equilibrium composition for dimethylhexanes is nearly eight times as great as actually found in isobutene alkylate; whereas, the trimethylpentane concentration of isobutene alkylate exceeds the equilibrium concentration by about eight times. When considering only the trimethylpentanes, the 2,2,4 content of isobutene alkylate is very near that for equilibrium (70.37 percent vs 69.77 percent). Agreement for the other three trimethylpentanes on this basis is poor. The conclusion is that the alkylation reactions are quite specific, and that isomerization of alkylation products is minor.

Dimethylhexane Formation. Dimethylhexane formation is believed to result largely from reactions of butene-I. This includes (I) codimerization of butene-I and isobutene, (2) dimerization of butene-I, and (3) dimerization of isobutene, and (4) isomerization of dimethylhexyl carbonium ions; each of these reactions is followed by abstraction of a hydride ion from isobutane. Reactions follow:

Codimerization

(12)

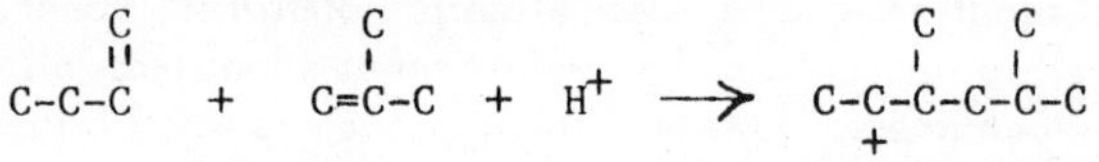

$$\underset{C-C-C}{\overset{\overset{\displaystyle C}{\parallel}}{}} \;+\; \underset{C=C-C}{\overset{\overset{\displaystyle C}{\mid}}{}} \;+\; H^+ \;\longrightarrow\; \underset{\underset{+}{C-C-C-C-C}}{\overset{\overset{\displaystyle C\;\;\;C}{\mid\;\;\;\mid}}{}}$$

(12-A)

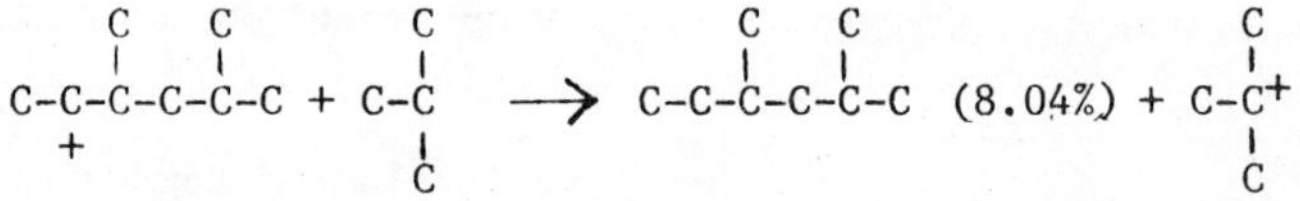

$$\underset{\underset{+}{C-C-C-C-C}}{\overset{\overset{\displaystyle C\;\;\;C}{\mid\;\;\;\mid}}{}} + \underset{\underset{\displaystyle C}{\mid}}{\overset{\displaystyle C-C}{}} \;\longrightarrow\; \underset{C-C-C-C-C-C}{\overset{\overset{\displaystyle C\;\;\;C}{\mid\;\;\;\mid}}{}}\ (8.04\%) + \underset{\underset{\displaystyle C}{\mid}}{C-C^+}$$

Dimerization

(13)

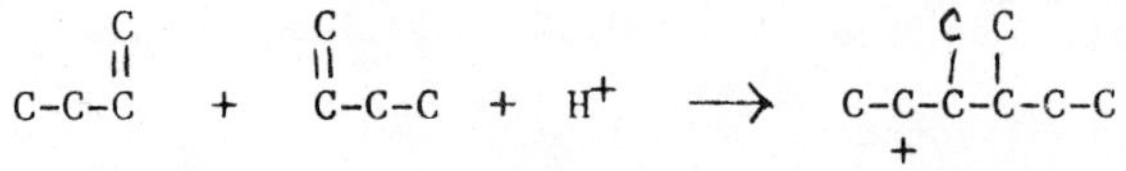

$$\underset{C-C-C}{\overset{\overset{\displaystyle C}{\parallel}}{}} \;+\; \underset{C-C-C}{\overset{\overset{\displaystyle C}{\parallel}}{}} \;+\; H^+ \;\longrightarrow\; \underset{\underset{+}{C-C-C-C-C-C}}{\overset{\overset{\displaystyle C\;C}{\diagup\;\mid}}{}}$$

(13-A)

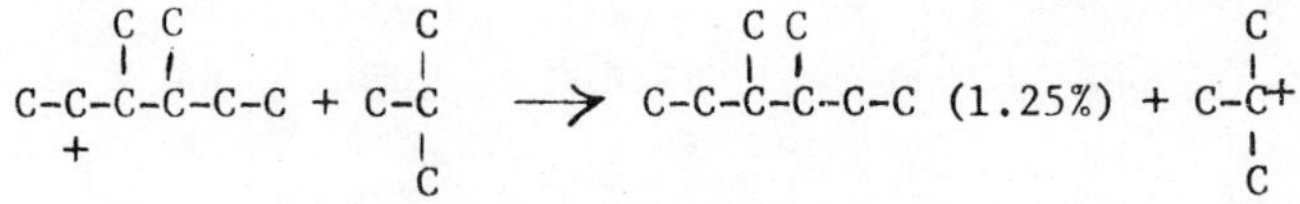

$$\underset{\underset{+}{C-C-C-C-C}}{\overset{\overset{\displaystyle C\;C}{\mid\;\diagup}}{}} + \underset{\underset{\displaystyle C}{\mid}}{\overset{\displaystyle C-C}{}} \;\longrightarrow\; \underset{C-C-C-C-C-C}{\overset{\overset{\displaystyle C\;C}{\mid\;\mid}}{}}\ (1.25\%) + \underset{\underset{\displaystyle C}{\mid}}{C-C^+}$$

(14)

$$\underset{C-C-C}{\overset{\overset{\displaystyle C}{\parallel}}{}} \;+\; \underset{C-C-C}{\overset{\overset{\displaystyle C}{\parallel}}{}} \;+\; H^+ \;\longrightarrow\; \underset{\underset{+}{C-C-C-C-C-C}}{\overset{\overset{\displaystyle C\;\;\;\;C}{\mid\;\;\;\;\mid}}{}}$$

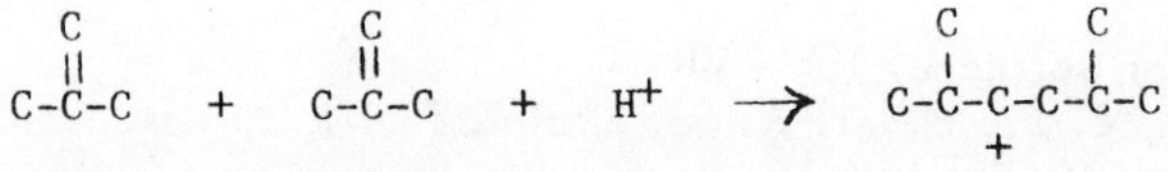

(14-A)

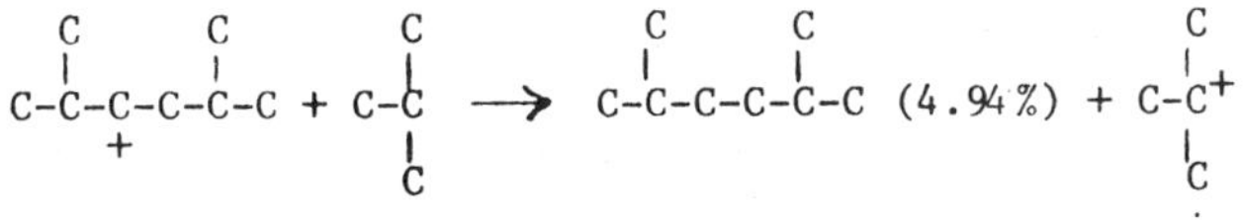

$$\underset{+}{\overset{\displaystyle C \qquad C \qquad\quad C}{C-C-C-C-C-C}} + C-\overset{\displaystyle C}{\underset{\displaystyle C}{C}} \longrightarrow \overset{\displaystyle C \qquad\quad C}{C-C-C-C-C-C} \ (4.94\%) + C-\overset{\displaystyle C}{\underset{\displaystyle C}{C^+}}$$

Carbonium Ion Isomerization

(15)

$$\underset{+}{\overset{\displaystyle C \quad C}{C-C-C-C-C-C}} \longrightarrow \underset{+}{\overset{\displaystyle C \quad C}{C-C-C-C-C-C}}$$

(15-A)

$$\underset{+}{\overset{\displaystyle C \; C}{C-C-C-C-C-C}} + C-\overset{\displaystyle C}{\underset{\displaystyle C}{C}} \longrightarrow \overset{\displaystyle C \; C}{C-C-C-C-C-C} \ (8.87\%) + C-\overset{\displaystyle C}{\underset{\displaystyle C}{C^+}}$$

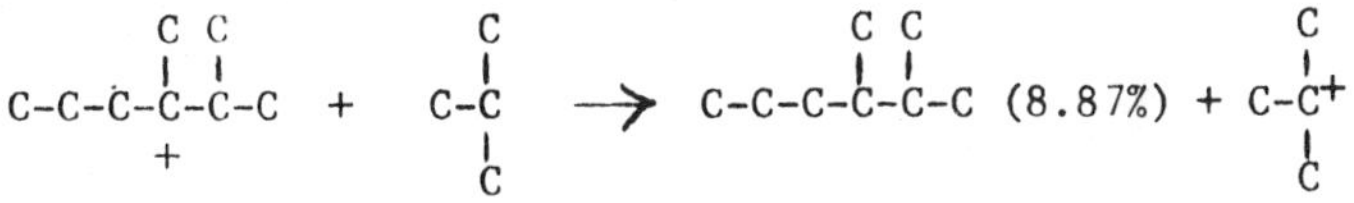

(The above percentage values in parentheses denote concentrations in normalized C_8 fractions as shown in Table VII for butene-1.)

Propylene Reactions. The following reaction mechanisms are gener-ally recognized as the principal ones occurring in propylene–isobutane alkylation with hydrofluoric acid catalyst (Ciapetta, 1945). In parenthe-ses are shown amounts of products from each mechanism; these are from Table VII for propylene:

Primary Alkylation (56.18%)

(16)

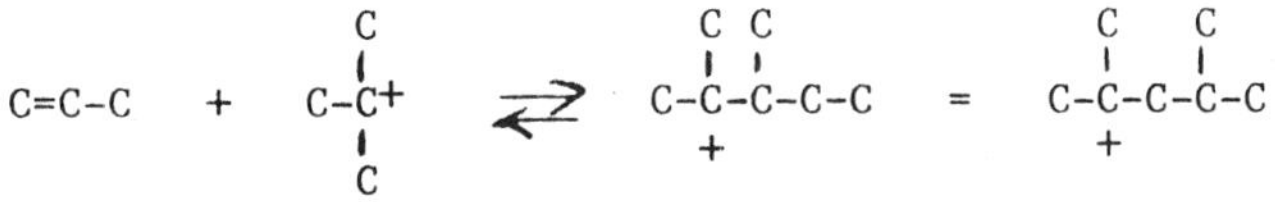

$$C=C-C + C-\overset{\displaystyle C}{\underset{\displaystyle C}{C^+}} \; \rightleftarrows \; \underset{+}{\overset{\displaystyle C \; C}{C-C-C-C-C}} = \overset{\displaystyle C \quad C}{C-C-C-C-C}$$

(16-A)

$$\underset{+}{\overset{\displaystyle C \; C}{C-C-C-C-C}} + C-\overset{\displaystyle C}{\underset{\displaystyle C}{C}} \longrightarrow \overset{\displaystyle C \; C}{C-C-C-C-C} + C-\overset{\displaystyle C}{\underset{\displaystyle C}{C^+}}$$

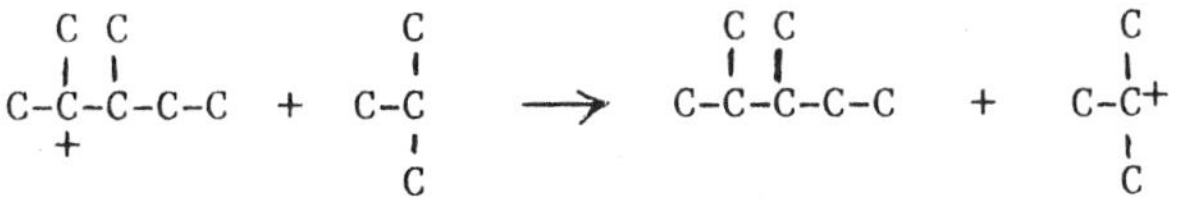

(A reaction similar to 16-A could take place with the 2,4-dimethylpentyl ion.) The overall reaction for primary alkylation may be written:

(16-B) $C_3H_6 + iC_4H_{10} = iC_7H_{16}$

Hydrogen Transfer (24.50%)

(17) $C=C-C$ + H^+ $\rightleftharpoons$ $\underset{+}{C-C-C}$

(17-A) $\underset{+}{C-C-C}$ + $\underset{\underset{C}{|}}{\overset{\overset{C}{|}}{C-C}}$ $\rightarrow$ $C-C-C$ + $\underset{\underset{C}{|}}{\overset{\overset{C}{|}}{C-C^+}}$

(17-B) $\underset{\underset{C}{|}}{\overset{\overset{C}{|}}{C-C^+}}$ $\rightleftharpoons$ $\underset{\underset{C}{|}}{\overset{\overset{C}{||}}{C-C}}$ + H^+

(17-C) $\underset{\underset{C}{|}}{\overset{\overset{C}{||}}{C-C}}$ + $\underset{\underset{C}{|}}{\overset{\overset{C}{|}}{C-C^+}}$ $\rightleftharpoons$ $\underset{\underset{C}{|}}{\overset{\overset{C}{|}}{C-C-C-\underset{+}{C}-C}}$

(17-D) $\underset{\underset{C}{|}}{\overset{\overset{C}{|}}{C-C-C-\underset{+}{C}-C}}$ + $\underset{\underset{C}{|}}{\overset{\overset{C}{|}}{C-C}}$ $\rightarrow$ $\underset{\underset{C}{|}}{\overset{\overset{C}{|}}{C-C-C-C-C}}$ + $\underset{\underset{C}{|}}{\overset{\overset{C}{|}}{C-C^+}}$

The overall reaction for hydrogen transfer may be written:

(17-E) C_3H_6 + $2iC_4H_{10}$ = C_3H_8 + iC_8H_{18}

Polymerization (6.92%)

(18) $2\ C=C-C$ + $\underset{\underset{C}{|}}{\overset{\overset{C}{|}}{C-C^+}}$ $\rightleftharpoons$ $iC_{10}H_{21}^+$

(18-A) $iC_{10}H_{21}^+$ + $\underset{\underset{C}{|}}{\overset{\overset{C}{|}}{C-C}}$ $\rightarrow$ $iC_{10}H_{22}$ + $\underset{\underset{C}{|}}{\overset{\overset{C}{|}}{C-C^+}}$

The overall polymerization reaction may be written:

(18-B) $2C_3H_6$ + iC_4H_{10} = $iC_{10}H_{22}$

Thus, the polymer molecule is made up of two (or more) molecules of propylene plus one molecule of isobutane. Cracking or scission of C_{10}

or larger carbonium ions, producing an olefin plus a smaller carbonium ion, can account for many additional compounds commonly seen in propylene alkylates in small amounts.

Disproportionation (Isopentane = 6.11%) (Only one example is shown.)

$$
(19) \quad
\begin{array}{c} C \\ | \\ C-C \\ | \\ C \end{array}
\;+\;
\begin{array}{c} C \quad C \\ | \quad\; | \\ C-C-C-C-C \end{array}
\;=\;
\begin{array}{c} C \\ | \\ C-C-C-C \end{array}
\;+\;
\begin{array}{c} C \\ | \\ C-C-C-C-C \end{array}
$$

Although this reaction is shown in summary form, it is believed to occur via a carbonium ion mechanism.

Dimerization (Isohexanes = 4.38%)

$$(20) \quad 2\,C_3H_6 \;+\; H^+ \;\rightleftharpoons\; C_6H_{13}{}^+$$

$$(20\text{-}A) \quad C_6H_{13}{}^+ \;+\; iC_4H_{10} \;\longrightarrow\; C_6H_{14} \;+\; iC_4H_9{}^+$$

$$(20\text{-}B) \quad iC_4H_9{}^+ \;\rightleftharpoons\; C_4H_8 \;+\; H^+$$

$$(20\text{-}C) \quad iC_4H_9{}^+ \;+\; C_4H_8 \;\rightleftharpoons\; C_8H_{17}{}^+$$

$$(20\text{-}D) \quad C_8H_{17}{}^+ \;+\; iC_4H_{10} \;\longrightarrow\; C_8H_{18} \;+\; iC_4H_9{}^+$$

The overall reaction for propylene dimerization may be written:

$$(20\text{-}E) \quad 2\,C_3H_6 \;+\; 2\,iC_4H_{10} \;\longrightarrow\; C_6H_{14} \;+\; C_8H_{18}$$

Thus, propylene dimerization could result in the formation of both isohexanes and isooctanes. However, little propylene dimerization is thought to take place since it should result in the formation of mainly 2,3-dimethylbutane; a small amount of this isohexane is found in propylene alkylate.

Cracking

$$(21) \quad 2\,C_3H_6 \;+\; iC_4H_9{}^+ \;\rightleftharpoons\; iC_{10}H_{21}{}^+$$

(21-A) $iC_{10}H_{21}+$ $\longrightarrow$ iC_5H_{10} + $iC_5H_{11}+$

Cracking or scission of large ions is believed to be of minor significance.
Both of the products (there are others) are subject to further reaction
under alkylation conditions and can account for a number of isoparaffins
commonly formed in propylene alkylate in minor amounts.

Examination of C_8 fractions from propylene and isobutene alkylates
(see Table I) shows a great degree of similarity. Data are summarized in
Table II.

TABLE II. SIMILARITY OF C_8 FRACTIONS IN ISOBUTENE

AND PROPYLENE ALKYLATES

	In Isobutene Alkylate	In Propylene Alkylate
2,2,4 TMP	62.63	64.94
2,5 DMH	4.07	2.88
2,4 DMH	4.45	3.16
2,2,3 TMP	1.34	1.08
2,3,4 TMP	13.35	16.54
2,3,3 TMP	11.67	11.04
2,3 DMH	2.22	0.18
2 M Heptane	-	0.04
3,4 DMH	0.26	0.11
3 M Heptane	-	0.04
	100.00	100.00

The similarity of these C_8 fractions is strong evidence that (1) propylene-
derived octanes come largely from hydrogen transfer reactions (self alkylation of isobutane) and (2) isobutene dimerization is probably an intermediate reaction step for production of 2,5-dimethylhexane in both propylene and butene alkylation. A small quantity of isooctane could also be

formed in the hydrogen transfer reactions associated with propylene dimerization. (see the reaction 20 group.)

Effects of Water in HF Catalyst. A number of investigators have pointed out that water has an important role in alkylation catalysts. Schmerling (1955) stated that the use of HF catalyst with one percent water produced a favorable result in propylene-isobutane alkylation, whereas, with
a catalyst containing ten percent water, isopropyl fluoride was the principal product and no alkylate was formed. (Both reactions were at 25C.)
Albright et al. (1972) found the water content of sulfuric acid to be
"highly important" in affecting the quality and yield of butene-isobutane
alkylate. They postulated that the water content of sulfuric acid controlled the level of ionization and hydride transfer rate in the catalyst
phase. It appears that dissolved water affects HF alkylation catalyst
similarly and also exerts further physical influence on the catalyst phase
such as reducing viscosity, interfacial tension, and isobutane solubility.

Alkylation tests were conducted in the pilot plant with a cat cracker
mixed olefin feed. As the water in HF was increased from 0.25 to about
2.8 percent, the alkylate composition changed dramatically. These
changes are summarized in Table III; detailed alkylate compositions are
given in Table VIII.

TABLE III.   EFFECT OF WATER ON ALKYLATE COMPOSITION[a]

| Water in | Alkylate Composition, weight per cent | | | | Residue + |
HF, wt %	C_5	C_6	C_7	C_8	C_9
0.25	10.39	4.94	17.76	48.75	18.16
0.43	5.51	3.83	19.43	58.30	12.87
0.6	4.57	2.89	19.16	60.60	12.79
0.9	3.62	2.63	24.29	60.39	9.07
1.4	2.89	2.35	23.29	62.83	8.63
2.0	3.51	2.54	26.35	60.10	7.50
2.8	3.72	2.88	25.48	61.43	6.49
4.4	4.28	3.44	28.40	56.56	7.32

(a) Feed composition, liquid % (paraffin-free basis): propylene 42.5,
butenes 57.2, pentanes 0.3.

Referring to Table III, the C_8's appear to have peaked when catalyst contained near 1.4 percent water. However, residue, which has an octane
number of only about 60 (RON + 3) exerted a strong influence on alkylate quality, and the best quality alkylate resulted when catalyst water
content was about 2.8 percent. As discussed previously, residue is be -
lieved to result from excessive olefin polymerization prior to hydride abstraction. Earlier work characterized residue as having a molecular weight
of about 178-196; it could therefore be considered a trimer or cotrimer
when alkylating isobutane and propylene-butene mixtures. Typically, an
overall reaction as described in an earlier paragraph would be: $2C_4H_8$ +
$iC_4H_{10} = iC_{12}H_{26}$. Thus, the effect of water in HF alkylation catalyst
appears to be that of slowing polymerization (trimerization) and reducing
residue; proper water control should be given serious consideration in
order to maximize alkylate quality and yield.

Isobutane-to-Olefin Ratio with Propylene Feed. The isobutane-to-
olefin ratio has long been recognized as an important process variable in
the alkylation of isobutane with either butenes or propylene (Phillips
Petroleum Company, 1946). By maintaining a sufficiently high concentration of isobutane in the reaction zone, the abstraction of hydride ions
from isobutane is favored over abstraction from product isoparaffins.
Even with propylene feed, a high isobutane-to-olefin ratio influences the
product toward predominantly C_8 hydrocarbons which have the highest
octane number and also improves yields. Thus, both alkylate quality and
yield are found to improve with increasing ratio and olefin dilution. In
Table IX, detailed propylene-isobutane alkylate composition data are
shown, where the volume ratio was increased from 4.6 to 126. For quick
reference, composition data are summarized in Table IV.

The C_7-fraction (direct propylene-isobutane alkylation product) decreased from 55.11 to 40.49 percent, while the C_8-fraction (hydrogen
transfer product) increased from 19.52 to 54.1 percent. At the same time,
the concentration of residue decreased from 15.69 to only 2.06 percent,
indicating a substantial reduction in the role of olefin polymerization as
olefin dilution with isobutane was increased.
        Considering the normalized C_8-fractions (Table IX), the trimethyl-
pentane content of the product increased from 86.5 to 95.9 percent as the
ratio was increased. Higher isobutane ratios suppressed the formation of
isopentane and isohexanes, which are believed to result from disproportionation or scission of large polymer ions and (to some extent) propylene
dimerization.

TABLE IV.
EFFECT OF ISOBUTANE-TO-OLEFIN RATIO ON ALKYLATE COMPOSITION

	Composition, wt %				Residue +
Ratio	C_5	C_6	C_7	C_8	C_9
4.6	5.09	4.55	55.11	19.52	15.69
13	3.56	2.95	48.78	32.23	11.95
16	4.82	2.90	45.79	34.01	10.99
22	3.69	2.63	48.84	36.22	8.47
52	2.22	2.08	45.50	45.13	3.57
109	2.24	2.02	42.58	49.54	3.32
126	1.36	1.72	40.49	54.19	2.06

Propylene-Isobutane Alkylation with Propane Added. The principal effect of adding a straight-chain paraffin to alkylation feed appears to be a decrease in the percentage of the olefin going into chain initiation. This, in turn, decreases the percentage of the alkylate formed via chain termination, thereby reducing the amount of trimethylpentanes formed. Specifically, when the feed concentration of isobutane is lowered by propane dilution, the addition of propylene to tertiary butyl carbonium ions to produce C_7 carbonium ions (precursors of isoheptanes) is favored over hydrogen transfer reactions.

The experimental studies reported here in Tables VII-XI indicate that as the propane-to-propylene ratio in alkylation feed was increased from 0 to 3.6/1, alkylate yield and isobutane consumption decreased significantly. The main effect on alkylate composition was an increase in the isoheptanes at the expense of the trimethylpentanes. Since, with propylene feed, trimethylpentane formation is a result mainly of hydrogen transfer reactions, the synthetic propane would be expected to decrease. This decrease was observed. A similar result occurred when normal butane was used as a diluent in alkylation feed.

HF versus H_2SO_4 Alkylates. The quality and yield advantages of HF alkylate over H_2SO_4 alkylate made from mixed olefin are attested in the literature (NPRA, 1973). HF alkylates typically contain higher concentrations of high-octane C_8's and lesser amounts of low-octane C_9^+ material. Volumetric yields based on olefin feed are generally higher for HF alkylates because of their lower density.

Marked differences are observed in component distribution of isobu-
tane-mixed olefin alkylates prepared with HF and H_2SO_4 catalysts. The
alkylate samples in Table V were prepared under typical operating conditions for the respective processes from similar feedstocks.

Significant composition advantages which HF alkylate shows over
H_2SO_4 alkylate are: (l) Lower residue (C_9^+) content (6.49 vs l2.85 percent). This is believed to be due to a lesser amount of polymerization
occurring with HF catalyst. (2) Higher C_8-fraction content (6l.43 vs
45.32 percent) and lower heptane content (25.48 vs 30.84 percent).   HF
catalyst favors a greater amount of hydrogen transfer reaction from propylene, resulting in high yields of isooctanes and a lesser amount of direct
propylene-isobutane alkylate. (3) Lesser amounts of C_5 and C_6 material
(6.60 vs l0.99 percent). Examination of the normalized C_8-fraction (see
Table IV) shows the HF and the H_2SO_2-catalyzed products to contain
about the same amounts of total trimethylpentanes (88.30 vs 86.40 percent). (4) Lesser amounts of dimethylhexanes (ll.6l vs l3.88 percent).
The octane advantage in favor of HF alkylate was approximately l.5 RON.

TABLE V.   COMPARISON OF HF- AND H_2SO_4-CATALYZED ALKYLATES

Olefin Feed Composition (liquid volume %)

Olefin	HF	H_2SO_4
C_3H_6	42.5	43.5
C_4H_8	57.2	54.9
C_5H_{10}	0.3	1.6
	100.0	100.0

Alkylate Composition (weight %)[(a)]

Fraction	HF	H_2SO_4
C_5	3.72	5.34
C_6	2.88	5.65
C_7	25.48	30.84
C_8	61.43	45.32
C_9^+	6.49	12.85
	100.00	100.00

(a) Detailed alkylate analyses are shown in Table X; compositions are
     summarized.

Hydrocarbon Disperson. Hydrocarbon dispersion (in HF catalyst) was found to be an important factor in producing alkylates of different composition and quality from the same olefin under the same reaction conditions. As the degree of dispersion was changed from poor to excellent by improving mass transfer, alkylate composition underwent drastic changes. These changes are summarized in Table VI. (Detailed alkylate compositions are in Table XI).

TABLE VI.   EFFECT OF DISPERSION ON ALKYLATE COMPOSITION

Dispersion	Alkylate Composition	
	Poor	Excellent
Isopentane	14.04%	7.31%
Isohexanes	7.65	3.61
Isoheptanes	22.30	22.15
2,2,5-Trimethylhexane	4.65	0.69
Trimethylpentanes	27.17	50.74
Residue	14.79	5.87

Thus, good dispersion or mass transfer favors olefin isomerization (to isobutene), isobutene dimerization, and maximizes hydrogen transfer and primary alkylation reactions, i.e., yielding the greatest amount of high-octane-number trimethylpentanes, and minimizing low-octane-number byproducts from secondary reactions such as excess polymerization.

There seems a strong possibility that olefin isomerization and dimerization could take place in the catalyst phase and that alkylation could take place in the hydrocarbon phase. When hydrocarbon dispersion is poor, i.e., droplets are large, a mass transfer limitation exists, and olefin isomerization and dimerization are reduced. This results in a product containing large amounts of residue, which may be due to secondary alkylation of, say, an isobutene molecule with an isooctyl carbonium ion to produce a dodecyl carbonium ion which can then undergo hydride transfer or scission. Under conditions of good hydrocarbon dispersion (small droplets with large amount of surface area), the reactions yield less heavies and scission products and more of the desirable trimethylpentanes.

TABLE VII
EFFECT OF OLEFIN FEED

| | Isobutene | | | Butene-1 | | |
	Wt %	Mol %	Normalized	Wt %	Mol %	Normalized
Isopentane	5.43	8.43	99.09	2.67	4.25	99.26
n-Pentane	0.05	0.08	0.91	0.02	0.03	0.74
2,2-DMB	0.0	0.0	0.0	0.0	0.0	0.0
2,3-DMB	2.02	2.63	64.95	1.07	1.43	62.94
2MPentane	0.86	1.12	27.65	0.48	0.64	28.24
3MPentane	0.23	0.30	7.40	0.15	0.20	8.82
n-Hexane	0.0	0.0	0.0	0.0	0.0	0.0
2,2-DMP	0.0	0.0	0.0	0.0	0.0	0.0
2,4-DMP	2.39	2.67	62.40	1.60	1.83	58.39
Triptane	0.0	0.0	0.0	0.0	0.0	0.0
3,3-DMP	0.0	0.0	0.0	0.0	0.0	0.0
2MHexane	0.13	0.15	3.39	0.14	0.16	5.11
2,3-DMP	1.18	1.32	30.81	0.80	0.92	29.20
3MHexane	0.13	0.15	3.39	0.20	0.23	7.30
2,2,4-TMP	50.12	49.18	62.63	39.67	39.86	46.34
2,5-DMH	3.26	3.20	4.07	4.23	4.25	4.94
2,4-DMH	3.56	3.49	4.45	6.88	6.91	8.04
2,2,3-TMP	1.07	1.05	1.34	1.26	1.27	1.47
2,3,4-TMP	10.68	10.48	13.35	13.33	13.39	15.57
2,3,3-TMP	9.34	9.16	11.67	11.37	11.42	13.28
2,3-DMH	1.78	1.75	2.22	7.59	7.63	8.87
2MHeptane	0.0	0.0	0.0	0.20	0.20	0.23
3,4-DMH	0.21	0.21	0.26	1.07	1.08	1.25
3MHeptane	0.0	0.0	0.0	0.0	0.0	0.0
2,2,5-TMH	1.31	1.14	100.00	0.40	0.36	100.00
Residue	6.25	3.50	100.00	6.88	3.95	100.00

Table VII (Cont.)

	cis-Butene-2			trans-Butene-2			Propylene	
Wt %	Mol %	Normalized	Wt %	Mol %	Normalized	Wt %	Mol %	Normalized
2.21	3.46	98.22	2.52	3.95	95.82	4.10	6.11	100.00
0.04	0.06	1.78	0.11	0.17	4.18	0.0	0.0	0.0
0.0	0.0	0.0	0.0	0.0	0.0	0.0	0.0	0.0
1.15	1.51	63.19	1.18	1.55	65.19	2.31	2.88	65.81
0.52	0.68	28.57	0.48	0.63	26.52	0.99	1.24	28.21
0.15	0.20	8.24	0.15	0.20	8.29	0.21	0.26	5.98
0.0	0.0	0.0	0.0	0.0	0.0	0.0	0.0	0.0
0.0	0.0	0.0	0.0	0.0	0.0	0.02	0.02	0.04
1.59	1.79	57.61	1.58	1.78	56.03	14.78	15.86	28.23
0.0	0.0	0.0	0.0	0.0	0.0	0.0	0.0	0.0
0.0	0.0	0.0	0.0	0.0	0.0	0.0	0.0	0.0
0.17	0.19	6.16	0.20	0.23	7.09	0.20	0.21	0.38
0.83	0.94	30.07	0.84	0.95	29.79	36.70	39.38	70.09
0.17	0.19	6.16	0.20	0.23	7.09	0.66	0.71	1.26
49.03	48.51	54.28	48.60	48.07	54.36	18.06	17.00	64.94
2.87	2.84	3.18	2.91	2.88	3.25	0.80	0.75	2.88
3.53	3.49	3.91	3.52	3.48	3.94	0.88	0.83	3.16
1.27	1.26	1.41	1.27	1.26	1.42	0.30	0.28	1.08
18.39	18.19	20.36	18.08	17.88	20.22	4.60	4.33	16.54
12.47	12.34	13.80	12.38	12.24	13.85	3.07	2.89	11.04
2.50	2.47	2.77	2.41	2.38	2.70	0.05	0.05	0.18
0.0	0.0	0.0	0.0	0.0	0.0	0.01	0.01	0.04
0.27	0.27	0.30	0.24	0.24	0.27	0.03	0.03	0.11
0.0	0.0	0.0	0.0	0.0	0.0	0.01	0.01	0.04
0.17	0.15	100.00	0.24	0.21	100.00	0.28	0.23	100.00
2.66	1.46	100.00	3.09	1.69	100.00	11.95	6.92	100.00

TABLE VIII
EFFECT OF WATER IN HF CATALYST
Mixed C_3-C_4 Olefins

Water in HF, Wt %

	0.25			0.43			0.6			0.9		
	Wt %	Mol %	Normalized	Wt %	Mol %	Normalized	Wt %	Mol %	Normalized	Wt %	Mol %	Normalized
Isopentane	10.35	15.70	99.62	5.51	8.50	98.92	4.56	7.10	99.78	3.60	5.54	99.45
n-Pentane	0.04	0.06	0.38	0.06	0.09	1.08	0.01	0.02	0.22	0.02	0.03	0.55
2,2-DMB	0.04	0.05	0.81	0.02	0.03	0.52	0.01	0.01	0.35	0.01	0.01	0.38
2,3-DMB	2.42	3.07	48.99	1.92	2.48	50.13	1.47	1.92	50.87	1.60	2.06	60.84
2MPentane	1.74	2.21	35.22	1.30	1.68	33.94	1.00	1.30	34.60	0.73	0.94	27.76
3MPentane	0.67	0.85	13.56	0.51	0.66	13.32	0.37	0.48	12.80	0.28	0.36	10.65
n-Hexane	0.07	0.09	1.42	0.08	0.10	2.09	0.04	0.05	1.38	0.01	0.01	0.38
2,2-DMP	0.01	0.01	0.06	0.01	0.01	0.05	0.01	0.01	0.05	0.01	0.01	0.04
2,4-DMP	6.52	7.12	36.71	7.53	8.36	38.75	7.58	8.50	39.56	8.08	8.96	33.26
Triptane	0.12	0.13	0.68	0.09	0.10	0.46	0.10	0.11	0.52	0.10	0.11	0.41
3,3-DMP	0.02	0.02	0.11	0.03	0.03	0.15	0.01	0.01	0.05	0.01	0.01	0.04
2MHexane	0.73	0.80	4.11	0.62	0.69	3.19	0.49	0.55	2.56	0.35	0.39	1.44
2,3-DMP	9.66	10.55	54.39	10.45	11.61	53.78	10.39	11.65	54.23	15.21	16.86	62.62
3MHexane	0.70	0.76	3.94	0.70	0.78	3.60	0.58	0.65	3.03	0.53	0.59	2.18
2,2,4-TMP	28.00	23.96	51.28	32.43	31.60	55.63	35.80	35.20	59.06	35.77	34.78	59.23
2,5-DMH	3.51	3.36	7.20	3.26	3.18	5.59	2.51	2.47	4.14	1.89	1.84	3.13
2,4-DMH	3.88	3.72	7.96	3.82	3.72	6.55	3.14	3.09	5.18	2.67	2.60	4.42
2,2,3-TMP	2.05	1.96	4.21	1.93	1.88	3.31	1.49	1.47	2.46	1.14	1.11	1.89
2,3,4-TMP	4.67	4.47	9.58	5.92	5.77	10.15	6.97	6.85	11.50	8.44	8.21	13.98
2,3,3-TMP	5.86	5.62	12.02	7.21	7.03	12.37	7.42	7.30	12.24	7.22	7.02	11.96
2,3-DMH	2.86	2.74	5.87	2.95	2.87	5.06	2.57	2.53	4.24	2.70	2.63	4.47
2MHeptane	0.25	0.24	0.51	0.16	0.16	0.27	0.15	0.15	0.25	0.08	0.08	0.13
3,4-DMH	0.44	0.42	0.90	0.44	0.43	0.75	0.44	0.43	0.73	0.42	0.41	0.70
3MHeptane	0.23	0.22	0.47	0.18	0.18	0.31	0.13	0.13	0.21	0.06	0.06	0.10
2,2,5-TMH	2.81	2.40	100.00	1.35	1.17	100.00	1.11	0.97	100.00	0.68	0.59	100.00
Residue	15.35	9.45	100.00	11.52	6.90	100.00	11.68	7.06	100.00	8.39	4.81	100.00

	1.4			2.0			2.8			4.4		
	Wt %	Mol %	Normalized	Wt %	Mol %	Normalized	Wt %	Mol %	Normalized	Wt %	Mol %	Normalized
	2.88	4.46	99.65	3.49	5.33	99.43	3.72	5.66	100.00	4.28	6.47	100.00
	0.01	0.02	0.35	0.02	0.03	0.57	0.0	0.0	0.0	0.0	0.0	0.0
	0.0	0.0	0.0	0.02	0.03	0.79	0.0	0.0	0.0	0.0	0.0	0.0
	1.42	1.84	60.43	1.62	2.07	63.78	1.94	2.47	67.36	2.33	2.95	67.73
	0.65	0.84	27.66	0.63	0.81	24.80	0.67	0.85	23.26	0.77	0.97	22.38
	0.24	0.31	10.21	0.24	0.31	9.45	0.26	0.33	9.03	0.32	0.41	9.30
	0.04	0.05	1.70	0.03	0.04	1.18	0.01	0.01	0.35	0.02	0.03	0.58
	0.0	0.0	0.0	0.01	0.01	0.04	0.0	0.0	0.0	0.0	0.0	0.0
	8.20	9.14	35.21	7.35	8.09	27.89	7.61	8.33	29.87	7.46	8.12	26.27
	0.08	0.09	0.34	0.12	0.13	0.46	0.08	0.09	0.31	0.11	0.12	0.39
	0.0	0.0	0.0	0.03	0.03	0.11	0.0	0.0	0.0	0.0	0.0	0.0
	0.32	0.36	1.37	0.26	0.29	0.99	0.29	0.32	1.14	0.28	0.30	0.99
	14.23	15.86	61.10	18.12	19.94	68.77	17.08	18.70	67.03	20.12	21.90	70.85
	0.46	0.51	1.98	0.46	0.51	1.75	0.42	0.46	1.65	0.43	0.47	1.51
	38.52	37.66	61.31	35.93	34.69	59.78	35.61	34.21	57.97	32.27	30.82	57.05
	1.78	1.74	2.83	1.43	1.38	2.38	1.64	1.58	2.67	1.47	1.40	2.60
	2.49	2.43	3.96	2.16	2.09	3.59	2.37	2.28	3.86	2.12	2.02	3.75
	1.09	1.07	1.73	0.94	0.91	1.56	0.99	0.95	1.61	0.86	0.82	1.52
	8.66	8.47	13.78	9.61	9.28	15.99	9.90	9.51	16.12	10.07	9.62	17.80
	7.52	7.35	11.97	7.11	6.86	11.83	7.74	7.43	12.60	7.04	6.72	12.45
	2.33	2.28	3.71	2.46	2.37	4.09	2.76	2.65	4.49	2.40	2.29	4.24
	0.08	0.08	0.13	0.06	0.06	0.10	0.02	0.02	0.03	0.05	0.05	0.09
	0.33	0.32	0.53	0.34	0.33	0.57	0.36	0.35	0.59	0.26	0.25	0.46
	0.03	0.03	0.05	0.06	0.06	0.10	0.04	0.04	0.07	0.02	0.02	0.04
	0.51	0.44	100.00	0.58	0.50	100.00	0.67	0.57	100.00	0.69	0.59	100.00
	8.12	4.66	100.00	6.92	3.86	100.00	5.82	3.19	100.00	6.63	3.66	100.00

TABLE IX
EFFECT OF ISOBUTANE TO OLEFIN RATIO ON ALKYLATE COMPOSITION
Propylene Feed

Ratio	4.6			13			16		
	Wt %	Mol %	Normalized	Wt %	Mol %	Normalized	Wt %	Mol %	Normalized
Isopentane	5.09	7.56	100.00	4.10	6.11	100.00	4.82	7.19	100.00
n-Pentane	0.0	0.0	0.0	0.0	0.0	0.0	0.0	0.0	0.0
2,2-DMB	0.0	0.0	0.0	0.0	0.0	0.0	0.0	0.0	0.0
2,3-DMB	3.06	3.81	67.25	2.31	2.88	65.81	2.67	3.33	68.46
2MPentane	1.12	1.39	24.62	0.99	1.24	28.21	0.90	1.12	23.08
3MPentane	0.36	0.45	7.91	0.21	0.26	5.98	0.33	0.41	8.46
n-Hexane	0.01	0.01	0.22	0.0	0.0	0.0	0.0	0.0	0.0
2,2-DMP	0.01	0.01	0.02	0.02	0.02	0.04	0.01	0.01	0.02
2,4-DMP	14.26	15.25	25.88	14.78	15.86	28.23	15.98	17.16	34.90
Triptane	0.0	0.0	0.0	0.0	0.0	0.0	0.20	0.21	0.44
3,3-DMP	0.01	0.01	0.02	0.0	0.0	0.0	0.0	0.0	0.0
2MHexane	0.45	0.48	0.82	0.20	0.21	0.38	0.50	0.54	1.09
2,3-DMP	39.74	42.50	72.11	36.70	39.38	70.09	28.40	30.49	62.02
3MHexane	0.64	0.68	1.16	0.66	0.71	1.26	0.70	0.75	1.53
2,2,4-TMP	12.43	11.66	63.68	18.06	17.00	64.94	21.73	20.46	63.89
2,5-DMH	0.76	0.71	3.89	0.80	0.75	2.88	1.16	1.09	3.41
2,4-DMH	0.99	0.93	5.07	0.88	0.83	3.16	1.20	1.13	3.53
2,2,3-TMP	0.25	0.23	1.28	0.30	0.28	1.08	0.27	0.25	0.79
2,3,4-TMP	2.34	2.20	11.99	4.60	4.33	16.54	5.33	5.02	15.67
2,3,3-TMP	1.86	1.74	9.53	3.07	2.89	11.04	3.22	3.03	9.47
2,3-DMH	0.69	0.65	3.53	0.05	0.05	0.18	0.80	0.75	2.35
2MHeptane	0.03	0.03	0.15	0.01	0.01	0.04	0.10	0.09	0.29
3,4-DMH	0.13	0.12	0.67	0.03	0.03	0.11	0.20	0.19	0.59
3MHeptane	0.04	0.04	0.20	0.01	0.01	0.04	0.0	0.0	0.0
2,2,5-TMH	0.04	0.03	100.00	0.28	0.23	100.00	0.49	0.41	100.00
Residue	15.69	9.50	100.00	11.95	6.92	100.00	10.99	6.35	100.00

	22			52			109			126		
Wt %	Mol %	Normalized	Wt %	Mol %	Normalized	Wt %	Mol %	Normalized	Wt %	Mol %	Normalized	
3.69	5.49	100.00	3.33	4.91	100.00	2.24	3.33	100.00	1.36	2.03	100.00	
0.0	0.0	0.0	0.0	0.0	0.0	0.0	0.0	0.0	0.0	0.0	0.0	
0.0	0.0	0.0	0.01	0.01	0.48	0.01	0.01	0.50	0.04	0.05	2.33	
1.71	2.13	65.02	1.06	1.31	50.96	0.89	1.11	44.06	0.68	0.85	39.53	
0.70	0.87	26.62	0.61	0.75	29.33	0.54	0.67	26.73	0.38	0.47	22.09	
0.22	0.27	8.37	0.22	0.27	10.58	0.22	0.27	10.89	0.16	0.20	9.30	
0.0	0.0	0.0	0.18	0.22	8.65	0.36	0.45	17.82	0.46	0.57	26.74	
0.02	0.02	0.04	0.03	0.03	0.07	0.04	0.04	0.09	0.04	0.04	0.10	
14.09	15.10	28.85	9.74	10.34	21.41	7.70	8.25	18.08	7.24	7.78	17.88	
0.10	0.11	0.20	0.08	0.08	0.18	0.08	0.09	0.19	0.06	0.06	0.15	
0.0	0.0	0.0	0.01	0.01	0.02	0.01	0.01	0.02	0.01	0.01	0.02	
0.40	0.43	0.82	0.27	0.29	0.59	0.30	0.32	0.70	0.28	0.30	0.69	
33.53	35.94	68.65	34.84	37.00	76.57	33.96	36.38	79.76	32.40	34.82	80.02	
0.70	0.75	1.43	0.53	0.56	1.16	0.49	0.52	1.15	0.46	0.49	1.14	
23.71	22.29	65.46	31.52	29.36	69.84	33.97	31.92	68.57	37.15	35.02	68.56	
0.93	0.87	2.57	0.57	0.53	1.26	0.62	0.58	1.25	0.63	0.59	1.16	
0.90	0.85	2.48	0.66	0.61	1.46	0.70	0.66	1.41	0.70	0.66	1.29	
0.15	0.14	0.41	0.48	0.45	1.06	0.60	0.56	1.21	0.63	0.59	1.16	
6.47	6.08	17.86	6.85	6.38	15.18	8.11	7.62	16.37	9.08	8.56	16.76	
3.16	2.97	8.72	4.20	3.91	9.31	4.65	4.37	9.39	5.08	4.79	9.37	
0.80	0.75	2.21	0.69	0.64	1.53	0.73	0.69	1.47	0.78	0.74	1.44	
0.0	0.0	0.0	0.05	0.05	0.11	0.05	0.05	0.10	0.05	0.05	0.09	
0.10	0.09	0.28	0.10	0.09	0.22	0.09	0.08	0.18	0.07	0.07	0.13	
0.0	0.0	0.0	0.01	0.01	0.02	0.02	0.02	0.04	0.02	0.02	0.04	
0.15	0.13	100.00	0.39	0.32	100.00	0.30	0.25	100.00	0.18	0.15	100.00	
8.47	4.69	100.00	3.57	1.84	100.00	3.32	1.73	100.00	2.06	1.08	100.00	

TABLE X

ALKYLATE COMPOSITIONS - MIXED OLEFIN FEED

	HF Catalyst		
	Wt %	Mol %	Normalized
Isopentane	3.61	5.56	98.90
n-Pentane	0.04	0.06	1.10
2,2-DMB	0.0	0.0	0.0
2,3-DMB	1.50	1.93	67.57
2MPentane	0.52	0.67	23.42
3MPentane	0.20	0.26	9.01
n-Hexane	0.0	0.0	0.0
2,2-DMP	0.0	0.0	0.0
2,4-DMP	6.21	6.89	25.60
Triptane	0.04	0.04	0.16
3,3-DMP	0.01	0.01	0.04
2MHexane	0.19	0.21	0.78
2,3-DMP	17.52	19.43	72.22
3MHexane	0.29	0.32	1.20
2,2,4-TMP	37.85	36.81	61.88
2,5-DMH	1.20	1.17	1.96
2,4-DMH	1.80	1.75	2.94
2,2,3-TMP	0.77	0.75	1.26
2,3,4-TMP	10.41	10.13	17.02
2,3,3-TMP	6.91	6.72	11.30
2,3-DMH	2.00	1.95	3.27
2MHeptane	0.01	0.01	0.02
3,4-DMH	0.20	0.19	0.33
3MHeptane	0.02	0.02	0.03
2,2,5-TMH	0.57	0.49	100.00
Residue	8.11	4.63	100.00

Table X (Cont.)

H₂SO₄ Catalyst		
Wt %	Mol %	Normalized
5.34	8.07	100.00
0.0	0.0	0.0
0.0	0.0	0.0
4.42	5.59	78.23
0.82	1.04	14.51
0.41	0.52	7.26
0.0	0.0	0.0
0.20	0.22	0.65
10.17	11.06	32.98
0.0	0.0	0.0
0.0	0.0	0.0
0.31	0.34	1.01
19.65	21.37	63.72
0.51	0.55	1.65
19.23	18.35	42.43
1.85	1.77	4.08
1.64	1.56	3.62
1.23	1.17	2.71
8.94	8.53	19.73
9.76	9.31	21.54
2.16	2.06	4.77
0.10	0.10	0.22
0.41	0.39	0.90
0.0	0.0	0.0
1.75	1.49	100.00
11.10	6.51	100.00

TABLE XI

EFFECTS OF DISPERSION

Mixed Olefin Feed

		Excellent	
	Wt %	Mol %	Normalized
Isopentane	7.31	10.92	100.00
n-Pentane	0.0	0.0	0.0
2,2-DMB	0.0	0.0	0.0
2,3-DMB	2.36	2.95	65.37
2MPentane	0.91	1.14	25.21
3MPentane	0.34	0.43	9.42
n-Hexane	0.0	0.0	0.0
2,2-DMP	0.02	0.02	0.09
2,4-DMP	9.26	9.96	41.81
Triptane	0.10	0.11	0.45
3,3-DMP	0.0	0.0	0.0
2MHexane	0.48	0.52	2.17
2,3-DMP	11.81	12.71	53.32
3MHexane	0.48	0.52	2.17
2,2,4-TMP	32.32	30.51	53.54
2,5-DMH	3.42	3.23	5.67
2,4-DMH	3.42	3.23	5.67
2,2,3-TMP	0.90	0.85	1.49
2,3,4-TMP	8.63	8.15	14.30
2,3,3-TMP	8.89	8.39	14.73
2,3-DMH	2.31	2.18	3.83
2MHeptane	0.0	0.0	0.0
3,4-DMH	0.48	0.45	0.80
3MHeptane	0.0	0.0	0.0
2,2,5-TMH	0.69	0.58	100.00
Residue	5.87	3.16	100.00

Table XI (Cont.)

Poor		
Wt %	Mol %	Normalized
14.04	20.58	98.53
0.21	0.31	1.47
0.0	0.0	0.0
5.23	6.42	68.37
1.74	2.13	22.75
0.68	0.83	8.89
0.0	0.0	0.0
0.03	0.03	0.13
8.90	9.39	39.91
0.10	0.11	0.45
0.0	0.0	0.0
0.48	0.51	2.15
12.31	12.99	55.20
0.48	0.51	2.15
15.66	14.50	43.09
3.19	2.95	8.78
3.19	2.95	8.78
0.64	0.59	1.76
5.45	5.04	15.00
5.42	5.02	14.91
2.31	2.14	6.36
0.0	0.0	0.0
0.48	0.44	1.32
0.0	0.0	0.0
4.65	3.83	100.00
14.79	8.73	100.00

References

Albright, L. F., Houle, L., Sumatka, A. M., Eckert, R. E.,"Ind. Eng.
    Chem. Process Des. Develop", II, 446 (1972).
Albright, L. F., Li, K. W., "Ind. Eng. Chem. Process Des. Develop.,"
    9, 447 (1970).
Ciaipetta, F. G., "Ind. Eng. Chem.", 37, 1210 (1945).
Cupit, C. R., Gwyn, J. E., Jernigan, E. C., "Petro/Chem. Eng.",
    33 (13), 203 (1961).
Doshi, B., Albright, L. F., "Ind. Eng. Chem. Process Des. Develop.",
    15, 53 (1976).
Grosse, A V., and Linn, C. B., "J. Org. Chem., 3, 26 (1938).
Hofmann, J. E., Schriesheim, A., "J. Amer. Chem. Soc.," 84, 957
    (1962).
Hutson, Jr., T., Logan, R. S., "Hydrocarbon Processing, 54" (9), 107
    (1975).
Miron, S., Lee, R. J., "J. Chem. Eng. Data," 8, 150 (1963).
NPRA's '72 Panel, "Hydrocarbon Processing, 52" (4), 143 (1973).
Phillips Petroleum Company (1946), Hydrofluoric Acid Alkylation,
    Bartlesville, Ok, Phillips, pp. 5, 117, 126, 170, 305-7, 343.
Schaad, R E (1955) in The Chemistry of Petroleum Hydrocarbons
    Vol. III, Brooks, B. T., Kurtz, S. S., Boord, C. E., Schmerling,
    L., Ed. New York, N.Y., Reinhold, Chapter 48.
Scharfe, G., "Hydrocarbon Processing", 52 (4), 171 (1973).
Schmerling, L., "J. Amer. Chem. Soc.," 68, 275 (1946).
Schmerling, L. (1955) in The Chemistry of Petroleum Hydrocarbons
    Vol. III, Brooks, B. T., Kurtz, S. S., Boord, C. E., Schmerling,
    L., Ed.,New York, N.Y., Reinhold, Chapter 54.
Sparks, W. J., Rosen, R., Frolich, P. K., "Trans. Faraday Soc", 35
    1040 (1939).
Whitmore, F. C., "Ind. Eng. Chem.", 26, 94 (1934).

Fluorosulfonic Acid Promoters in HF Alkylation

ROBERT A. INNES

Gulf Research and Development Co., P. O. Drawer 2038, Pittsburgh, PA 15230

High octane blending stock for gasoline is produced by the strong acid catalyzed alkylation of C_3 to C_5 olefins with isobutane. Alkylate typically comprises 10 to 15 percent of the gasoline pool. Although alkylation has been an important refinery process for more than thirty years, substantial improvements can still be made. In particular, refiners would like to increase the octane ratings of their alkylate to help compensate for the removal of lead from gasoline. This can be accomplished by controlling the many side reactions which accompany alkylation. For example, a typical refinery produces 92 RON alkylate from a C_3-C_5 feedstock. It is theoretically possible to produce 95 RON alkylate from the same feedstock by eliminating those side reactions which are detrimental to alkylate quality and encouraging those which are beneficial. Such fine tuning of the alkylation process requires modification of the acid catalyst. This paper describes how the catalytic properties of HF may be enhanced by the addition of minor amounts of trifluoromethanesulfonic acid (CF_3SO_3H) or fluorosulfonic acid (FSO_3H).([1,2])

Experimental Methods

Apparatus and Procedure. Blends of C.P. Grade isobutane with various olefinic feedstocks were reacted in the continuous flow apparatus shown in Figure 1. The reactor was a small stainless steel autoclave equipped with a magnetically driven stirrer. A polyethylene-lined sight glass was employed as a settling vessel for the separation of acid catalyst from hydrocarbon product. Acid blends were stored in a stainless steel bomb. The system was pressurized to 50 psig with nitrogen prior to the start of each run. A portion of the acid blend was then transferred to the reactor using the sight glass to gauge the amount. The acid level was adjusted so that the autoclave would hold approximately equal volumes of acid and hydrocarbon under reaction conditions.

58   INDUSTRIAL AND LABORATORY ALKYLATIONS

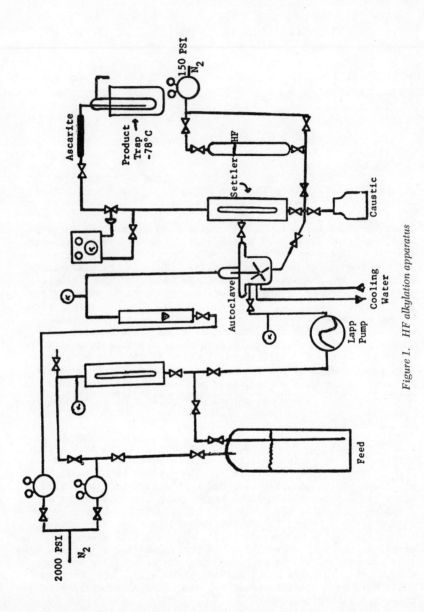

Figure 1. HF alkylation apparatus

The reactor was then pressurized to 200 psig and the run
begun. Isobutane-olefin feed was pumped into the autoclave.
The reactor was stirred at 1600 RPM, intimately mixing the acid
and hydrocarbon phases. Contact times (half the reactor volume
divided by the hydrocarbon feed rate) ranged from 0.3 to 3.0
minutes. The temperature inside the reactor was controlled by
passing water or antifreeze solution from a controlled temperature reservoir through the jacket of the autoclave. The reaction temperature was monitored by a sheathed thermocouple inserted into the autoclave just below the stirrer. A stream of
acid-hydrocarbon emulsion passed continuously from the autoclave
to the settler. The acid catalyst settled to the bottom and
returned to the reactor by gravity flow. The hydrocarbon product passed out the top of the settler through a pressure control valve which maintained the reactor at 200 psig.
     After passage through a bed of Ascarite and alumina beads
to remove dissolved HF, the alkylation product was collected
at -78°C. The collection of the products reported herein was
not begun until the volume of feed pumped equaled 15 times the
reactor volume. At this point the product composition was
changing very slowly, if at all. The duration of a run was
three hours.

     Analytical Methods. Liquefied samples of the feed and re-
actor effluent were analyzed by gas chromatography. All gas
chromatographs were tied to a chromatographic data processing
system which determined peak areas and calculated sample compositions. Sample components were identified on the basis of
their retention times. Response factors were determined experimentally, using synthetic blends resembling actual alkylation feeds and products. Except in those cases where RON was
to be determined on a test engine, n-hexane was added to the
samples as an internal standard.
     All samples were analyzed using a gas chromatograph equip-
ped with a flame ionization detector and a 200' x 0.01"
squalane-coated capillary column. Analyses were also made on
conventional packed columns in chromatographs equipped with
thermal conductivity detectors. Feeds were analyzed on a
silver nitrate-benzyl cyanide column. Products were analyzed
on a 10' x 1/4" column packed with 25% hexatriacontane on
Chromosorb R. After the C$_8$'s had been eluted, the hexatriacontane column was backflushed through the detector to determine the heavies content of the sample.

     Determination of Octane Numbers. In some cases Research
and Motor octane numbers were estimated from alkylate compositions, using the formula,

$$N_{est} = \frac{\Sigma_i W_i N_i}{\Sigma_i W_i}$$

where W_i is the weight of component i and N_i is the ASTM (clear) octane number ($\underline{3},\underline{4}$) of component i in its pure form. The octane ratings of the "heavies" portion of the alkylate were taken as 87 RON and 84 MON.

Otherwise, RON was determined experimentally on a test engine. The reactor effluent was distilled to remove isobutane then submitted to our Product Evaluation Division for determination of RON by the standard ASTM method.

Results and Discussion

Refiners normally charge a mixture of olefins to their alkylation units. However, in studying alkylation, it is often instructive to alkylate individual olefins. Accordingly, to demonstrate the effect of temperature on alkylate composition, we ran the HF alkylation of various pure olefin feeds at 4°C and 45°C. The results are shown in Tables I and II. The contact time was 3.0 minutes for the runs at 4°C and 1.0 minute for those at 45°C. The isobutane-olefin molar ratio was 24 when alkylating diisobutene and 12 when alkylating other olefins. The product names are abbreviated: IP for isopentane, 23DMB for 2,3-dimethylbutane, 224TMP for 2,2,4-trimethylpentane, 23DMH for 2,3-dimethylhexane, and so on.

If alkylation were a selective process, one would expect to obtain 23DMP from propylene, 23DMH from 1-butene, 224TMP from isobutene and diisobutene, and a mixture of TMP's from cis- or trans-2-butene. ($\underline{5},\underline{6},\underline{7}$) These are the products which predominated at 4°C. The many other products listed in Tables I and II are the result of the various side reactions which accompany alkylation. At 45°C, the yield of primary alkylation products was greatly reduced. Alkylation yielded increased amounts of 24DMP and 224TMP from propylene, mixed TMP's from 1-butene, DMH's from the other C_4 olefins, and heavy and light ends from all feedstocks. Thus, as the reaction temperature was increased, side reactions became increasingly important.

From an octane number standpoint TMP's are the most desirable alkylation product, while the formation of DMH's and heavy and light ends should be avoided. When alkylating propylene, it is important to note that 23DMP has a higher octane rating than 24DMP. Accordingly, most side reactions are detrimental to alkylate quality. Two exceptions to this rule are the hydrogen transfer and butene isomerization reactions. Hydrogen transfer converts propylene and isobutane to propane and isobutylene,

Table I

ALKYLATE PRODUCED FROM PURE OLEFIN FEEDS AT 4°C

Olefin	C_3H_6	C_4H_8-1	cis C_4H_8-2	trans C_4H_8-2	iso C_4H_8	244-TMP-1
Component, Wt %						
IP	3.9	1.0	1.7	2.7	4.7	3.7
23 DMB	1.7	0.5	0.8	1.0	1.7	1.5
2 MP	0.9	0.7	0.3	0.3	0.6	0.5
3 MP	0.2	0.1	0.2	0.2	0.3	0.2
23 DMP	56.5	0.2	0.5	0.5	1.2	0.9
24 DMP	5.4	0.5	0.9	1.0	1.7	1.5
Other C_7	0.1	0.3	0.2	0.2	0.3	0.2
223 TMP	0.2	0.9	1.5	1.4	1.2	0.8
224 TMP	11.5	18.7	45.0	42.6	54.7	59.7
233 TMP	1.6	5.5	13.8	13.5	6.5	6.4
234 TMP	2.0	10.2	25.2	25.6	9.6	10.7
23 DMH	0.4	37.7	3.0	2.4	1.3	1.1
24 DMH	0.5	5.9	1.9	2.0	2.7	2.5
25 DMH	0.4	1.1	1.0	1.1	1.7	1.7
34 DMH	0.1	2.7	0.6	0.4	0.2	0.2
225 TMH	0.6	0.2	0.4	0.8	1.8	1.5
Heavies	13.6	13.3	3.2	4.2	9.7	6.8
Est. RON	91.3	84.3	98.6	98.4	96.1	96.9
Est. MON	89.0	85.8	95.9	95.6	94.6	95.4

Table II

ALKYLATE PRODUCED FROM PURE OLEFIN FEEDS AT 45°C

Olefin Component, Wt %	C_3H_6	C_4H_8-1	cis C_4H_8-2	trans C_4H_8-2	iso C_4H_8	244-TMP-1
IP	5.0	4.1	5.6	5.1	6.6	6.6
23 DMB	2.5	1.2	1.9	1.8	2.2	2.1
2 MP	1.2	0.9	0.8	0.8	0.9	0.9
3 MP	0.4	0.3	0.3	0.3	0.3	0.3
23 DMP	33.6	1.0	1.2	1.2	1.3	1.3
24 DMP	19.7	2.1	2.3	2.3	2.5	2.4
Other C_7	1.9	0.6	0.5	0.5	0.5	0.5
223 TMP	0.5	1.2	2.1	2.3	1.8	1.8
224 TMP	20.9	35.4	42.3	44.3	46.4	46.3
233 TMP	2.8	6.0	10.7	11.0	9.1	8.9
234 TMP	3.6	7.2	12.5	12.1	9.8	10.6
23 DMH	0.8	11.8	2.9	3.1	2.4	2.5
24 DMH	1.2	9.7	4.4	4.6	4.1	4.1
25 DMH	1.0	5.0	3.7	3.6	3.5	3.5
34 DMH	0.2	1.4	0.6	0.6	0.5	0.5
225 TMH	0.7	0.8	1.6	1.4	2.0	1.8
Heavies	3.6	11.3	5.8	4.6	5.4	5.7
Est. RON	90.6	88.4	94.6	94.8	94.6	94.6
Est. MON	89.1	88.4	92.6	92.9	92.9	92.9

$$C-C=C + H^+ \longrightarrow C-\overset{+}{C}-C$$

$$C-C-\overset{+}{C} + C-\overset{C}{\underset{C}{C}}-C \longrightarrow C-C-C + C-\overset{+}{\underset{C}{C}}-C$$

$$C-\overset{+}{\underset{C}{C}}-C \longrightarrow C-\overset{C}{\underset{C}{C}}=C + H^+$$

while butene isomerization converts 1-butene to 2-butene.

$$C-C-C=C + H^+ \longrightarrow C-C-\overset{+}{C}-C$$

$$C-C-\overset{+}{C}-C \longrightarrow C-C=C-C + H^+$$

The product olefins yield TMP's when alkylated rather than the lower octane 23DMP or 23DMH, expected from the starting olefins.

A comparison of Tables I and II demonstrates the importance of side reactions in determining alkylate quality. The quality of the alkylates produced from isobutene, diisobutene, cis-2-butene, and trans-2-butene was greatly improved when side reactions were reduced by lowering the reaction temperature to 4°C. At 45°C, the alkylates obtained from these olefins were remarkably similar in composition. The estimated RON's ranged from 94.6 to 94.8. Reducing the reaction temperature to 4°C inhibited the formation of DMH's and increased the yield of TMP's, boosting octane ratings 1.5 to 4.0 RON.

Reducing side reactions does not always mean greatly improved octane ratings. Propylene alkylate was only slightly improved at the low temperature, because the hydrogen transfer reaction was inhibited along with undesirable side reactions. A decrease in the yield of 24DMP and other undesirable products was offset by a decrease in the yield of 224TMP. The quality 1-butene alkylate at low temperatures was very poor because the desirable butene-1 isomerization reaction was inhibited. At 45°C most 1-butene isomerized to 2-butene prior to alkylation and TMP's were the major product; but, at 4°C, less than half the 1-butene isomerized and 23DMH predominated.

The same effects were seen with mixed olefin feeds. The refinery stream described in Table III was blended with C.P. Grade isobutane to obtain a 9.0-to-1.0 isobutane-to-olefin molar ratio. This feed was alkylated at temperatures ranging from 4°C to 45°C. The contact time was held constant at 1.0 minutes. The results are shown in Table IV. The alkylate compositions include pentanes derived from the feed, but only

Table III

C_3–C_5 REFINERY STREAM

Component	Vol %
Ethane	0.06
Propane	13.26
Propene	27.34
Isobutane	19.55
N-butane	4.71
Butene-1	4.44
Isobutene	4.64
Trans-butene-2	5.20
Cis-butene-2	3.40
Isopentane	12.28
N-pentane	0.61
Pentene-1	0.60
Cis and trans pentene-2	1.26
2-methylbutene-1	1.35
2-methylbutene-2	0.90

Table IV

HF ALKYLATION OF C3-C5 REFINERY FEED WITHOUT PROMOTER

Temperature, °C	4	10	12	15	20	27	32	38	45
Yield, Wt %	172	179	185	198	193	206	207	208	204
RON	92.3	92.4	92.4	92.6	92.6	92.5	92.4	92.1	91.8
Alkylate Composition, Wt %									
IP	18.3	18.3	18.2	17.3	16.9	18.0	18.0	18.1	18.2
P	0.8	0.9	0.8	0.9	0.8	0.8	0.8	0.8	0.8
23 DMB	2.0	2.2	2.0	1.8	1.8	1.8	1.9	2.0	2.3
2 MP	0.6	0.7	0.7	0.6	0.6	0.7	0.7	0.8	1.0
3 MP	0.2	0.2	0.2	0.2	0.2	0.3	0.3	0.3	0.4
23 DMP	30.5	28.4	27.7	25.9	24.9	18.9	18.0	17.5	16.6
24 DMP	2.3	2.5	2.7	3.0	3.8	5.1	6.4	8.1	9.8
Other C7	0.1	0.2	0.2	0.2	0.2	0.4	0.5	0.8	1.0
223 TMP	0.3	0.4	0.5	0.5	0.5	0.5	0.5	0.5	0.6
224 TMP	14.7	15.4	16.3	20.1	20.0	22.5	22.0	21.0	19.9
233 TMP	3.2	3.2	3.4	3.7	3.9	4.0	4.1	4.1	3.9
234 TMP	6.3	6.7	6.7	6.9	7.1	6.5	6.5	6.8	5.4
23 DMH	4.9	4.6	4.6	3.9	4.1	3.4	3.0	2.0	2.2
24 DMH	1.3	1.6	1.5	1.6	1.8	2.4	2.3	2.4	2.7
25 DMH	0.6	0.7	0.7	0.7	0.8	1.1	1.2	1.6	1.7
34 DMH	0.5	0.5	0.5	0.5	0.5	0.4	0.4	0.4	0.4
Other C8	0.1	0.1	0.1	0.1	0.2	0.2	0.2	0.2	0.3
225 TMH	2.7	2.8	2.9	2.5	2.8	3.0	3.1	3.0	3.3
Heavies	10.6	10.1	11.0	9.5	9.2	10.1	10.2	10.2	9.6

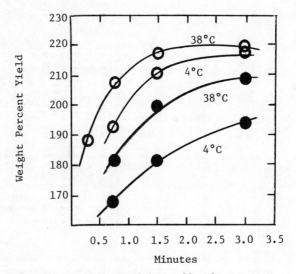

Figure 2. Variation of alkylate yield with contact time.
○, *6.5 wt % CF₃SO₃H added;* ●, *no CF₃SO₃H added.*

the net increase in these materials was considered in calculating
the yield of alkylate.

As with the pure olefin feeds, lowering the reaction temperature inhibited side reactions. The yield of primary alkylation products such as 23DMH and 23DMP increased, as the formation of other DMH isomers and 24DMP was inhibited. TMP's are
both a primary product and a byproduct. Thus, TMP's from isobutene and 2-butene increased as TMP's from propylene and 1-
butene decreased. The total TMP yield was optimized at 27°C.
The net effect on alkylate quality was that RON increased from
91.8 to 92.6, then fell back to 92.3.

The theoretical yield of alkylate based on olefin in the
feed was 220 wt %. Below 27°C, olefin conversion was incomplete.
Unreacted olefins were detected in the reactor effluent and
alkylate yields were considerably below the theoretical value.
From 27°C to 45°C, the olefin feed was completely reacted. In
this temperature range the olefin yield averaged 206 wt %.

Small amounts of added CF_3SO_3H dramatically changed the
catalytic properties of the HF phase. Tables V, VI, and VII
show the effect at low temperatures with propylene, 1-butene,
and mixed-olefin feeds, respectively. The first two feeds contained 95 wt % isobutane and 5 wt % olefin. The third feed,
prepared from C.P. Grade isobutane and the C_3-C_5 refinery stream
shown in Table III, had an isobutane-olefin molar ratio of 12.
With HF alone, the yield of primary alkylation products was very
high. The addition of CF_3SO_3H selectively restored the hydrogen
transfer and butene isomerization reactions. TMP yields were
enhanced at the expense of 23DMP and 23DMH without unduly increasing undesirable byproducts such as 24DMP. As a result,
substantial increases in alkylate quality were recorded for all
three feedstocks. Most importantly, the octane rating of the
alkylate produced from the C_3-C_5 refinery stream was boosted
from 92.1 to 93.9 RON.

Furthermore, reaction rates in the presence of CF_3SO_3H
were several times those obtained with HF alone. The increased
alkylate yields resulting from CF_3SO_3H were due in part to increased olefin conversions. In an extension of the experiments
in Table VII, contact time was varied at several temperatures
and promoter levels. The alkylate yields were plotted against
contact time as in Figure 2. An estimate of the relative reaction rates was made by comparing the contact times required
to achieve a 195 wt % yield of alkylate. The Figure shows
that the enhancement of reaction rate by CF_3SO_3H was more than
enough to compensate for a reduction in temperature from 38°C
to 4°C.

The effect of FSO_3H on HF alkylation was very much like
that of CF_3SO_3H. The 9-to-1 blend of isobutane with refinery
olefins, alkylated earlier with HF alone (Table IV), was used
to study the catalytic properties of HF-FSO_3H blends. Table
VIII gives the results of runs made at 4°C and 1.0 minute

Table V

EFFECT OF CF3SO3H ON PROPYLENE ALKYLATION

Wt % CF3SO3H	0	3.8	6.6
Contact Time, Minutes	3.0	3.0	3.0
Temperature, °C	2	3	4
Wt % Yield	176	199	230
Est. RON	90.8	92.8	93.5

Alkylate Composition, Wt %

C5	1.7	2.5	2.5
C6	1.3	2.2	2.4
23 DMP	77.0	51.8	41.6
24 DMP	3.9	9.9	12.2
TMP's	9.5	30.3	37.5
DMH's	0.8	1.5	1.8
225 TMH	0.1	0.1	0.2
Heavies	6.2	2.0	1.8

Table VI

EFFECT OF CF_3SO_3H ON BUTENE-1 ALKYLATION

Wt % CF_3SO_3H	0	3.6	7.0
Contact Time, Minutes	3.0	3.0	3.0
Temperature, °C	2	3	4
Wt % Yield	188	198	203
Est. RON	83.8	90.8	94.5
Alkylate Composition, Wt %			
C_5	0.9	1.6	1.6
C_6	1.2	1.5	2.2
C_7	0.9	1.2	1.2
TMP's	40.2	62.3	75.0
DMH's	48.9	31.1	18.2
Other C_8	0.3	0.2	0.3
225 TMH	0.1	0.5	0.5
Heavies	7.5	1.2	1.2

Table VII

EFFECT OF CF_3SO_3H WHEN ALKYLATING
A C_3–C_5 REFINERY STREAM AT 4°C

Wt % CF_3SO_3H	0	2.7	6.5	8.6
Contact Time, Minutes	1.5	1.5	1.5	1.5
Wt % Yield	182	205	211	217
RON	92.1	92.9	93.9	93.7

Alkylate Composition, Wt %

C_5	19.1	18.8	18.9	17.1
C_6	2.3	2.4	2.3	2.0
23 DMP	28.5	23.7	21.5	16.8
24 DMP	2.3	2.9	3.1	4.4
Other C_7	0.1	0.1	0.2	0.2
TMP's	24.3	32.5	36.8	38.7
DMH's	7.3	5.5	4.5	4.4
Other C_8	0.2	0.2	0.1	0.1
225 TMH	2.7	3.0	3.0	3.0
Heavies	13.1	11.0	9.6	13.3

Table VIII

EFFECT OF ADDED FSO$_3$H ON THE ALKYLATION OF C$_3$–C$_5$ REFINERY FEED AT 4°C

Catalyst Composition, Wt %									
FSO$_3$H	--	3.3	6.6	9.7	15.7	21.8	32.4	50.0	100
HF	100.0	96.7	93.4	90.3	84.3	78.2	67.6	50.0	--
Yield, Wt %	172	191	206	208	215	210	213	204	204
RON	92.3	93.0	93.6	93.9	93.6	93.2	92.6	91.9	90.1
Alkylate Composition, Wt %									
IP	18.3	18.0	18.8	17.2	17.8	18.7	19.0	20.1	25.5
P	0.8	0.8	0.8	0.8	0.9	0.9	0.9	1.0	1.0
23 DMB	2.0	1.9	1.9	1.6	1.4	1.7	2.1	2.2	2.7
2 MP	0.6	0.6	0.6	0.6	0.8	1.1	1.2	1.4	2.4
3 MP	0.2	0.2	0.2	0.2	0.3	0.5	0.4	0.5	0.9
23 DMP	30.5	24.9	21.2	21.3	10.5	8.6	8.2	7.4	5.8
24 DMP	2.3	3.3	4.1	4.2	5.8	5.8	6.3	6.9	5.2
Other C$_7$	0.1	0.1	0.1	0.3	0.7	0.6	0.9	1.3	1.9
223 TMP	0.3	0.3	0.2	0.5	1.4	1.7	2.4	2.3	2.8
224 TMP	14.7	21.1	25.1	24.0	28.6	26.6	24.7	21.0	14.2
233 TMP	3.2	4.1	4.8	5.6	6.3	5.9	6.0	6.1	3.7
234 TMP	6.3	6.3	6.0	6.6	4.7	4.2	3.7	3.5	2.3
23 DMH	4.9	4.0	2.9	2.3	1.9	2.0	1.9	1.8	1.4
24 DMH	1.3	1.1	0.8	1.0	1.7	1.8	2.2	2.3	2.7
25 DMH	0.6	0.5	0.5	0.4	1.2	1.5	1.9	2.0	2.6
34 DMH	0.5	0.4	0.4	0.3	0.3	0.4	0.4	0.4	0.4
Other C$_8$	0.1	0.1	0.0	0.0	0.1	0.2	0.2	0.2	0.6
225 TMH	2.7	2.8	3.1	3.0	3.8	4.6	5.4	6.0	8.7
Heavies	10.6	9.6	8.2	9.9	11.6	13.1	12.0	13.6	15.4

contact time. The percentage of FSO3H in the catalyst blend
was varied from 0 to 100 wt %. The highest quality alkylate
(93.9 RON) was obtained with a catalyst blend containing 9.7
wt % FSO3H. Once again the promoter selectively restored the
hydrogen transfer and butene isomerization reactions, boosting
TMP yields at the expense of 23DMP and 23DMH. Above the optimum promoter level, alkylate quality deteriorated because
DMH's, 24DMP, and heavy and light ends were increasingly formed.
With 50% or 100% FSO3H, alkylate octane ratings fell below those
obtained with HF alone. Reaction in the absence of promoter was
relatively slow, so olefin conversion was incomplete. The addition of FSO3H markedly increased reaction rates, and alkylate
yields approached the theoretical value of 220 wt %. Only
6.6 wt % FSO3H was required to achieve complete conversion, so
the enhancement of reaction rates was of the same magnitude
achieved with CF3SO3H.

In Table IX alkylation was carried out at various temperatures with a catalyst blend containing 9.7 wt % FSO3H. As the
reaction temperature was raised, undesirable side reactions
increased. A comparison of Tables IX and IV shows that above
25°C the addition of 9.7 wt % FSO3H was detrimental to alkylate
quality.

Conclusion

A commercial HF alkylation unit operating at 45°C produces
92.1 RON alkylate from the C3-C5 refinery stream used in these
experiments. We have shown that 93.9 RON alkylate can be obtained from the same feedstock by lowering the reaction temperature to 4°C and adding an optimum amount of CF3SO3H or FSO3H
promoter. Reducing the reaction temperature increases the
selectivity for primary alkylation products, but in the absence
of promoter this has only a small effect on RON because both
desirable and undesirable side reactions are inhibited. The
addition of promoter restores hydrogen transfer and butene
isomerization without markedly increasing undesirable side reactions. Consequently, alkylate quality is improved. The
addition of CF3SO3H or FSO3H also greatly accelerates the rate
of alkylate production. This means that throughputs for a modified HF alkylation process employing one of these promoters
would be at least as high as in conventional HF alkylation,
despite the lower reaction temperature.

The utilization of CF3SO3H or FSO3H as promoters in HF alkylation will depend on the value assigned by refiners to an
incremental octane number. The added value of the alkylate must
be high enough to justify the installation and operation of the
required refrigeration equipment. At present most refiners are
still able to increase the quality of their gasoline pool by
less costly methods such as by operating their reforming units
at higher severity. However, if the demand for high-octane

Table IX

HF ALKYLATION OF C$_3$–C$_5$ REFINERY FEED PROMOTED BY 9.7 WT % FSO$_3$H

Temperature, °C	4	7	10	16	20	25	32	38	45
Yield, Wt %	208	205	222	210	210	213	212	207	203
RON	93.9	93.7	93.5	93.2	92.8	92.5	92.1	91.7	91.3
Alkylate Composition, Wt %									
IP	17.2	16.6	16.9	17.1	17.4	17.8	18.2	18.5	19.1
P	0.8	0.8	0.8	0.8	0.8	0.9	0.9	0.9	0.9
23 DMB	1.6	1.7	1.8	1.9	2.2	2.3	2.3	2.2	2.2
2 MP	0.6	0.7	0.7	0.9	1.2	1.2	1.4	1.4	1.5
3 MP	0.2	0.2	0.3	0.4	0.4	0.4	0.4	0.5	0.5
23 DMP	21.3	15.4	15.3	13.1	12.0	10.9	10.1	9.1	8.5
24 DMP	4.2	6.2	6.4	7.3	7.7	7.5	7.9	8.3	8.7
Other C$_7$	0.3	0.6	0.7	0.8	1.1	1.0	1.2	1.3	1.5
223 TMP	0.5	1.0	1.1	1.3	1.5	1.4	1.5	1.5	1.5
224 TMP	24.0	27.7	27.0	26.6	25.1	25.0	23.3	22.4	19.9
233 TMP	5.6	6.0	5.9	6.0	5.8	5.8	5.3	5.0	4.9
234 TMP	6.6	5.2	5.1	4.9	4.6	4.7	4.0	3.8	3.7
23 DMH	2.3	2.1	2.2	2.1	2.2	2.5	2.6	2.8	3.1
24 DMH	1.0	1.5	1.6	1.9	2.1	2.2	2.3	2.6	2.9
25 DMH	0.4	1.0	1.1	1.3	1.5	1.7	2.0	2.3	2.5
34 DMH	0.3	0.4	0.4	0.4	0.4	0.4	0.5	0.6	0.7
Other C$_8$	0.0	0.1	0.2	0.1	0.2	0.2	0.3	0.3	0.4
225 TMH	3.0	3.3	3.4	3.7	4.1	3.7	3.9	4.2	4.5
Heavies	9.9	9.2	9.1	9.2	9.4	10.4	11.9	12.3	13.0

unleaded gasoline increases significantly, the use of these
promoters could be justified.

Literature Cited

(1) Innes, R. A., U.S. Patent Application Serial No. 501,664,
    August 29, 1974.
(2) McCaulay, D. A., U.S. Patent 3,928,487 (December 23, 1975).
(3) "Knocking Characteristics of Pure Hydrocarbons", ASTM
    Special Publication No. 225, American Society for Testing
    Materials, Philadelphia (1958).
(4) ASTM Manual for Rating Motor, Diesel, and Aviation Fuels,
    pp 16 & 37, American Society for Testing Materials,
    Philadelphia (1971).
(5) Schmerling, L., J. Amer. Chem. Soc., 67, 1778 (1945).
(6) Schmerling, L., "Alkylation of Saturated Hydrocarbons",
    Chemistry of Petroleum Hydrocarbons, III, 363, Reinhold,
    New York (1955).
(7) Schmerling, L., "Alkylation of Saturated Hydrocarbons",
    Friedel Crafts and Related Reactions, 2, 1075,
    Interscience, New York (1964).

Isoparaffin/Olefin Alkylation over Resin/Boron Trifluoride Catalysts

T. J. HUANG

Mobil Research and Development Corporation, P. O. Box 1025, Princeton, NJ 08540

S. YURCHAK

Mobil Research and Development Corporation, Paulsboro Laboratory,
Paulsboro, NJ 08066

Conventional sulfuric acid and HF alkylation processes (1,2,3) employ liquid-liquid catalytic systems which are expensive and troublesome because of such problems as maintaining an acid/hydrocarbon emulsion, product separation and waste disposal (H2SO4 process only). A solid catalyst should eliminate many of these problems. In view of their high activity, zeolites have been used by a number of workers (4,5,6,7) to catalyze isoparaffin/olefin alkylation with varying degrees of success. Zeolites appear to have limited attractiveness because they age rapidly and cannot perform effectively at high olefin space velocities.

Macroreticular acidic ion exchange resins represent a class of materials available with a rigid pore structure. While they have shown good low-temperature activity for a number of acid-catalyzed reactions, they are ineffective for isoparaffin/olefin alkylation since they lack hydride transfer capability. It may, however, be possible to increase the acidity of ion exchange resins by forming a complex between the acid groups of the resin and a Lewis acid. This approach was taken by Kelly (8) who found that isoparaffin/olefin alkylation could be catalyzed by a gel-type acidic ion exchange resin in the presence of BF3. The efficiency of this catalyst system seems to have been impaired by the non-swelling nature of gel-type resins in hydrocarbons. Such limitations are not imposed on macroreticular ion exchange resins.

In this paper, we present some of our data on the use of macroreticular ion exchange resins to catalyze the alkylation of isobutane with butene in the presence of BF3.

Experimental

In conducting isoparaffin/olefin alkylation, particular attention must be given to minimizing the chief side reaction -- olefin polymerization. In the laboratory, this can be conveniently done by using a semi-batch reactor with olefin being charged continuously to a pool of isobutane, catalyst and product. The olefin feed rate is kept low enough so that it does not accumulate in the reactor. In this manner, the bimolecular polymerization reaction rate is minimized. The experiment is continued for a time sufficient to give the desired external isoparaffin/olefin ratio. This ratio bears some similarity to the isoparaffin/olefin feed ratio to a continuous reactor employed in commercial alkylation systems.

In accordance with the above, most runs were performed in a 300 ml, Type 316 stainless steel stirred autoclave equipped for semi-batch operation with olefin. Reactor pressure was sufficient to maintain the reactants in the liquid phase. In the standard procedure, 7 grams of dry resin (Rohm and Haas, see Table I) was placed in the reactor and 93 grams of isobutane (Instrument Grade, Matheson) was charged into the reactor by N_2 pressure. Then about 6-7 grams of BF_3 (Pfaulty and Bauer) was charged and the contents stirred at 1800 rpm. After the desired temperature was reached, butene (C.P., Matheson) was fed in slowly for the duration of the run. An on-line chromatograph, which was equipped with a flame-ionization detector and a digital integrater, was used to monitor the course of alkylation. Details on operating conditions are given in the tables.

At the end of the run, the product was discharged under N_2 flow into a metal bomb which was kept at -73 °C. The product was warmed to room temperature and transferred to the atmospheric pressure weathering system which consisted of two BF_3-scrubbers, a solenoid valve and a ten-liter gas collector which was equipped with an automatic pressure controller. Both the weathered liquid and the weathered gas were analyzed on a Scot Pak column.

Resins were pretreated in the following manner: elution with methanol; distilled water wash; exchange with 4 percent NaOH, distilled water wash; exchange with 15 percent H_2SO_4, distilled water wash until acid free. Treated runs were stored in distilled water.

The pretreated resin was dried in vacuo for three hours at 120°C prior to use. All resins were ground to pass through 100 mesh unless otherwise specified.

TABLE I

PHYSICAL PROPERTIES OF VARIOUS MACRORETICULAR RESINS

Resin	Amberlyst-15	Amberlyst-XN-1005	Amberlyst-XN-1010	Amberlyst-XN-1011	Amberlyst-XN-1008	Amberlite-200
Skeletal structure	Styrene-DVB	"	"	"	"	"
Ionic functionality	RSO_3H	"	"	"	"	"
Hydrogen ion concentration meq/g dry	4.9	3.4	3.3	4.2	4.5	4.3
Porosity, %	32	42	47	24	–	–
Avg. pore diameter, Å	200–600	80–90	40–50	–	400–800	–
Maximum Temperature, °C	150	~150	~150	~150	~150	150
Cross-linkage, % divinylbenzene	~20	–	–	–	–	~20
Surface Area, m2/g dry	40–50	100–120	550–600	28	30–40	40–50

Results and Discussion

The C_5^+ hydrocarbon yield together with chromatographic data and bromine number has been used to assess
the effectiveness of a catalyst for catalyzing iso-
paraffin/olefin alkylation. This yield is defined as
the grams of C_5^+ hydrocarbon produced per gram of olefin
converted. For butene feed, if only alkylation occurs,
the C_5^+ yield would be 2.04 g C_5^+/g $C_4^=$ converted; however, if only polymerization occurs, the C_5^+ yield
would be 1.0 g C_5^+/g $C_4^=$ converted. The $C_4^=$ WHSV is
defined as g $C_4^=$/g catalyst/hour.

I. Amberlyst-15 Resin

Preliminary experiments were performed using
Amberlyst-15 resin/BF_3 catalyst to explore the stoichiometric requirements necessary to obtain a good alkylation catalyst. As the results in Table II show, BF_3
alone, Amberlyst-15 alone and Amberlyst-15/BF_3 catalyst with a 1:1 equivalent ratio of BF_3 to $-SO_3H$ groups
in the resin are not effective alkylation catalysts.
However, in the presence of a large excess of BF_3 (BF_3/
$-SO_3H$ > 2 equivalent/equivalent), the catalyst is
active for alkylation as indicated by the C_5^+ yield
being equal to 1.92 g C_5^+/g $C_4^=$ converted (Run 4, Table
II), or 1.95 g C_5^+/g $C_4^=$ converted (Run 6, Table II).
Amberlyst-15 in the Na^+ form (prepared by exchange
with NaOH) was used to determine if the H^+ on the $-SO_3H$
group participates in the alkylation. The results are
shown below (conditions: 40°C, 2.6 $C_4^=$ WHSV, and
i-C_4/$C_4^=$ = 5):

	H^+	Na^+
Cation in resin		
Olefin conversion %	100	91
C_5^+ yield, $=$ g C_5^+/g C_4 converted	1.95	0.93

The difference in C_5^+ yield clearly demonstrates
that the sulfonic acid group is essential for isobutane
alkylation.
Also shown in Table II is the effect of olefin
space velocity. Comparison of Runs 4 and 6 shows that
the Amberlyst-15/BF_3 catalyst can alkylate isobutane
with butene in good yield at an olefin WHSV of 2.6 g
olefin/g resin-hour. The alkylate yields are slightly
lower than the theoretical value of 2.04 due to removal
of some of the reactor contents via the on-line sampling system. The yields shown are based on the liquid

TABLE II

ISOBUTANE/TRANS-2-BUTENE ALKYLATION OVER AMBERLYST-15/BF_3 CATALYST AT 40°C

Run No.*	Catalyst	Equiv. Ratio BF_3/SO_2H	Olefin Conv. %	C_5^+ Yield	Wt % of C_8 in C_5^+	Bromine No. of Product	Comment
1	Amberlyst-15 alone	0.0	No Product for Analysis				No reaction but trace of olefin polymerization
2	BF_3 alone	–	96.9	1.16	47.8	25.5	Primarily poly-merization
3	Amberlyst-15/BF_3	1.0	100.0	1.21	56.6	–	Primarily poly-merization
4	Amberlyst-15/BF_3	2.1	100.0	1.92	79.2	< 0.1	Alkylation
5	Amberlyst-15/BF_3	1.7	99.2	1.02	50.3	–	Primarily poly-merization
6	Amberlyst-15/BF_3	2.4	100.0	1.95	52.4	< 0.1	Alkylation

*Operating conditions: $i-C_4/C_4^= = 7.5$ and $C_4^=$ WHSV $= 0.2$ for Runs 1-4;

$i-C_4/C_4^= = 5.1$ and $C_4^=$ WHSV $= 2.6$ for Runs 5-6.

recovered from the reactor at the end of the experiment.

It is noteworthy that good performance was achieved at high olefin space velocities with low external isobutane/butene ratio. For comparison purposes, sulfuric acid alkylation operates at an olefin space velocity of approximately 0.2-0.4, and an external isobutane/olefin ratio of about 5. HF alkylation, while permitting operation at higher olefin space velocities than H_2SO_4 alkylation, requires external isobutane/olefin ratio of about 15.

It is well known that olefin space velocity and external isobutane/olefin ratio has a pronounced effect on alkylate quality with both H_2SO_4 and HF alkylation. A similar situation was obtained with resin/BF_3 catalyst, as shown below (conditions: Amberlyst-15/BF_3 and 40°C):

$i-C_4/C_4=$	$C_4=$ WHSV	Wt. % of TMP in C_5^+
10	0.7	60
10	1.4	54
5	1.4	45
5	2.6	41

Clearly, the alkylate quality, as measured by the trimethylpentane (TMP) content of the alkylate, deteriorates as olefin space velocity is increased. In addition, at equal olefin space velocity, reducing the external isobutane/olefin ratio also lowers the alkylate quality.

For H_2SO_4 and HF catalysts, three main process variables affect alkylate quality: temperature; olefin space velocity; and external isobutane/olefin ratio. For resin/BF_3 catalyst, the above process variables also affect alkylate quality. However, with the resin/BF_3 catalyst, the surface area of resin in addition to the functional group of the resin, may also play an important role in directing alkylation. Some results illustrating the effect of the resin's surface area on alkylate quality are shown in Table III. Clearly, increasing the resin's surface area improves the alkylate quality both in terms of the fraction of trimethylpentanes in the C_5^+ alkylate and the clear research octane number (RON) of the C_5^+ alkylate.

While this study of the effect of the resin's surface area was performed by using different macroreticular resins, it is felt that the primary variable being changed is surface area, since all the

TABLE III

THE RESULTS OF SCREENING TESTS ON ISOBUTANE/TRANS-2-BUTENE ALKYLATION OVER VARIOUS RESIN/BF$_3$ COMPLEXES AT 40°C AND 2.6 C$_4$ = WHSV*. (Effect of Surface Area of Resin).

Type of Resin	Surface Area m^2/g dry	Wt. % of TMP in C$_5^+$	Wt. % of TMP in C$_8$	Wt. % of C$_9^+$ in C$_5^+$	C$_5^+$ Yield	RON C$_5^+$ Alkylate
Macro-reticular:						
Amberlyst XN-1011	28	45.9	75.5	19.3	1.88	92.3
Amberlyst XN-1008	35	39.4	73.1	21.9	1.86	92.4
Amberlite-200	45	52.0	80.9	22.0	1.91	93.7
Amberlyst-15	45	52.2	79.1	19.2	1.89	93.6
Amberlyst XN-1005	110	56.5	86.4	20.7	1.76	95.6
Amberlyst XN-1010	570	69.7	88.3	11.6	1.88	96.5

*100% olefin conversion for all resins.

resins are composed basically of a sulfonated styrene-divinylbenzene copolymer.

II. Amberlyst XN-1010 Resin

Since the Amberlyst XN-1010/BF_3 catalyst produced an alkylate with quality superior to the other resins tested, additional work was performed with this resin/ BF_3 catalyst to explore the effects of particle size and temperature, and sensitivity to olefin type. Olefin space velocity was kept at 2.6 grams $C_4^=$/gram resin-hour for all the results reported below.

A. Effect of Particle Size. The high reactivity of Amberlyst XN-1010/BF_3 catalytic system suggested that a diffusional limitation may be imposed in this system. Therefore, the effect of particle size on alkylation was investigated in order to define a proper particle size range for the subsequent work. The results on the effect of particle size are summarized below (40°C, i-C_4/$C_4^=$-2 = 5.1, and 100% olefin conversion):

Run Number	7	8	9	10
Particle size, mesh	30-40	100-200	100$^+$	325$^+$
Wt. % of TMP in C_5^+	62.2	68.5	69.7	69.9
Wt. % of TMP in C_8	85.4	88.6	88.3	87.7
Wt. % of C_9^+ in C_5^+	16.1	13.6	11.6	9.6
RON clear of C_5^+ alkylate	94.6	95.9	96.5	95.9

For the resin with a particle size of 30-40 mesh, less TMP and more C_9^+ were produced in the C_5^+ resulting in an alkylate with a lower RON+O. With particles smaller than 100 mesh, there is almost no difference in alkylate quality, although the C_9^+ content seems to increase only slightly with increasing particle size. It is concluded that particles with 100$^+$ mesh are sufficiently small to overcome the diffusion limitation problem.

B. Effect of Olefin Feedstock. Different butenes were used as feedstocks to determine if the Amberlyst XN-1010/BF_3 catalyst discriminates among them. The results are shown below (40°C, i-C_4/$C_4^=$ = 5.1 and 100% olefin conversion):

Run Number	9	11	12
Olefin feedstock	trans-2-Butene	1-Butene	Isobutene
C_5^+ yield, g C_5^+/ g $C_4^=$ converted	1.88	2.06	1.82
Wt. % of TMP in C_5^+	69.7	64.3	59.0
Wt. % of TMP in C_8	88.3	83.2	87.4
Wt. % of C_9^+ in C_5^+	11.6	14.2	18.4
RON clear of C_5^+ alkylate	96.5	93.4	97.7

The unusual finding was that isobutene gave lower TMP in C_5^+, but with a higher octane number for the C_5^+ alkylate (normally, the octane number of alkylate parallels to the TMP content in C_5^+).

By comparison with conventional HF and H_2SO_4 alkylation catalysts as shown below, Amberlyst XN-1010/BF_3 showed relatively much less difference among these three olefin isomers. HF alkylation discriminates strongly against 1-butene while H_2SO_4 alkylation gives relatively poor results with isobutene (2,3). This suggests that the resin/BF_3 system is unique and not related to conventional systems.

	Wt. % of TMP in C_5^+		
	HF*	H_2SO_4**	Resin/BF_3***
2-Butene	74	72	70
1-Butene	22	70	64
Isobutene	59	52	59

*At -10°C and 5 min. contact time (see Ref. 2).
**At 7°C and 0.2 $C_4^=$ WHSV (see Ref. 2).
***At 40°C, i-C_4/$C_4^=$ = 5 and 2.6 $C_4^=$ WHSV.

C. Effect of Temperature. The effect of temperature on isobutane/trans-2-butene alkylation over Amberlyst XN-1010/BF_3 catalyst was studied at 0°, 20°, 40°, and 60°C. The results are summarized below (i-C_4/$C_4^=$ = 5.1 and 100% olefin conversion):

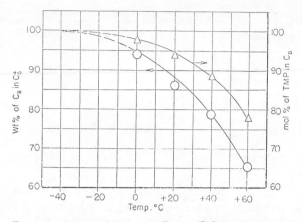

Figure 1. Percent of C_8 isoparaffins in alkylate as function of temperature (Amberlyst XN-1010/BF$_3$ catalyst)

Run Number	13	14	9	15
Temperature, °C	0	20	40	60
C_5^+ yield, g C_5^+/ g $C_4^=$ converted	1.96	1.99	1.88	1.81
Wt. % of TMP in C_5^+	90.3	80.6	69.7	50.6
Wt. % of TMP in C_8	97.1	93.5	88.3	77.5
Wt. % of C_9^+ in C_5^+	4.2	9.2	11.6	20.7
RON clear of C_5^+ alkylate	101.9	99.1	96.5	96.0

    As shown, the selectivity for TMP and the RON of C_5^+ alkylate increased sharply with decreasing temperature. An alkylate with a RON clear of 101.9 was produced at 0°C and an olefin space velocity of 2.6.
    The plots of wt. % of C_8 in C_5^+ and wt. % of TMP in C_8 vs. temperature are shown in Figure 1, while the plot of TMP distribution vs. temperature is given in Figure 2. It is speculated that the reverse trends of 2,2,4-TMP and 2,3,4-TMP in TMP distribution at 60°C (Figure 2) could be due to increased degree of self-alkylation and butene dimerization, followed by rapid hydride and methyl group transfers. Based on the extrapolations from these two figures, the RON of C_5^+ alkylate at -40°C was calculated to be 102.7. The RON

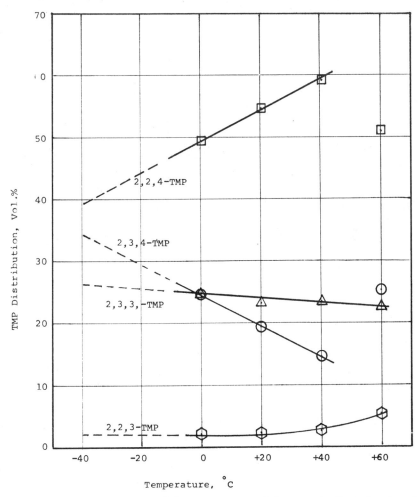

Figure 2. Distribution of trimethylpentanes in trimethylpentane family as function of temperature (Amberlyst XN-1010/BF₃ catalyst)

of C_5^+ alkylate is plotted against temperature in Figure 3. The RON-temperature curve appears to approach a plateau at about 0°C, indicating that lower temperature ($\leq$0°C) would give relatively little further improvement in octane number.

The details on C_5^+ composition at 0°C are given in Table IV.

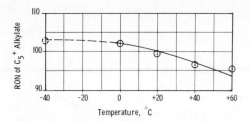

Figure 3. RON of alkylate as function of temperature (Amberlyst XN-100/BF₃ catalyst)

TABLE IV

COMPOSITION OF C_5^+ PRODUCT FROM ISOBUTANE/
TRANS-2-BUTENE ALKYLATION OVER THE
AMBERLYST XN-1010/BF_3 CATALYST AT
0°C, i-C_4/$C_4^=$ = 5.1 AND 2.6 $C_4^=$ WHSV.

Component	Mol %	Wt %
2-Methyl Butane	1.0	0.6
2,3-DM-C_4 + 2-M-C_5	0.6	0.5
3-Methyl Pentane	0.1	0.0
2,4-Dimethyl Pentane	0.4	0.3
2,3-Dimethyl Pentane	0.4	0.3
2,2,4-Trimethyl Pentane	44.0	43.9
2,5-Dimethyl Hexane	0.8	0.8
2,4-Dimethyl Hexane	1.9	1.9
2,2,3-Trimethyl Pentane	1.8	1.8
2,3,4-Trimethyl Pentane	22.7	22.6
2,3-DM-C_6 + 2M-3E-C_5 + 2,3,3-TMP	23.0	23.0
2,2,5-Trimethyl Hexane	0.4	0.4
2,2,4-Trimethyl Hexane	0.2	0.2
Other Nonanes	0.1	0.1
Decanes +	2.8	3.5

Summary

Three phenomena were observed in this catalytic system:

1. BF_3 alone or resin alone is ineffective for alkylation.

2. The H^+ of the sulfonic acid group of the resin does participate in alkylation.

3. Polymerization predominates when the equivalent ratio of total BF_3 to $-SO_3H$ is less than 2.0, while alkylation prevails over polymerization when the equivalent ratio is greater than 2.0.

These above three facts strongly suggest that the active species for hydride transfer, and possibly also methyl group migration, may involve one sulfonic acid group and two BF_3 molecules.

Among the resins tested, the Amberlyst XN-1010/BF_3 catalyst system appeared to be the most selective one, producing the best alkylate quality. This catalyst showed much less difference among different butene feedstocks than conventional HF and H_2SO_4 alkylation catalysts. In addition, its alkylate quality was improved as the temperature was decreased, reaching a plateau at about 0°C at which an alkylate with a RON clear of 101.9 in high yield (1.96 grams alkylate/gram olefin converted) was produced at a $C_4^=$ WHSV of 2.6 (g $C_4^=$ /g resin/hour) and an isobutane/butene external ratio of 5 to 1. The selectivity and productivity of this solid catalytic alkylation system appear to be superior to conventional HF and H_2SO_4 alkylation processes.

Acknowledgment

The authors are indebted to Drs. H. Heinemann and A. J. Silvestri for their constant interest in this project, helpful discussions, and valuable comments on the original manuscript; and to Mr. C. L. Tatsch for his technical assistance.

Abstract

Liquid phase isobutane/butene-2 alkylation was conducted at 0°-60°C using a stirred-tank reactor in the presence of a catalyst comprising of a macroreticular acid cation exchange resin and boron trifluoride. Neither BF_3 nor resin alone was effective for alkylation. For the sulfonic acid resin/BF_3 sys-

tem, an equivalent ratio of total BF_3 to $-SO_3H$ of exceeding 2 is essential for effective isoparaffin alkylation. The H^+ of the sulfonic acid group of the resin
does participate in alkylation.

Among the resins tested, the Amberlyst XN-1010/BF_3
catalyst appeared to be the most selective one, producing the best alkylate quality. This catalyst showed
much less difference among various butene feedstocks,
than conventional HF and H_2SO_4 alkylation catalysts.
In addition, the alkylate quality was improved as the
temperature was decreased, reaching a plateau at about
0°C at which an alkylate with a RON clear of 101.9 in
high yield (1.96 g alkylate/g olefin converted) was produced at an olefin WHSV of 2.6 (g $C_4^=$/g resin/hr.) and
an isobutane/butene external ratio of 5 to 1.

Literature Cited

(1) Ewing, R. C., Oil & Gas J., 61 (1971).
(2) Cupit, E. R., Gwyn, J. E., and Jernigan, E. C.,
    Petro-Chem. Eng., 43 (1961); and 49 (1962).
(3) Gorin, M. H., Kuhn (Jr.), C.S., and Miles, C. B.,
    Ind. Eng. Chem., 38, 795 (1946).
(4) Garwood, W. E., Leaman, W. K., Myers, C. G., and
    Plank, C. J., U. S. Patent 3,251,902 (1966).
(5) Kirsh, F. W., Potts, J. D., and Barmby, D. S.,
    French Patent 1,598,716 (1968).
(6) Kirsh, F. W., and Potts, J. D., Reprint, Div. Pet.
    Chem. A.C.S., 15, A109 (1970).
(7) Yang, C. L., U. S. Patent 3,893,942 (1975).
(8) Kelly, J. T., U. S. Patent 2,843,642 (1958).

5

Isoparaffin–Olefin Alkylation over Zeolite Catalysts

KH. M. MINACHEV, E. S. MORTIKOV, S. M. ZEN'KOVSKY,
N. V. MOSTOVOY, and N. F. KONONOV

N. D. Zelinsky Institute of Organic Chemistry, Academy of Sciences,
Leninsky pr. 47, Moscow, USSR

Alkylation processes with strong acid catalysts are fraught with many difficulties. Problems involved are the handling of highly corrosive and costly acids and the necessity of pretreatment of alkylates aimed at removal of traces of acids, harmful sulphate esters, and some others.

Accordingly the search for a novel highly active and selective catalyst for isoparaffin-olefin alkylation is still an objective of numerous recent patents on the subject (1).

In the present paper, catalytic activity of a number of zeolitic catalysts in alkylation of isobutane with various olefins have been investigated and some considerations of ethylene alkylation have been made.

Experimental

Olefins used in the alkylation were ethylene, propylene, isobutene and 2-methyl-butene-2, all 98-99% pure; butene-1, 98% pure but containing some isobutane and traces of butene-2; and a butene-1 and -2 mixture with a component ratio of about 1:2 respectively. Isobutane had a purity of 96% with 4% of n-butane. The experiments were performed isothermally in a stainless steel tube. Optimum reaction conditions of 80-100°C, 20 atm, LHSV=1hr^{-1}, and an isobutane-olefin ratio of 15-20 were determined by mathematical planning of experiments and regression analysis (2).

Composition of the liquid products was determined by gas chromatography, using a 100m capillary column with squalane, gas products were analyzed over vaseline oil. The catalysts were prepared by the ion exchange of the sodium form of zeolite Y with metal cations (3).

Butylamine titration and ammonia thermodesorption methods were used to determine surface acidity of the catalysts.

89

Results and Discussion

Use of Catalyst Containing Rare Earths. A number of zeolite catalysts containing polyvalent metals cations such as Ca^{2+} and rare earth cations (RE^{3+}) have been prepared and investigated. The total exchange ratio was about 90% (Table I).

Table I
Yields and Compositions of Alkylates Obtained over
Zeolite Y Catalysts (Containing Rare Earths and Hydrogen Ions)
$100°C$, 20 atm, 1 hr^{-1}, $iC_4:C_4^= =20:1$

Hydrocarbons	Catalyst				
	CaY	82 REY	3.9 Ca: 89 REY	16.6 Ca: 64.2 REY	35.8 Ca: 57.6 REY
C_5	1.2	2.8	7.7	2.1	2.8
C_6	4.4	4.3	6.5	2.1	3.1
C_7	5.5	5.8	7.6	5.1	6.3
C_8	57.6	71.8	59.9	77.7	70.2
C_9 and higher	31.3	15.3	18.3	13.0	17.6
Yield based on olefins reacted, wt %	22.5	86.0	77.5	89.5	48.0

Sodium forms of zeolites X and Y are known to be inactive for alkylation. Calcium introduction (catalyst 1) has resulted in a catalyst with some activity. Selectivity of the sample was not high; about 57% of the alkylate were octanes with a ratio of TMP to DMH of 2:1. The yield and quality of the alkylate were improved, if Na^+ cations were replaced with cations of rare-earth elements (catalysts 2 and 3). Product yield for catalysts 2 and 3 were 86.0% and 77.5% respectively with a TMP content in C_8-fraction of about 85%. Unfortunately the stabilities of these two catalysts were rather low in both cases, and alkylates yields and quality declined after 3 or 4 runs. For example the percentage of unsaturated hydrocarbons in the hydrocarbon product for catalyst 3 increased from 18 up to 30%, and TMP concentration decreased to 35% after several runs. Catalyst 4 has proved to be the most active and stable catalyst and the yields and quality of alkylates obtained over it have been the same even after many reaction-regeneration cycles. Further increase of calcium content in the catalyst (catalyst 5) deteriorated its catalytic properties.

We believe that the difference in catalytic activity and stability of the catalysts tested may be due to the different locations of Ca^{2+} and RE^{3+}. Calcium cations are located mainly on S_I sites of the zeolite, thus preventing RE^{3+} migration from their S_{II} sites to S_I sites (4) and also increasing the stability of the catalyst.

These considerations have been confirmed for CaREY zeolite by the direct correlations between the heat of adsorption of benzene, on the one hand, and acidity and the catalytic activity of the catalyst for ethylene-benzene alkylation, on the other hand (5).

Isobutane-olefin alkylation is accepted to proceed via carbonium-ion mechanism. The close similarity (Table II of the product obtained over a zeolite catalyst (Catalyst 4 of Table I) and that produced in the presence of H_2SO_4 contributes to the point.

TABLE II
Composition of Alkylates Obtained over
CaREY (16.6:64.2) Zeolite and H_2SO_4

Hydrocarbon	Catalyst	
	zeolite	H_2SO_4 (6)
isopentane	1.9	
n-pentane	0.2	
C_5	2.1	8.9
2,3-dimethylbutane	-	4.7
2-methylpentane	1.6	1.1
3-methylpentane	0.5	0.4
C_6	2.1	6.2
2,2,3-trimethylbutane	-	0.2
2,2-dimethylpentane	0.2	0.2
2,4-dimethylpentane	2.4	3.4
2-methylhexane	0.6	}0.3
3-methylhexane	}1.9	
2,3 dimethylpentane		2.3
C_7	5.1	6.4
2,2,4-trimethylpentane	16.5	24.3
2,2,3-trimethylpentane	3.5	1.2
2,3,3-trimethylpentane	29.6	12.3
2,3,4-trimethylpentane	18.5	13.0
2,2-dimethylhexane	-	0.2
2,3-dimethylhexane	tr	3.0
2,4-dimethylhexane	}5.7	}6.6
2,5-dimethylhexane		
3,4-dimethylhexane	3.9	0.4
C_8	77.7	61.0
C_9 and higher	13.0	17.5

Table III indicates alkylates composition depending on the olefin used; zeolite Y containing 16-20% of Ca^{2+} and 60-80% of RE^{3+} was used.

Table III
Alkylates Obtained Using Various Olefins as Alkylating Agents

100°C, 20 atm., 1 hr.$^{-1}$, $iC_4:C_4^= $ =15-20:1; CaREY

Hydrocarbons	Alkylation Agent					
	pro-pylene	butene-1	mixture of n-butenes	iso-butene	iso-amylene	commerc. bbf
1	2	3	4	5	6	7
C_5	3.2	10.9	9.9	3.6	8.9	4.8
C_6	8.7	8.5	6.5	4.4	10.3	5.1
C_7	35.6	8.1	7.2	5.8	15.0	5.0
C_8	23.1	55.8	60.6	45.4	19.7	44.4
C_9 and higher	29.4	16.7	15.8	40.8	46.1	40.7
Distribution of C_8-hydrocarbons						
TMP		78.7	81.3	83.7	44.0	81.8
DMH		21.3	18.7	16.3	56.0	18.2
Distribution of TMP						
2,2,4-TMP	45.5	62.5	71.7	70.3	36.2	54.5
2,2,3-TMP		5.7	3.0	1.9		6.0
2,3,4-TMP	} 54.5	13.4	10.5	11.0	} 63.8	14.0
2,2,3-TMP		18.4	14.2	16.8		25.5

The products from butene-1 and n-butenes mixture containing mainly butene-2 were of comparable quality, probably because of isomerization of the double bonds of the butenes, as in the case of H_2SO_4 alkylation. This was confirmed by a special experiment where butene-1 diluted with ten-fold n-butane was passed over the catalyst employed under the conditions of alkylation. In the product obtained both butene-1 and butene-2 isomers were present in thermodynamic quantities. Moreover unexpectedly we have found that n-butane had been alkylated with butene-1 resulting in a liquid product composed mainly of DMH. 72% of the DMH fraction was 3,4 DMH - the product of the direct interaction of n-butane and butene-1. This reaction in the presence of conventional mineral acids is not known and is very interesting from a theoretical standpoint.

If isobutene (or commercial b.b.f.) was used as alkylating agent, up to 40% of heavy hydrocarbons (C_9 and higher) were formed, as a result of an increased tendency to polymerize the isobutene.

Alkylation of isobutane with propylene or 2-methylbutene-2 resulted in products containing about 20% of isooctanes (Table III, columns 2 and 6). Similarly to the sulfuric acid alkylation, C_8 paraffins formation may be explained by self-alkylation of isobutane.

Use of Catalysts Containing Transition Metal Cations. Ethylene being alkylated over certain zeolite catalysts reacts specifically. Ethylene can not, however, be alkylated with isobutane in the presence of H_2SO_4, because of the formation of stable ethylsulphates. We examined the isobutane - ethylene alkylation over crystalline aluminosilicates and found that those catalysts containing RE^{3+} and/or Ca^{2+} in combination with transition metal cations were most active. The alkylation has resulted in not hexanes as would be expected, but an alkylate containing octane isomers as the major product (about 80%). Moreover, the product composition was similar to that obtained from n-butene over CaREY. The TMP-to-DMH ratios were 7.8 and 7.1 respectively.

The explanation of the experimental results is that the alkylation proceeds in two steps - first ethylene dimerization takes place (7) and then n-butene (or its precursor) formed alkylated with isobutane as follows:

(Step 1) $2CH_2=CH_2 \longrightarrow CH_3CH=CHCH_3 +$ (dimerization)

(Step 2) $CH_3CH=CHCH_3 + H^+ \longrightarrow CH_3CH_2CHCH_3$ (alkylation)

$CH_3CH_2CHCH_3 + iC_4H_{10} \longrightarrow nC_4H_{10} + tC_4H_9^+$

$tC_4H_9^+ + CH_3CH=CHCH_3 \longrightarrow$ alkylation products.

This scheme was confirmed by the presence of n-butenes in the reaction gases, and the close similarity of the products obtained from ethylene and n-butenes over zeolite CaMEY, as indicated in Table IV. ME represents a transition metal; nickel, chromium, and cobalt were all found effective for ethylene alkylations. Nickel was the metal used for experiments reported in Tables IV and V.

Dimerization presumably takes place on the transition metal-containing sites, and alkylation on the acidic sites of zeolitic surface. The sodium form of zeolite exchanged with transition metal cations is capable of dimerization (and further polymerization), but does not practically exhibit alkylating capacity. This explains the composition of the product obtained from ethylene and isobutane over this catalyst (Table V, column 3).

Table IV
Composition of the Products of Alkylation of Isobutane
with Ethylene and n-butenes over CaMEY (ME is Nickel)
80°C, 20 atm, LHSV-1 hr^{-1}

Hydrocarbons	Ethylene, %	n-butenes, %
C_5	0.7	0.4
C_6	3.2	1.2
C_7	6.0	5.6
C_8	78.2	71.6
C_9 and higher	11.9	21.2

Table V
Ethylene-Isobutane Alkylation over Various
Zeolitic Catalysts (ME is Nickel)
80°C, 20 atm, LHSV 1 hr^{-1}

Hydrocarbons	NaY %	NaMEY %	CaMEY %	CaY %
1	2	3	4	5
C_5		0.1	0.2	
C_6	no	1.8	1.6	no
C_7	product	0.7	2.0	product
C_8	formed	14.8	94.5	formed
C_9 and higher		83.3	1.7	

Ca^{2+} cation introduction leads to the catalyst containing both
kinds of catalytic sites, and the ethylene-isobutane interaction
over this catalyst proceeds through dimerization to the alkylation
step, yielding high quality alkylate (Table V, column 4). The
alkylate yield was 120%. The reaction does not occur unless
transition metal cations are present in the catalyst even though
the latter may contain acidic sites (CaY for example). It is
understood that no butenes are formed in this case, and ethylene
does not interact directly with isobutane under the conditions
of this experiment.

Transition metal cations are known to be responsible for
dimerization (8). If this is true for ethylene-isobutane interaction, then deactivation of the active sites of dimerization
occurs through the reduction of the cations into metal as follows:

Me^{2+} ———→ $Me°$ will lead to the loss of the catalyst activity. This reduction was conducted in a hydrogen atmosphere for 12 hours at 450°C and 100 atm. The treatment resulted in the catalyst as inactive for ethylene alkylation as zeolyte CaY. The experiment contributes additionally to the proposed scheme of ethylene alkylation.

Miscellaneous results. Runs have been conducted for four hours before the catalyst was regenerated. To restore the catalytic activity of the catalysts, the usual procedure of oxidative regeneration was used, where dried air was passed over the deactivated catalyst at 500°C for 3-4 hours. After the first regeneration, the catalytic activity decreased slightly in comparison with the initial activity; but the activity did not change with further reaction-regeneration cycles (up to at least 20 cycles).

Conclusions

Further development of zeolitic catalysts to the point of their commercial application would help to eliminate many problems of the conventional methods of high quality gasoline production. A number of important problems such as creation of a catalyst of higher activity and stability, development of more simple regeneration procedure, etc. must be solved before zeolites can be used commercially.

Literature Cited

1. US Pat. 3.647.916, 7.3.72; 3.706.814, 19.12.72; 3.795.714, 5.5.74.

2. С.М.Зеньковский, Е.С.Мортиков, А.Г.Погорелов, Н.В.Мостовой, А.М.Доценко, Х.М.Миначев, Нефтехимия, 15, №4, 516 (1975).

3. Х.М.Миначев, Е.С.Мортиков, Н.В.Мостовой, С.М.Зеньковский, Н.Ф.Кононов и др. Авт.свид. СССР №507350, 20.I.1972г., Откр.,изобр., пром.обр., тов.зн., №II, 1976.

4. D.H.Olson, G.T.Kokotailo, J.Charnell, Nature,215, 271,1967.

5. Е.С.Мортиков, А.С.Леонтьев, А.А.Маслобоев, Н.В.Мирзабекова, Н.Ф.Кононов, Х.М.Миначев. Изв.АН СССР, сер.хим. № 3, 1976г.

6. C.A.Zimmermann, J.T.Kelly, J.C.Dean, Ind.Eng.Chem.Prod.Res.Dev. 1, n2,124 (1962).

7. H.Uchida, H.Imai, Bull.Chem.Soc. Japan., 40, 321,1967.

8. Х.М.Миначев, Я.И.Исаков, Г.В.Антошин, В.П.Калинин, Е.С.Шпиро, Изв.АН СССР, Сер.хим., 25-27 (1973).

6

Two-Step Alkylation of Isobutane with C₄ Olefins: Reactions of C₄ Olefins with Sulfuric Acid

LYLE F. ALBRIGHT, BHARAT DOSHI,[1] MARTIN A. FERMAN,[2] and ADA EWO

School of Chemical Engineering, Purdue University, West Lafayette, IN 47907

When isobutane-olefin mixtures are contacted with sulfuric acid at alkylation conditions of commercial interest, the olefins often, if not always, disappear from the hydrocarbon phase at a faster rate than the isobutane (1,2). Subsequently, the isobutane reacts to produce alkylate which is often predominantly trimethylpentanes (TMP's). Other isoparaffins formed to a large extent during the initial stages of alkylation include dimethylhexanes (DMH's), C_5-C_7 isoparaffins referred to as light ends (LE's), C_9 and heavier isoparaffins often called heavy ends (HE's), and acid-soluble hydrocarbons sometimes referred to as conjunct polymer or red oil. The relatively rapid disappearance of the olefins from the hydrocarbon phase is undoubtedly caused in part by the relatively high solubility of the olefins in the acid phases.

The question that has not yet been satisfactorily answered is what reactions occur as the olefins enter the acid phase and how does the isobutane eventually react in order to produce the final wide range of isoparaffins. The mechanism (3,4) that has been widely accepted and that involves a chain reaction in which isobutane and the olefin react in consecutive reactions to form both TMP's and DMH's does not explain all the phenomena noted in the alkylation mechanism. Hofmann and Schriesheim (5) for example, suggested that the acid-soluble hydrocarbons provided a source of hydride ions that entered into the reaction. More recently, Albright and Li (6) suggested that at least some olefins react in the acid phase with acid-soluble hydrocarbons. Goldsby (7,8) has shown that propylene often reacts with sulfuric acid to form sec-propyl sulfates. These sulfates then can be reacted with isobutane to form alkylate.

An objective of the present investigation was to clarify the reactions between butenes and sulfuric acid. In subsequent and companion papers, reactions to produce alkylate will be described

Present address: 1 Worldwide Construction Co., Wichita, Kansas
2 General Motors Co., Detroit, Michigan

in which the isobutane reacts with butyl sulfates and other products obtained from olefin reactions (9); in addition, a comprehensive mechanism will be proposed for the production of alkylate-
type hydrocarbons under a wide variety of conditions including
alkylation conditions (10).

Experimental Details

The reactor similar to the one used earlier (11,12) was a
large glass test tube that had an internal volume of about
150 ml. Four vertical stainless-steel baffles were spaced inside the reactor 90° apart, and three square-pitch impellers
mounted on an agitator shaft provided vigorous agitation to the
contents of the reactor. The reactor was partly immersed in an
acetone bath which could be cooled with dry ice or with a cold-
finger refrigeration unit to any desired temperatures down to
-50°C.

A large cork provided with four holes was fitted into the
top opening of the reactor. The holes in the cork were used
for the agitator shaft that could be rotated up to 1600 rpm,
thermistor probe connected to a temperature-control device, and
feed lines for the acid and chilled hydrocarbon liquids. The
drain line at the bottom of the reactor was provided with a
stopcock.

In most runs, a hydrocarbon mixture of isobutane and a C₄
olefin was first added to the reactor, and this mixture was
cooled to the desired temperature usually -30°, -20°, or -10°C.
Agitation was started, and sulfuric acid was dripped in from a
glass burrette.

Exothermocities were noted as the acid was added, but the
rate of additon was adjusted to prevent temperature rises over
2°C. The desired amounts of acid were usually added within
1-5 minutes. Agitation was maintained until a sample was to be
taken. Then agitation was stopped generally for a minute or
less allowing partial separation of the phases. A chilled micro-
syringe was used to sample the top hydrocarbon phase; this sample
was quickly injected in the gas chromatographic equipment. Agitation was then started and continued until the next sample was
to be taken.

The chromatographic equipment was able to analyze major isoparaffins in the C₄ to C₉ range. In the first portion of this
investigation, an isothermal column was used, and the analyses
of the HE's were less accurate than those for subsequent runs in
which a temperature-programmed column was used. At least 18
compounds were separated and measured in the C₉ to C₁₂ range.

The isobutane used was 95% pure having n-butane as the
major impurity and with a trace of propane. 1-Butene was over
99% pure; trans-2-butene was about 92% pure with cis-2-butene
and n-butane as major impurities; cis-2-butene was about 80% pure
with 18-20% trans-2-butene as the major impurity; and isobutylene

was more than 99% pure but contained traces of 2-butenes. 2,2,4-trimethylpentene-1 obtained from Phillips Petroleum Co. had greater than 99.8% purity.

95-96% sulfuric acid, fuming sulfuric acid, water, and used (or spent alkylation) sulfuric acid obtained from both Amoco Oil Co. and Sohio Oil Co. were blended as needed to produce the desired concentrations of sulfuric acid. These used acids had 88-89% acidities and contained 5-7% dissolved hydrocarbons.

Additional experimental details are reported by Doshi (11, 12).

Results with n-Butenes

Sulfuric acid reacted readily with the n-butenes in liquid hydrocarbon mixtures of isobutane and either 1-butene, cis-2-butene, or trans-2-butene at -30°C to -10°C. The olefin in all cases was extracted from the hydrocarbon phase which upon completion of the reaction was essentially pure isobutane. Although no attempt was made to prove specifically that the product was sec-butyl sulfate, all evidence as will be reported later, indicate that it was the major, if not exclusive, product after a small excess of acid had been added. At these conditions, little or no isobutane reacted. The resulting butyl sulfate was soluble in the acid phase.

In this investigation, acid was rather slowly added to the isobutane-olefin mixture. By the time sufficient acid was added to obtain an acid/olefin (A/O) molar ratio of about 1.2, all olefins had been removed from the hydrocarbon phase. While the acid was being added, acid droplets were well dispersed because of vigorous agitation throughout the hydrocarbon phase. Acids tested were 92.5 to 98% fresh acid (water being the only impurity) and both Amoco and Sohio used alkylation acids. When 98% fresh acid was used, part of the acid initially froze and deposited on the walls of the reactor. Gradually, the frozen acid disappeared presumably because it reacted to form butyl sulfates.

As the acid was dripped into the hydrocarbon phase, a small temperature rise occurred. This rise was in all cases 2°C or less depending on the isobutane-to-olefin (I/O) ratio and on the rate of acid addition. I/O ratios of 1.5 to 20 were investigated. As the acid was contacted with the hydrocarbon phase, the volume of the acid phase increased. The total volume of the two phases remained essentially constant.

A series of runs were made at -20°C in which the acid was added in three to five increments to obtain a final A/O ratio of about 1.2. Figure 1 indicates how the composition of the hydrocarbon phase changed after each addition of 96.5% fresh acid until a 1.2:1 ratio of acid to 1-butene was obtained. After each addition of acid, part of the 1-butene isomerized to cis- and especially trans-2-butene, and part formed butyl sulfate. The sulfate formation was followed by the decrease in the C_4 olefin

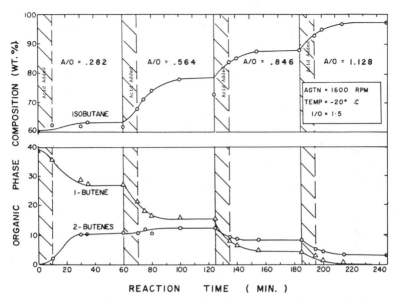

Figure 1. Reaction between sulfuric acid and mixture of isobutane and 1-butene

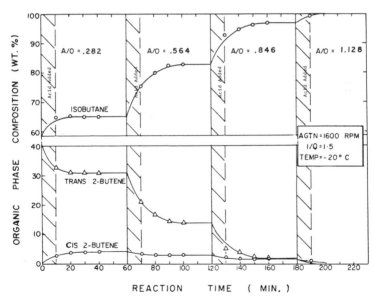

Figure 2. Reaction between sulfuric acid and mixture of isobutane and trans-2-butene

content of the hydrocarbon phase. After three additions of acid
(at an A/O ratio of about 0.85), 2-butenes became the predominant
olefins. Both the isomerization and sulfate-forming reactions
continued for about 20-40 minutes after each increment of acid
was added; as butyl sulfate was produced, the concentration of
free acid decreased resulting in an increased ratio of water to
unreacted sulfuric acid. Below a certain acidity, both isomerization and sulfate formation ceased.

Figure 2 shows the results of a similar run made using a
mixture of isobutane and trans-2-butene. Butyl sulfate formed
in a manner very similar to that shown in Figure 1 (for 1-butene).
Part of the trans-2-butene isomerized to form cis-2-butene. If
any 1-butene was produced, it was in quantities too small to be
detected with the gas chromatographic column used.

When fresh acids were used with A/O ratios from 0.3 up to 1,
a third "phase" formed at the bottom of the reactor when agitation
was suspended. This layer was unstable when brought to room
temperature. It is thought to be an "ice" layer that became
insoluble in the acid layer as the sulfuric acid reacted and as
butyl sulfate was formed. When more acid was added and as the
acidity of the acid layer increased at A/O ratios greater than 1,
the third phase dissolved again in the acid phase. Except for
this whitish layer, the acid and hydrocarbon phases remained
essentially colorless when fresh acid was used and when temperatures of -20°C or less were used. For a run at -10°C, however,
some side reactions occurred as indicated by a slight color
formation in the acid phase.

A run was also made in which used acid was added in several
increments to a mixture of isobutane and cis-2-butene. The
reactions were similar to those using fresh acid. Trans-2-butene
was the only C_4 isomer formed, and eventually its concentration
was higher than that of cis-2-butene. A small amount of the
colored materials in the acid phase were extracted by the hydrocarbon phase.

Figure 3 shows a plot of olefin conversions after equilibrium
had been obtained versus A/O ratios in the feed. The conversion
results were essentially identical for all n-butenes and for all
sulfuric acids tested. More than 10-30% excess sulfuric acid
was needed to produce butyl sulfate (and also to cause olefin
isomerization).

Stability of Butyl Sulfate

Acid mixtures containing dissolved butyl sulfate and relatively
small amounts of acid were stable at temperatures from -30° to
-10°C. There was no indication of instability for samples stored
several days at -20°C or several hours at -10°C. Furthermore,
subsequent reactions of these stored butyl sulfates with isobutane,
as will be discussed later (9), resulted in identical products
as compared to freshly prepared butyl sulfate samples.

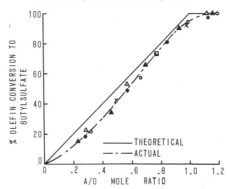

● R-14(1-Butene,96.4%Acid,-20°C)

▲ R-17(2-Butene,96.4%Acid,-20°C)

○ R-59(1-Butene,Used Acid(AMOCO),-20°C)

△ R-130(2-Butene,Used Acid(SOHIO),-20°C)

× R-149(1-Butene,95.5%Acid,-10°C)

▫ R-150(1-Butene,Used Acid(SOHIO),-10°C)

Figure 3. First-step reactions with n-butenes: percent olefin conversion to sulfate vs. A/O ratio

At 0°C and higher, mixtures of butyl sulfate and acid were however, unstable resulting in the formation of both a hydrocarbon phase and acid-soluble hydrocarbons. In these tests, the original mixture was allowed to warm up without agitation. When the mixture was warmed to room temperature, the relatively clear acid phase darkened in color within about one minute; furthermore, a colorless hydrocarbon phase began to form. After 8-10 minutes, no further visual changes were noted. The hydrocarbon phase was analyzed and found to contain in each case 5-10% isobutane, and the remainder was a mixture of all isoparaffins found in alkylates (see Table I); it was essentially a low-quality (about 87-90 research octane number) hydrocarbon mixture. The acidity of the final acid indicated that 3-4% acid-soluble hydrocarbons were present in it; calculations indicated that approximately 20-40% of the initial olefin (used to produce the butyl sulfate) had reacted to form acid-soluble hydrocarbons.

At 0°C, acid solutions containing butyl sulfate reacted slowly. No observable changes were noted for 15-20 minutes. Slowly the color of the acid darkened, and after 1.75 hours a hydrocarbon phase began to form. Calculations indicated that more "alkylate" and less acid-soluble hydrocarbons were formed at 0°C as compared to higher temperatures.

Production of isobutane, the same isoparaffins (as found in alkylates), and acid-soluble hydrocarbons undoubtedly occurred

Table I

Composition of Acid-Insoluble Hydrocarbons
Formed by Decomposition of Butyl Sulfate at
Various Operating Conditions

RUN	123A	123B	125A	125B	126A	126B	127A	127B
Feed Acid H_2SO_4	95.5	95.5	97.0	97.0	97.0	97.0	88.2	88.2
C.P.	0.0	0.0	0.0	0.0	0.0	0.0	8.8	8.8
Temperature $^{\circ}$C	0	25	5	26	2	25	2	20
(A/O) Ratio	2	2	1.5	1.5	3	3	3	3
RON	90.0	87.9	89.0	88.1	89.9	89.1	92.3	91.0
Composition (by weight) of Acid-insoluble Hydrocarbons								
Isopentane	2.3	2.1	6.0	11.6	9.1	10.6	4.0	4.0
2,3-DMB	3.6	4.9	5.5	9.0	6.6	7.6	4.9	5.3
2-MP	0.8	1.5	1.4	2.6	1.5	1.9	0.9	0.8
2,4,-DMP	2.5	3.2	2.9	3.5	3.2	3.5	2.2	2.2
2,2,3-TMB	1.1	1.1	1.0	1.0	0.8	1.0	0.8	1.0
2-MH	1.0	2.5	1.3	2.7	1.2	1.8	0.6	0.8
2,3-DMP	2.6	3.9	2.9	4.9	2.7	3.4	3.1	2.9
2,2,4-TMP	17.9	11.0	11.6	9.3	15.9	13.7	16.5	16.0
DMH's	6.3	8.0	6.5	6.9	6.5	6.7	5.6	6.4
2,2,3-TMP	1.4	1.3	1.4	1.1	1.3	1.2	1.3	1.3
2,3,4-TMP	12.2	8.4	7.6	8.2	7.5	6.3	15.6	13.2
2,3,3-TMP	9.1	8.4	7.7	7.2	7.7	7.0	12.2	10.5
2,2,5-TMH	18.8	10.5	16.7	13.3	19.1	18.5	18.0	17.7
Residue	20.3	24.2	24.4	16.0	13.6	13.8	11.4	14.5
Total L.E.'s	13.9	19.3	21.0	35.3	25.2	29.8	16.5	17.0
Total Octanes	47.0	37.1	38.0	35.9	42.1	37.8	54.2	50.8
Total H.E.'s	39.1	43.6	41.0	29.0	32.7	33.3	29.4	32.2

because of many complicated reaction steps. The initial step likely was the decomposition of butyl sulfate into sulfuric acid and n-butenes. Based on thermodynamic considerations, the n-butenes probably contained over 99% 2-butenes. The subsequent reactions that produced isoparaffins and acid-soluble hydrocarbons will be considered later (10).

Table I shows the results at 0°C and higher for several decomposition runs for acid mixtures containing butyl sulfate. These results indicate the relative comparisons including the effect of various operating conditions. The quality or research octane number (RON) of the hydrocarbon phase obtained in these butyl sulfate decomposition runs increased either with reduced temperature, with higher ratios of acid to butyl sulfate, or when more dilute acids were used to produce the butyl sulfate. More dilute acid were obtained by using either more water or more acid-soluble hydrocarbons in the acid. Increased RON in general meant more TMP's, but lesser amounts of LE's, DMH's, HE's, and acid-soluble hydrocarbons. Considering just the TMP, LE, or DMH family in the hydrocarbon phase, the composition of each family was very similar in each of the following hydrocarbon mixtures:

(a) Product formed in this investigation by degradation of of butyl sulfate.
(b) Alkylate formed during conventional alkylations with sulfuric acid used as the catalyst (2).
(c) Alkylate formed during the two-step alkylation product to be discussed later (9).

This finding is considered to be highly significant relative to the alkylation mechanism also to be considered later (10). For the TMP family, 2,2,4-TMP, 2,3,4-TMP, and 2,2,3-TMP were always the major TMP's; generally their relative importance is in the above order of listing.

Results with Isobutylene

Reactions between isobutylene and sulfuric acid were very different than those between n-butene and sulfuric acid. At least three types of reactions occurred. These were dimerization to form C₈ olefins, formation of heavy ends including C₉ to C₁₄ hydrocarbons (many of which are olefins), and acid-soluble hydrocarbons. These latter hydrocarbons probably included conjunct polymers and perhaps t-butyl sulfate. As will be reported later (9), some of the acid-soluble hydrocarbons react under select conditions in the presence of isobutane to form alkylate.

When the acid was added to isobutane-isobutylene mixtures, large exothermocities were noted; higher exothmocities were noted when the acid was added to pure isobutylene. When sufficient acid to produce an A/O ratio of 0.2 or 0.4 was added quickly to mixtures of hydrocarbons originally at -30°C, temperature rises of 4.5 to 20°C were noted. When, however, the acid was added in

Table II. Heavy End Hydrocarbons Produced in Reactions Between Isobutylene and
Sulfuric Acid

Hydrocarbon	Predicted Boiling Point °C	Probable Carbon Number	Run 28B[1]	Run 29B[2]	Run 32[3]	Run 34[4]
19	161	C_9	0.0	0.0	0.0	0.0
20			0.0	0.0	0.0	0.0
21	181	C_{10}	2.8	0.7	0.0	0.0
22	199		1.9	0.6	0.0	1.2
23	201		0.0	0.3	0.0	2.4
24	209	C_{12}	29.3	26.3	20.7	26.2
25	215	C_{12}	22.0	22.7	24.0	23.6
26	218	C_{12}	6.8	3.6	8.8	8.9
27	224	C_{12}	4.8	3.1	6.8	5.0
28			2.8	1.7	1.1	3.6
29			0.0	2.3	0.0	0.0
30	244	C_{13}	13.3	13.7	13.3	14.9
31	245	C_{13}	2.3	2.8	2.9	2.4
32			2.1	5.8	5.5	1.5
33			2.0	4.8	4.3	1.3
34			2.5	2.6	3.1	0.0
35			0.0	1.1	2.0	1.0
36	265	C_{14}	0.0	0.0	0.0	0.0
Total 19-36			96.2	93.6	92.5	92.0

[1] --- , 95.6% fresh acid, -30°C, I/O = 10 and A/O = 7
[2] --- , 95.6% fresh acid, -30°C, I/O = 9, and A/O = 10
[3] 0.1 hour, 95.6% fresh acid, -30°C, I/O = 5, and A/O = 0.5
[4] 0.16 hour, 95.6% fresh acid, -30°C, I/O = 10, and A/O = 1.0

Note: Residual hydrocarbons were C_5 - C_9's

relatively small increments (equivalent to an A/O ratio of about
0.04:1), temperature rises of only 1-2°C were noted. The acid
phase at the end of the reaction was generally reddish-yellow
indicating the presence of conjunct polymer and other acid-
soluble hydrocarbons.

In the present investigation, 60-90 ml of liquid mixtures of
isobutane and isobutylene were first added to the reactor and then
cooled to either -30° or -20°C. Sulfuric acid was then dripped
into the hydrocarbon mixture that was vigorously agitated. The
amount of isobutylene that disappeared from the hydrocarbon phase
(mainly because of reactions) depended on the amount of acid
added and on the temperature. At -30°C, all of the isobutylene
had reacted by the time an A/O ratio of 0.3 was obtained. At
-20°C, a lower ratio of perhaps 0.2 resulted in complete disappearance of the isobutylene. There was no evidence at these
low A/O ratios and low temperatures that any isobutane had
reacted. The acid-insoluble hydrocarbons produced as the acid
was first added were dimers of isobutylene and to a lesser extent
trimers. As more acid was added, the dimers reacted forming a
large number of compounds in the C₉ and higher range, as will be
discussed in more detail later. In addition, part of the isobutylene was dissolved in the acid phase as indicated by an increase
in the volume of the acid phase. Furthermore when only small
amounts of acid were added, the acid phase became rather whitish
and relatively viscous; the increased viscosity was noted when
agitation was started after the phases were allowed to separate.

When sufficient acid was added to result in A/O ratios of
about 1.0 or higher, a large number of acid-insoluble hydrocarbons
were formed. At least 18 hydrocarbons that were C₉'s or higher
were present. These hydrocarbons designated in Table II as hydrocarbons 19 to 36* have not yet been identified. The approximate
boiling points of these hydrocarbons as shown in Table II were
determined using the following technique. The retention times
in the chromatographic column of several known C₉ through C₁₄
hydrocarbons were plotted versus their boiling points. It was
assumed that this plot was applicable to the unknown hydrocarbons
obtained from isobutylene. Hydrocarbons 24, 25, and 30 were
produced in two runs in the 13-24% range. Some alkylate had however been produced as indicated by the presence of 2,2,4-TMP;
hence it is not clear whether the values for these three hydrocarbons are representative of values at the end of the first-
effect reactions.

One run was made in which a 1:1 mixture of isobutylene and
2-butenes was combined with isobutane to produce a 5:1 mixture of
I/O. A used alkylation acid was then slowly added to the mixture

* Based on the chromatographic analysis, hydrocarbons 1 through
4 are isobutane and C₄ olefins; hydrocarbons 5 through 17 are
major C₅ through C₈ isoparaffins found in alkylate; and hydrocarbon 18 is 2,2,5-trimethylhexane.

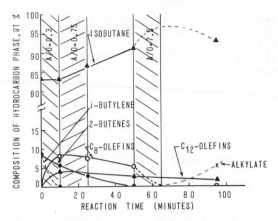

Figure 4. First-step reactions with a 50:50 mixture of
isobutylene and 2-butenes at $-20°C$

maintained at -20°C. At an A/O ratio of 0.3 (see Figure 4), all
isobutylene had reacted but 2-butenes were still present. C_8
and heavier olefins were formed. Material balance calculations
indicated that part of the 2-butenes had reacted to form acid-
soluble hydrocarbons and also butyl sulfate. The next addition
of acid forming an A/O ratio of 0.75 resulted in the disappearance of all of the 2-butenes and part of the C_8 olefins. There
was an increase in the volume of the acid phase because of dissolved or reacted olefins. When an excess amount of acid (A/O =
7.5) was added, alkylation occurred as will be discussed later (9).
    One run was made using a hydrocarbon mixture of isobutylene
and n-pentane; this n-paraffin presumably was relatively inert.
When the acid was added, isobutylene reacted as it did in the
presence of isobutane. Of interest, no isobutane was noted in the
reaction product.
    At lower temperatures and lower A/O ratios, isobutane was
unreactive or at least only slightly reactive with either the
acid-soluble or acid-insoluble hydrocarbons; when reactions did
occur, isoparaffins produced initially were LE's (particularly
isopentane and 2,3-dimethylbutane) and/or 2,2,4-TMP. Both higher
temperatures and higher A/O ratios promote alkylation reactions
to produce isobutane and all isoparaffins normally found during
commercial alkylations.
    The approximate conditions needed to obtain alkylate is
summarized below for various A/O ratios:

(a)   At an A/O ratio of 0.5, some LE's were first noted when
      the reaction mixture was heated to -10°C and maintained
      at this temperature for 30 minutes.
(b)   At an A/O ratio of 1.0, some LE's were first noted upon
      heating to -20°C.
(c)   At an A/O ratio of 2, no TMP's were formed until the
      temperature was raised to about -10°C.
(d)   At an A/O ratio of 7, alkylation occurred slowly at
      -30°C.
These alkylation reactions will be discussed later (9).

Results with 2,2,4-Trimethylpentene-1

A series of runs were made in which sulfuric acid was added
in small increments to a 4:1 mixture of isobutane and 2,2,4-
trimethylpentene-1 at -30°C or -20°C.   Increased amounts of the
feed isoolefin reacted as more acid was added.   Almost no
reaction products were noted in the hydrocarbon phase at A/O
ratios of 0.25 or less.   When the A/O ratio was increased to 0.5,
C_9 - C_{20} hydrocarbons were formed during a 1.5 hour period during
which time most of the original isoolefin reacted.

Discussion of Results

Although no specific analysis was made to identify sec-butyl
sulfate in the acid phase, there seems little doubt that this
sulfate was the reaction product obtained in the n-butene runs.
First, the n-butenes and acid reacted on essentially a 1:1 molar
basis; only a slight excess of acid was needed in all cases.   Of
interest, formation of butyl sulfates has also been reported to
occur in conventional alkylation reactors (13).   Second, the
reaction product was similar in many respects as compared to
sec-propyl sulfate; the product was unstable at higher temperatures and it reacted with isobutane in the presence of sulfuric
aicd to form alkylate (9).   Third, the reaction products from all
three n-butenes alkylated in identical manners as will be
discussed later (9).   Fourth, McCauley (14) has shown that butyl
fluorides can be formed and are quite stable at low temperatures
such as used here; he had contacted HF with n-butenes.
No evidence was found for the formation of di-sec-butyl
sulfates.   Di-sec-propyl sulfates are produced when propylene
and sulfuric acid are contacted.
If acid-olefin reactions which occurred in this investigation
at -10°C or lower also occur during conventional alkylations at
10°C or higher, some if not all of the significant differences
in the type of alkylations with n-butenes and with isobutylene
can be explained, as will be discussed later (10).

Literature Cited

1) Shlegeris, R. J. and Albright, L. F., Ind. Eng. Chem. Process Des. Dev. 8, 92 (1969).

2) Li, K. W., Eckert, R. E. and Albright, L. F., Ind. Eng. Chem. Process Des. Dev. 9, 434 (1970).

3) Schmerling, L., Ind. Eng. Chem., 45, 1447 (1953).

4) Schmerling, L., in "Friedel Crafts and Related Reactions in Alkylation and Related Reactions." Olah, G. A. (ed.), Interscience Publishers (1964).

5) Hoffman, J. E. and Schriescheim, A., J. Am. Chem. Soc., 84, 953 (1962).

6) Albright, L. F. and Li, K. W., Ind. Eng. Chem. Process Des. Dev. 9, 451 (1970).

7) Goldsby, A. R., U.S. Pat. 3,234,301 (Feb. 6, 1966).

8) Goldsby, A. R., U.S. Pat. 3,422,164 (Jan. 14, 1969).

9) Albright, L. F., Doshi, B. M. Ferman, M. A., and Ewo, A., "Two-Step Alkylation of Isobutane with Products of First-Step Reactions", This book, Chapter 7,1977.

10) Albright, L. F., "Mechanism for Alkylation of Isobutane with Light Olefin", This book, Chapter 8, 1977.

11) Doshi, B. H. and Albright, L. F., Ind. Eng. Chem. Process Des. Dev. 15, 53 (1976).

12) Doshi, B. M., Ph.D. Thesis, Purdue University (1975).

13) Goldsby, A. R. and Gross, H. H., U.S. Pat. 3,083,247 (March 26, 1973).

14) McCauley, D. A., U.S. Pat. 3,280,211 (1966).

Two-Step Alkylation of Isobutane with C$_4$ Olefins: Reaction of Isobutane with Initial Reaction Products

LYLE F. ALBRIGHT, BHARAT M. DOSHI, MARTIN A. FERMAN, and ADA EWO

School of Chemical Engineering, Purdue University, West Lafayette, IN 47907

As has been indicated earlier (1,2), C$_4$ olefins react to a significant, and maybe even predominant, extent before isotutane reacts during alkylations using sulfuric acid as a catalyst. In the previous paper (3) of this series the reactions that occur when n-butene and isobutylene were contacted with relatively small amounts of sulfuric acid in the temperature range of about -30° to -10°C were investigated. These reactions defined as first-step reactions in the two-step alkylation process were found to be as follows:

(a) n-Butenes react with sulfuric acid to form sec-butyl sulfates that dissolve in the acid phase.

(b) Isobutylene reacts forming both acid-insoluble and acid-soluble hydrocarbons. Acid-soluble hydrocarbons probably include both t-butyl sulfates and conjunct polymers. The acid-insoluble hydrocarbons are mainly in the C$_8$ and higher range; a large fraction of these hydrocarbons are olefins.

The second step of a two-step alkylation process involves reactions of the isobutane with the products of the first-step reactions. Although two-step processes were first reported many years ago (4-6), relatively little is known about the factors affecting the second step and the quality of the alkylate produced. Most investigators have largely ignored the second-step reactions, perhaps because only relatively poor quality alkylates had been produced in these earlier processes.

A modification of the second-step process has been employed recently though as a means of reducing acid consumption in alkylation plants using sulfuric acid as a catalyst (7,8). Propyl sulfates formed using propylene are reacted with isobutane in conventional reactors; the resulting alkylates contain appreciable amounts of dimethylpentanes. Some butyl sulfates have also been reportedly formed in conventional processes when n-butenes are in the feed. Furthermore, fairly high molecular weight olefins, up to at least C$_{20}$'s, react with isobutane in the presence of sulfuric acid to form fairly high quality alkylates (9).

In the present investigation, extensive information was
obtained on the second-step reactions of the two-step alkylation
process. Alkylate (that was mainly C_5 - C_{12} isoparaffins) was
the main reaction product in all cases. The theoretical importance of the present findings will be considered in the next paper
of this series (10).

Experimental Details

The reaction system and the analytical procedures were
identical with those used earlier for the first-step reactions
(3). Considerably more acid had to be added to the reaction
mixture however in order to obtain second-step reactions as
compared to first-step reactions. Operating procedures and
methods of obtaining hydrocarbon samples for analysis were
essentially identical to those used in the first-step investigation. Yields and research octane numbers (RON) were calculated
based on the analysis of the hydrocarbon product as reported
earlier (2).

Second-Step Runs with Butyl Sulfates

More than 30 runs were made in which sec-butyl sulfate was
reacted with isobutane in the presence of sulfuric acid. Variables investigated included the following:

Temperature   -30, -20, -10°C, and 0°C
Ratio of acid to olefin (A/O): 2 to 20
Composition of feed acid used: 1:15 to 8.0% water and 0 to
                                    8.8% acid-soluble hydrocarbons
Agitation: 1000 to 2400 rpm

The acid composition and ratio of acid to olefin reported
here were those of the first-step reactions. Since the acid and
n-butene reacted on a 1:1 basis, the ratio of free acid to butyl
sulfate at the start of the second-step reaction was less by 1.0
than the A/O ratio, i.e. the ratios of free acid to butyl sulfate
investigated varied from an initial value of about 1 to 19. At
A/O ratios up to about 8, the emulsions formed between the acid
and hydrocarbon phases were in general hydrocarbon-continuous.
At ratios of 10 or greater, the emulsions were generally acid-
continuous depending to some extent on the amount of excess isobutane used.

Kinetics of Butyl Sulfate Runs. In all runs, plots of the alkylate yield versus time indicates S-shaped curves such as shown in
Figure 1. The rates of alkylate production increased with time
immediately after the start of the batch runs. The rates passed
through maxima when approximately half of the theoretical amount
of alkylate was produced, and then decreased toward zero as the

reactions approached completion (i.e. at theoretical alkylate yields of 204 grams of alkylate per 100 grams of feed C_4 olefin). The butyl sulfate formed from 1-butene or the two 2-butenes reacted in all cases in identical manners. Such a conclusion was, of course, expected since all three n-butenes produce sec-butyl sulfate. Storage of two mixtures of butyl sulfate and sulfuric acid for several days at -20°C had no effect on the subsequent alkylation reactions. Identical results were obtained as compared with freshly prepared mixtures.

At -10°C and at A/O ratios less than 2, the kinetics of second-step reactions were always low, generally too low to measure. The kinetics increased significantly both as the A/O ratios increased and at higher temperatures. In the glass reactor (operated at atmospheric pressure), the highest temperature that could be investigated was about -10°C; higher temperatures resulted in excessive vaporization.

For runs at -10°C and for A/O ratios greater than 4, the reaction was usually completed within 1.5 to 6 hours. The rates of alkylate production decreased significantly as the temperature decreased from -10°C to -30°C.

Two runs were made at 0°C in a stainless steel reactor. In these runs, isobutane and mixtures of butyl sulfate and acid were first cooled to approximately -30°C and were then added to a stainless steel reactor having a volume of 50 ml. This reactor was a nipple closed at both ends with steel caps. The reactor was then immersed in an ice-water mixture, and the reactor was shaken at the rate of 250 oscillations per minute. Although the level of agitation in the steel reactor was undoubtedly quite different (probably less) than those in the glass reactor used in the other runs, the results of these runs clearly indicate the faster rates of alkylate production at 0°C as compared to -10°C. Figure 1 is a comparison of results at -20°C, -10°C, and 0°C.

Comparative runs indicated that the rates of alkylate formation increased when the following occurred:
(a) More concentrated feed acids were used in preparing the mixtures of acid and butyl sulfate, i.e. less water and/or less acid-soluble hydrocarbons were present in the feed acid. Comparative runs were made with 98.5% fresh acid, 95.5% fresh acid, and used alkylation acids containing several percent of both water and acid-soluble hydrocarbons.
(b) As the A/O ratio (for preparing the mixture of acid and butyl sulfate) increased from 2 to 8 which is the range in which the emulsions were hydrocarbon-continuous. At higher ratios the emulsions were acid-continuous, and changes in the ratio then had little effect on the kinetics of alkylation production.
(c) Higher ratios of isobutane to olefin (that resulted in larger amounts of excess isobutane). The initial rates

during a run were essentially the same regardless of
the ratio of isobutane to olefin; at the start of all
runs, the hydrocarbon phases were always essentially
pure isobutane. As the run progressed, the rates became as high as 10-15% greater for a run with a ratio
of 5 as compared to one with a ratio of 2.
(d) Higher levels of agitation. Increasing the level of
agitation from 1000 to 2400 rpm increased the kinetics
by about 30% for the one comparison made.

Modelling of Kinetic Results. To model the kinetic data for
reactions between butyl sulfate and isobutane, two possibilities
were considered. Either the main reactions occurred in the acid
phase or the main reactions occurred at the interface between
the acid and hydrocarbon phases. For both cases, the following
assumptions were made:
(a) The rate of alkylate formation was directly proportional to the sulfate concentration in the acid phase.
(b) The rate was directly proportional to the isobutane
concentration in the hydrocarbon phase.
(c) The rate was proportional to the square of H_0, the
Hammett acidity of the acid phase. As the batch run
progressed, additional acid was formed as the butyl
sulfate reacted; hence the Hammett acidity increased
during the run.
Second-order dependency relative to the Hammett acidity
is based on the results for the degradation of TMP's in the
presence of sulfuric acid (11). The acidity is, of course,
decreased because of dissolved acid-soluble hydrocarbons including conjunct polymers and butyl sulfates and because of
dissolved water. The term, $-H_0 - m(BS) - 0.065(CP)$, was employed
to predict the effective acidity of the acids used. It is an
expansion of the expression used earlier (11) for acids containing dissolved conjunct polymers.
Equations tested for representing the experimental data were
as follows:
For reactions occurring in the acid phase

$$R_{alk} = K \cdot V [-H_0 - m(BS) - 0.065 (CP)]^2 (C_{iso,A})(C_{BS,A}) \quad (1)$$

For reactions occurring at the interface

$$R_{alk} = K \cdot S [-H_0 - m (BS) - 0.065(CP)]^2 (C_{iso, HC})(C_{BS,A}) \quad (2)$$

where
R_{alk} = rate of alkylate production, moles alkylate/hr
K = reaction rate constant
V = volume of acid phase
S = interfacial area between phases
H_0 = Hammett acidity of fresh acid with same water content
     as acid used
BS = weight percent butylsulfate in acid

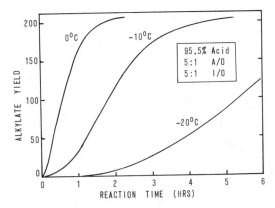

Figure 1.   *Effect of temperature on the rate of second-step alkylation reactions*

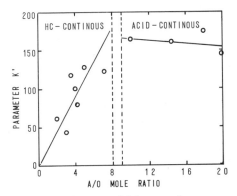

Figure 2.   *Kinetic model for second-step reactions: variation of K' (or interfacial surface area) with A/O ratio for isobutane-to-olefin feed ratio of 5:1*

CP = weight percent acid-soluble hydrocarbons in acid

m = constant for given run

$C_{iso,HC}$ = concentration of isobutane in hydrocarbon phase

$C_{iso,A}$ = concentration of isobutane dissolved in acid phase

$C_{BS,A}$ = concentration of butyl sulfate in acid phase

To solve Equation 1, solubility date are needed to determine the dissolved concentration of isobutane in the acid phase. Experimental measurements at equilibrium conditions indicated low solubilities that were hard to measure accurately. It was then assumed that $K \cdot (C_{iso,A}) = K'$, and hence:

$$R_{alk} = K' \cdot V [-H_0 - m(BS) - 0.065 (CP)]^2 (C_{BS,A}) \qquad (3)$$

No data are available for the interfacial area (S) in Equation 1. It was assumed that $K'' = K \cdot S$ was a constant for a given run so that:

$$R_{alk} = K'' [-H_0 - m(BS) - 0.065(CP)]^2 (C_{iso,HC})(C_{BS,A}) \qquad (4)$$

A non-linear weighted regression program was used to calculate values of K', K'', and m for the above equations. Based on values of the correlation coefficient that were in general 0.90 or higher, both Equations 3 and 4 gave reasonably good correlations of the experimental data of a given run. Equation 3 was, however, discarded for the following reason. If the reactions occurred in the acid phase (as postulated via Equation 3), isobutane would need to be transferred from the hydrocarbon phase and dissolved in the acid phase. The faster rates of alkylate formation when higher rates of agitation were provided would in such a case indicate that transfer of the isobutane was a rate-controlling step. Yet calculations indicated that sufficient agitation was available so that the transfer step for isobutane was not rate-controlling. To make these calculations, both the equilibrium solubility of isobutane in the acid phase and the mass transfer coefficients were needed. The equilibrium solubility was approximated in a separate experiment at 0.00015 moles/cc, and the value of K_0A of 30-1000 cc/sec as reported by Rushton et al (12) was used. The actual solubility of isobutane in the acid during a run was calculated to be 0.98 to 0.999 times the equilibrium solubility. The finding that transfer of isobutane was not rate controlling is not surprising since relatively high rates of agitation were provided and since the rates of alkylate production were quite slow (at least 1 to 1.5 hours were needed to complete the reactions).

The rates of alkylate formation can be explained readily by Equations 2 and 4. Increased agitation would result in higher interfacial surface areas, S, and hence in higher K'' values. Higher interfacial surface areas and hence higher K'' values would also occur in the hydrocarbon-continuous emulsions as the A/O ratios increased in the range from 2 to 7 (see Figure 2 and Table I). When, however, the emulsion became acid-continuous,

increased A/O ratios would result in no further significant
increase in surface areas or in K" values. With acid-continuous
emulsions, the level of agitation was probably fairly low
because of the high viscosities of the acids. Values of m for
Equation 4 increased rather linearly with increased A/O ratios
as indicated by Table I. Such an increase cannot be explained
on a theoretical basis.

Table I
K" and m Values for Various Second-Effect Runs

A/O	Feed Acid Concentration	Isobutane to Olefin ratio	K"	m	Correlation Coefficient
5	95.5	5	128	0.34	0.97
5	95.5	5	118	0.33	0.93
5	95.5	5	154	0.34	0.97
10	95.5	5	166	0.63	0.93
20	89.3	5	145	1.12	0.83
2	98.8	5	61	0.23	0.99
18	95.5	5	175	1.18	0.90
3	95.5	5	40	0.23	0.90
4	95.5	5	78	0.32	0.92
5	89.3 (used)	5	120	0.53	0.90
15	89.3 (used)	5	165	1.0	0.99
3.5	98.8	5	118	0.31	0.93
4	94	5	100	0.29	0.97
5	95.5	2	234	0.34	0.94
10	89.3 (used)	2	300	0.64	0.91

Although Equation 4 correlates the data for batch runs
reasonably well, there were several simplifying assumptions.
Equation 4 is at best only a semi-theoretical equation. Improved ways of measuring the acidity of an alkylation acid need
to be developed. Reliable values for the interfacial surface
area and the amount of isobutane dissolved in the acid phase
are also needed.

Characteristics of Alkylate Produced in Butyl Sulfate Runs. The
alkylate (produced from butyl sulfate and isobutane) contained
the identical C_5 to C_9 isoparaffins found in alkylates obtained
in conventional one-step alkylation processes. The isoparaffins
in the heavy end (H.E.) fraction, i.e. C_9's and higher, were also
presumably identical. Unfortunately the gas chromatographic
analyses for some hydrocarbon samples resulted in heavy end
analyses that were too low. Most analyses were made using an
isothermal two-column unit in which the second column was
designed to analyze the heavy-ends. This column generally
indicated only 1 to 5% heavy ends whereas subsequent analyses
indicated that values up to 15 to 20% sometimes occurred. These
latter analyses were made employing a temperature-programmed

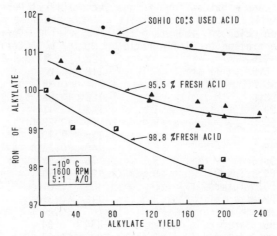

Figure 3.   *Effect of acid composition on the RON of the alkylate*

column and were also confirmed by an analytical group at Phillips Petroleum Co. The analytical results as obtained by the two analytical techniques for C_5 to C_9 isoparaffins were, however, found to be in excellent agreement.

Although the isothermal gas chromatographic column did not measure the heavy ends correctly, the results obtained with it give good comparisons of the composition and quality of the alkylates on essentially a heavy-end-free basis. These qualities are reported as the research octane number (RON) that was calculated (2) based on the composition of the alkylates as determined analytically. On essentially a heavy-end-free basis, RON values of 98-101 were common. Reanalyses of the alkylate in order to determine the heavy ends more correctly indicated that the above RON values were in general too high by 3-4 units. RON values of 94.5 to 98 are thought to have occurred.

The composition of the alkylate and especially of the trimethylpentane (TMP) family varied significantly as a batch run progressed (and as alkylate was produced). The first alkylate produced during a run generally contained a relatively large fraction of 2,3,4-TMP. In two cases, over 50% of the initial alkylate was 2,3,4-TMP. As a run progressed, the ratio of 2,2,4-TMP to 2,3,4-TMP increased. Of interest, the amounts of 2,2,4-, 2,3,3-, and 2,3,4-TMP were generally similar at the end of all batch runs (see Table II). The change of the composition of the TMP family was a major cause of the lower RON values for the alkylate as the run progressed. Figure 3 shows the decrease in RON as three batch runs progressed.

Alkylates produced in runs made at different operating conditions often were found to have relatively different amounts of the four main groups (or families) of isoparaffins, namely TMP's, LE's, DMH's, and HE's. As a general rule, increased RON values resulted because of increased fractions of TMP's and decreased fractions of the other three. Variables that had a significant effect on the alkylate quality were as follows:

(a)   Feed acid composition. Acids with lower acidities produced higher quality alkylates, as shown in Figure 3. The alkylate qualities at the end of a run varied from about 101 when used alkylation acid was employed to 93 with 98.8% fresh acid. Little or no differences in quality were noted, however, for the alkylates produced using 95.5% and 92% acids.

(b)   Ratio of isobutane to n-butene used in the feed to the first effect. The calculated RON decreased by perhaps 1 to 2 as the ratio decreased from 5 to 1.5; relatively little change, however, occurred in the 5 to 20 range.

(c)   Temperature. Temperature had a major effect on both the relative amounts of TMP's and LE's in the alkylate and also the compositions of these two families; temperature was the only variable that had such an effect. Tables II and III indicate the effect of temperature on

Table II

Effect of Temperature on the Compositions of Trimethylpentanes
and Light Ends Produced Using n-Butenes As the Olefins

	-20°C Two-Step Process	-10°C Two-Step Process	0°C Two-Step Process	(0°-25°C) Sulfate Stability	10°C Conventional Process	20°C Conventional Process	30°C Conventional Process
TMP Family							
2,2,4-TMP	26	31	37	--	50	52	54
2,2,3-TMP	1.5	2	3	--	5	6	5.5
2,3,4-TMP	45	37	34	--	17	16	13
2,3,3-TMP	27.5	30	26	--	27	26	26.5
L.E. Family							
Isopentane	21	26	30	(34)*	41	41	42
2,3-DMB	44	39	35	(26)	27	26.5	27
3-MP	3.5	3.8	4	(6)	4.5	4	5
2,4-DMP	13.5	13	11	(12)	14	15	14
2,2,3-TMB	--	1.7	3	(3)	1.4	1.0	1
2-MH	--	1.5	1	(1)	3	2.5	3
2,3-DMP	13	16	16	(16)	10	10	9

* Average values.

Table III

Effect of Temperature on the Compositions of Trimethylpentanes
and Light Ends Produced Using Isobutylene as the Olefin

	-20°C Two-Step Process	10°C Conventional Process	20°C Conventional Process	30°C Conventional Process
TMP Family				
2,2,4-TMP	63	59	58	57
2,2,3-TMP	2	4.5	5.0	5.3
2,3,4-TMP	13	13	13	14.0
2,3,3-TMP	22	23.5	24	24
L.E. Family				
Isopentane	29	36	43	45
2,3-DMB	39	26	26	24.4
3-MP	2	3.5	4	4.5
2,4-DMP	15.4	19	13	12
2,2,3-TMB	2	1.5	1.0	1.6
2-MH	0.6	0.8	2.5	1.5
2,3-DMP	11	12.5	10.5	9.5

the compositions of both TMP and LE groups produced by
second-step reactions using n-butenes and isobutylene
as feed olefins respectively. For LE's, the ratio of
isopentane to 2,3-dimethylbutane increased significantly
with temperature.

(d) C_4 olefin: As shown in Tables II and III, alkylations
with n-butenes and isobutylene resulted in significantly
different compositions for TMP's as will be discussed
later.

The compositions of the TMP's and LE's as determined for this
investigation showed important differences as compared to the
compositions calculated for alkylates produced in conventional
processes operated at 10 to 15°C. The alkylates of this investigation produced from sec-butyl sulfates contained larger relative
amounts of 2,3,4- and 2,3,3-TMP's but lesser amounts of 2,2,4-TMP.
The ratios of 2,3-dimethylbutane to isopentane were much higher
in the alkylates of the present investigation.

Second-Step Runs with Reaction Products from Isobutylene

More than 20 runs were made to investigate the reaction
between isobutane and the products of first-effect reactions with
isobutylene. Alkylate was not produced until excess acid was
used; and larger amounts of excess acid were needed for runs at
-30°C as compared to runs at -10°C (3). An acid-to-olefin ratio
of about 1.0 was required at -10°C to obtain TMP's whereas a 7:1
ratio was needed at -30°C.

In the present investigation, additional acid was generally
added to the agitated acid-hydrocarbon emulsion resulting from
the first-step reactions of isobutylene. The rate of alkylate
formation was generally highest at the start of the run and
decreased toward zero as alkylate yields of 80-120 were approached.
Appreciable amounts of acid-soluble hydrocarbons remained in the
acid at the end of all two-step runs using isobutylene, and these
hydrocarbons caused alkylate yields to be much less than the
theoretical value of 204. The yields of alkylate depended on the
operating conditions investigated, as will be discussed later.
In several runs, the kinetics of alkylate formation were slightly
less at the start of the run and then increased to a maximum at
intermediate conversions.

The rates of alkylate production increased when the following
occurred:

(a) The A/O ratio increased. Such a conclusion is based on
several runs at -10°C with ratios between 1.0 and 9;
the emulsions formed were hydrocarbon-continuous. For
a run at a ratio of 1.0, 4.5 hours resulted in an
alkylate yield of 52. With a ratio of 9, a yield of 98
was obtained in 4 hours.

(b) Increased temperatures in the -30° to -10°C range.

The rates of alkylate formation were affected to only a small, and as yet undetermined, degree by changes of the ratio of isobutane to isobutylene in the 5 to 11 range and by changes in the feed acids used. Comparative runs were made with 96.5% fresh acid and with used alkylation acid.

All alkylates produced by second-step reactions in which isobutylene was the feed olefin had relatively low RON's and contained relatively large amounts of LE's, DMH's, and HE's as compared to alkylates produced from butyl sulfate. The LE content of the alkylate generally varied between 23 to 28%, and the HE content between 30 to 40% (based on analyses with the temperature-programmed unit). The calculated RON values of the alkylate varied from about 88-91. The compositions of the TMP family were considerably different for the isobutylene runs (see Table III) as compared to those of the n-butene runs (see Table II). For runs using isobutylene, 2,2,4-TMP accounted for about 60% of the TMP family. The LE's had similar compositions, however, at a given temperature for both n-butene and isobutylene runs.

Several samples of the hydrocarbon phase were obtained during the course of each batch run to determine how the composition was changing. As a run progressed, LE's, DMH's, and TMP's, were the major compounds produced. Several heavy end hydrocarbons decreased toward zero as the run progresses: hydrocarbons 24, 25, and 30 as defined earlier (3) in particular disappeared as a result of reactions with isobutane. These reactive hydrocarbons were presumably C_{12} or heavier olefins. There were however large amounts of unreactive heavy ends, which are heavy isoparaffins. Calculations indicated that the heavy ends in the alkylate were produced for the most part during the first-step reactions.

Slightly higher quality alkylates were produced using isobutylene at lower temperatures and at higher I/O ratios. The improvement in quality was, however, less for isobutylene runs as compared to those with n-butenes.

Two runs were made using the products of first-step reactions in which mixtures of isobutylene and 2-butene had been used. The second-step reactions in such cases resulted in products that were intermediate between those products for runs with pure isobutylene and for runs with pure n-butenes. The RON of the alkylates obtaining using isobutylene, a 1:1 mixture of isobutylene and 2-butene, and 2-butene were 89.7, 93.0, and 100.8 respectively. The composition of the TMP family obtained with the mixed olefins also appeared to be intermediate in nature.

The alkylation mechanism for second-step reactions with isobutylene as the olefin have been significantly clarified by two runs, each conducted as follows. The acid and hydrocarbon phases produced by first-step reactions involving isobutylene and sulfuric acid were separated. The acid phase which contained some acid-soluble hydrocarbons or reaction products such as perhaps t-butyl sulfate was then contacted with fresh isobutane. This resulting mixture of reactants was designated as A below. The hydrocarbon phase for each run was contacted with fresh 96.5% acid; this mixture was designated as B.

[A]

$$\left(\begin{array}{l}\text{Isobutane} \\ + \\ \text{Isobutylene}\end{array}\right) + \left(\begin{array}{l}96.5\% \\ \text{sulfuric} \\ \text{acid}\end{array}\right)$$

→ $\left(\begin{array}{l}\text{Acid Phase} \\ \text{Including} \\ \text{Acid-soluble} \\ \text{hydrocarbons}\end{array}\right) + \left(\begin{array}{l}\text{Fresh} \\ \text{Isobutane}\end{array}\right)$ →Alkylate A

↘ $\left(\begin{array}{l}\text{Hydrocarbon} \\ \text{Phase;} \\ \text{Isobutane and} \\ \text{heavy ends}\end{array}\right) + \left(\begin{array}{l}96.5\% \\ \text{sulfuric} \\ \text{acid}\end{array}\right)$ [B] →Alkylate B

A summary of the amount of reactants and of the results for these two runs are shown in Table IV. In both runs, approximately the same amount and quality of alkylate was produced by mixtures A and B, and the calculated RON of the alkylates were in the 88.6 to 91 range. The qualities of the alkylates in the second run were somewhat higher than those of the first run, but it is now known if the differences are significant. The higher rates of reaction for mixtures A are thought to be caused by the larger amounts of acid used.

When isobutane was contacted in route A with the acid phase, some colored materials were extracted from the acid phase. Furthermore, apparently some isoparaffins were also extracted based on the analysis of the first sample of the hydrocarbon phase; more isoparaffins were present in the hydrocarbon phase than would seem probable based on the expected amount formed by initial alkylation reactions. Tentatively it is concluded that some lighter isoparaffins had been dissolved in the acid phase.

Second-Step Runs with Reaction Products from 2,2,4-Trimethylpentene-1

Several runs were made using the reaction products obtained when sulfuric acid was slowly dripped into a mixture of isobutane and 2,2,4-trimethylpentene-1. The yield, quality, and composition of the alkylates produced were similar to those for runs using isobutylene.

Discussion of Results

The results of this and a companion investigation (3) report significant information on a two-step alkylation process which has been investigated at Purdue University since 1970 and which appears to be of considerable commercial interest. This two-step process requires a much more selective choice of operating variables as compared to earlier processes (4-6). The current

Table IV

Mechanistic Investigation of Second-Step Reactions
Using Isobutylene and 96.5% Sulfuric Acid

RUN	1	2
First-Step Reactions		
Isobutane, ml	43	47
Isobutylene, ml	9	10
Acid, ml	16	25
Temperature, °C	-30	-30
Second-Step Reactions (Part A)		
Isobutane Added, ml	52	47
Temperature, °C	-10	-10
Yield of Alkylate	30-40	45-60
Time, Hr.	1.5	1.5
Second-Step Reactions (Part B)		
Acid Added, ml	10	15
Temperature, °C	-10	-10
Yield of Alkylate	30-40	55-60
Time, Hr.	3	2.5

process, however, offers several key advantages that will be
discussed later. The proposed new process has the following
unique features. First, hydrocarbon-continuous emulsions have
proved to be most satisfactory and probably are preferred. Yet
acid-continuous emulsions have been shown in the past (2) to be
highly preferred for conventional (one-step) alkylation processes.
Second, lower temperatures are preferred. Although past investigators have thought that such temperatures might offer advantages,
no one had realized that both hydrocarbon-continuous systems and/
or longer residence times in the reactor were then required in
order to operate satisfactorily at low temperatures.

Although further investigations are required in order to
determine the specific chemical steps of the second-step reactions
with butyl sulfate, butyl sulfate perhaps first decomposes to
release n-butene as follows:

$$\text{sec-}C_4H_9OSO_3H \longrightarrow \text{n-butene} + H_2SO_4$$

Thermodynamic information indicates that 2-butenes would be the
predominant olefins released. The resulting 2-butenes presumably
react in this process with a t-butyl cation to produce a trimethylpentyl ion. Hydride transfer from isobutane or more likely an
acid-soluble hydrocarbon would result in the production of a
trimethylpentane.

The relatively small amount of LE's and DMH's produced in
reactions with butyl sulfates were probably because of the high
ratios of isobutane to n-butenes in the reaction zone. Such a

high ratio would be expected because of the relatively slow rate of 2-butene formation as butyl sulfate decomposed at the low temperatures used. High isobutane to 2-butene ratios would result in the production of only small amounts of C_{12}^+ or higher ions; these ions are thought to be precursors for both LE's and DMH's.

The relatively large amount of HE's formed suggests that HE's were formed as a result of polymerization reactions in the acid phase. The n-butenes were presumably present primarily in the acid phase when they were released from butyl sulfate; such a conclusion is based on the fact that butyl sulfate is dissolved in the acid phase.

For the experiments in which the first-step reaction products of isobutylene were employed, reactions between high molecular weight olefins and isobutane were of major importance. Presumably a first step in the overall process was the protonation of the heavier olefins to form heavier isoalkyl cations. These cations apparently fragmented to a large extent to form mainly C_4-C_9 cations and olefins. The latter olefins also quickly protonated forming cations. Hydride transfer from isobutane or acid-soluble hydrocarbons resulted in the production of C_4 to C_9 isoparaffins. Heavy ends and conjunct polymers were, of course, produced to some extent during the second-step reactions, but most of these compounds were probably produced during the first-step reactions.

The results of the present investigation support the hypothesis that the main alkylation reactions occurred at the acid-hydrocarbon interface. Although such information does not apply directly to the phenomena occurring in industrial alkylation units, it is of interest that Kramer (13) supports this hypothesis in such units, and Doshi et al (11) have found that the interface is the location of degradation reactions for C_8 isoparaffins. One might expect the location of the main reactions to be the same in all three cases.

Two-step alkylation of isobutane with C_4 olefins not only helps clarify the mechanism of alkylation, but may be of commercial importance. Some of the advantages of the new process as compared to the conventional processes are as follows:

  (a) Quality of alkylate that is at least comparable when butyl sulfates are used in the second step of the process. In considering the alkylate quality, some thought has to be given to the best way for reporting the quality, either as research octane number (RON) or motor octane number (MON). 2,3,4-TMP and 2,2,3-TMP both have higher RON values than 2,2,4-TMP; the reverse is true for MON values. This point is of interest since the relative amounts of these three TMP's differ significantly in the alkylates produced in the two-step and the conventional processes. One of the favorable features of the alkylate produced by the new two-step process is the low amount of LE's; unfortunately the HE content has been

high in all experiments conducted to date.
(b) Isobutane-to-olefin ratios in the feed streams to the reactor can be low, perhaps as little as 1.5 and certainly no higher than 5. Cost of recovering and recycling the excess isobutane would be drastically reduced.
(c) The amount of acid that needs to be recycled will be low because of the low acid-to-hydrocarbon ratios used. The acid recycle costs would be decreased as compared to conventional processes.
(d) Agitation costs should be relatively low since hydrocarbon-continuous emulsions have fairly low viscosities.
Based on the present investigation, current disadvantages of the two-step alkylation process are as follows:
(a) Relatively poor quality alkylates are obtained when isobutylene is the feed olefin. Such low quality alkylates are produced in both the two-step process and the conventional alkylation process. There is hope that major improvements in quality can, however, be obtained for the two-step process. If t-butyl sulfates can be produced in high yields from isobutylene, then much improved alkylates likely would result. Tests to date indicate that the alkylate produced from mixed C_4 olefins is at least similar in quality to alkylates produced in commercial units. It is recommended that different composition acids and different temperatures be tested in future runs using isobutylene in the feedstock. Acids containing relatively large amounts of conjunct polymers, butyl sulfates (such as obtained using n-butenes), and/or water may result in improved performance.
(b) Large amounts of HE's in the alkylate. Further investigations need to be made to learn how HE production can be reduced. It is not known during what time period of batch runs using butyl sulfates that HE production is greatest. Future investigations with analytical techniques that will accurately measure the HE content should help develop improved methods of minimizing the formation of HE's.
(c) Refrigeration to provide rather low temperatures will be needed. The energy demands for refrigeration may be relatively moderate, however, since lesser amounts of isobutane and acid will need to be recycled.
Although residence times in the reactor will be relatively long, perhaps 1.0 - 1.5 hours, the volume of the reactor may not be greater than those of many current reactors. Such a conclusion is based on the considerable reductions of the volumes of excess isobutane used and the low ratios of acid to hydrocarbons in the reactor.
Further investigations of the two-step alkylation processes are recommended. There is a chance that alkylate can be produced cheaper by the two-step process and/or be of better quality.

Although the Purdue work has been restricted only to runs with
sulfuric acid, similarly promising processes can probably also
be developed with both HF and $AlCl_3$-type catalysts. Possibly
HF would even be a preferred catalyst since it is not an oxidizing
catalyst such as sulfuric acid often is.

The results reported by McCaulay (14) strongly support the
postulate that a two-step alkylation process will be feasible
with HF as the catalyst. Although he had emphasized the advantages
of intermittently adding the olefin to HF cataysts, he reports
two features of particular interest relative to the two-step
process.

(a)   Butyl fluorides are desired intermediates in his process.
(b)   Alkylates produced from butyl fluorides at low tempera-
      tures (-18° and -15°C) had RON values of 100.1 to 101.9.

For commercial development of a two-step process, provisions
must be provided so that high yields of the desired products are
obtained in both the first and the second steps of the process.
In some cases at least, it may be necessary to provide a different
reactor for each step in order to obtain the desired operating
conditions. McCaulay had apparently not considered such a system
for alkylation. Furthermore, he had apparently not succeeded in
producing butyl fluorides with either 1-butene or isobutylene
since the qualities of the alkylates obtained using these olefins
were significantly lower than those obtained using 2-butenes.

The results of Roebuck and Evering (15) who used $AlCl_3$-type
catalysts and temperatures down to -20°C can also be explained
by the production of esters with both 2-butene and ethylene. They
had found that improved alkylate qualities occurred as the reaction temperature was decreased, but yields of alkylate were
significantly reduced at lower temperatures. Probably they would
have obtained in all cases high yields of alkylate if they had
provided longer residence times for some of the lower temperature
runs; presumably sufficient time had not always been allowed for
complete reaction of all the esters.

The present investigation clearly indicates that in the range
of temperatures from at least -30 to 0°C a two-step alkylation
sequence is the predominant one. A logical question is, does a
similar two-step sequence occur at higher temperatures such as
employed in commercial reactors? Since some butyl sulfate and
some butyl fluorides are known to be formed in commercial reactors, it seems obvious that the two-step process is of at least
some importance at higher temperatures. The mechanism for alkylation will be considered in further detail in the next paper of
this series (10).

Literature Cited

1) Shlegris, R. J. and Albright, L. F., Ind. Eng. Chem. Process
    Des. Dev. 8, 92 (1969)
2) Li, K. W., Eckert, R. E., and Albright, L. F., Ind. Eng. Chem.

Process Des. Dev. <u>9</u>, 434 (1970)
3) Albright, L. F., Doshi, B. M., Ferman, M. A., and Ewo, A.,
"Two-Step Alkylation of Isobutane with C₄ Olefins: Reactions
of C₄ Olefins with Sulfuric Acid", This book, Chapter 6 (1977)
4) Linn, C. B., U.S. Pat. 2,307,799 (Jan. 12, 1943)
5) Matuszak, M. P., U.S. Pat. 2,387,162 (Oct. 16, 1945)
6) Goldsby, A. R., U.S. Pat. 2,420,369 (May 13, 1947)
7) Goldsby, A. R., U.S. Pat. 3,234,301 (Feb. 6, 1966)
8) Goldsby, A. R., U.S. Pat. 3,422,164 (Jan. 14, 1969)
9) Esso Research and Engineering Co. (by J. R. Lawley, et.al.)
Fr. 1,334,799 (Aug. 9, 1963)
10) Albright, L. F., "Mechanism for Alkylation of Isobutane with
Light Olefins", This book, Chapter 8 (1977)
11) Doshi, B. M., and Albright, L. F., Ind. Eng. Chem. Process
Des. Dev. <u>15</u>, 53 (1976)
12) Rushton, J. H., Nagata, S., and Rooney, T. B., AIChE Journal
10, 298 (1964)
13) Kramer, G. M., J. Org. Chem. <u>30</u>, 2671 (1965); U.S. Pat.
3,231,633 (Jan. 25, 1966)
14) McCaulay, D. A., U.S. Pat. 3,280,211 (Oct. 18, 1966)
15) Roebuck, A. K. and Evering, B. L., Ind. Eng. Chem. Prod. Res.
Dev. <u>9</u>, 76 (1970)

8

Mechanism for Alkylation of Isobutane with Light Olefins

LYLE F. ALBRIGHT

School of Chemical Engineering, Purdue University, West Lafayette, IN 47907

Considerable information has been obtained in the last several years at Purdue University that provides new insights to the mechanism of alkylation. Such information has been obtained by investigating conventional alkylation processes using sulfuric acid as a catalyst and several related reactions that employ sulfuric acid. These reactions include the following:

1) Conventional alkylations at 4 to 25°C employing either 1-butene, 2-butenes, isobutylene, mixtures of C_4 olefins, propylene, and trimethylpentenes (1,2,3,4).

2) Two-step alkylations at -30 to 0°C employing either n-butenes, isobutylene, mixed C_4 olefins, or 2,2,4-trimethylpentene-1 (5,6).

3) Degradation reactions that occur when trimethylpentanes, dimethylhexanes, or isobutane are contacted with sulfuric acid at -10° to 25°C (7).

4) Decomposition reactions of sec-butyl sulfate (when dissolved in sulfuric acid) at 0° to 25°C (5).

5) Reactions of C_4 olefins when contacted with sulfuric acid at 10°C (1,3).

When the results for these five reactions were considered in their entirety, two important observations were made. First the same isoparaffins were produced in each case. A total of 17 C_5-C_8 isoparaffins were detected in each product. Additional isoparaffins were probably also produced in minor quantities, but limitations with the analytical equipment did not permit conclusive evidence. Isobutane was also produced in each case based on the alkylation results of Hofmann and Schriesheim (8) who used C^{14}-tagged olefins and on the results for reactions 3,4, and 5 listed above. Within the limits of analytical precision, the same heavy end isoparaffins were also present in each hydrocarbon product. At least 18-20 isoparaffins in the C_9 and higher range were usually detected. The acid-soluble hydrocarbons were also apparently identical in all cases.

Dividing the hydrocarbon products into five families or groups indicated another important piece of information. These families

128

are trimethylpentanes (TMP's), light ends (LE's or C_5-C_7 isoparaffins), dimethylhexanes (DMH's), heavy ends (HE's), and acid-
soluble hydrocarbons (conjunct polymers, ester, red oil, acid
sludge, etc.). It was found that the composition of each family
was essentially a function of temperature only and was similar
for all five reactions. This later point was tested for especially
TMP's and LE's (6).

The relative importance of each family in the reaction
products often varied over wide ranges depending on the specific
reactions being tested and on the operating conditions employed.
At preferred operating conditions, up to 90% of the product was
TMP's for alkylates obtained by both conventional and two-step
alkylation processes. Much smaller fractions of TMP's were
produced by decomposition of butyl sulfate or by reaction of C_4
olefins in contact with sulfuric acid. Larger amounts of the
other four families were produced in such cases. As a general
rule, whenever the fraction of TMP's in the product decreased,
the relative importance of the other four families increased.

Based on these findings, it has been concluded that each of
the above families is formed from a common intermediate(s). Such
a conclusion is of importance since the combined results for the
five reactions outlined above can be employed to clarify and
develop an improved mechanism for alkylation. The mechanism
proposed in the paper is considered to be applicable for all five
reactions; special attention will, however, be given to conventional alkylation.

General Comments on Past Mechanisms

Production of TMP's has frequently been reported to be primarily by the following simple chain reaction (9,10):

$$t\text{-}C_4H_9^+ + \begin{array}{c} \text{2-butene} \\ \text{or} \\ \text{isobutylene} \end{array} \rightarrow TMP^+$$

$$TMP^+ + isobutane \rightarrow TMP + t\text{-}C_4H_9^+$$

where TMP^+ and TMP represent one of several trimethylpentyl cations
and trimethylpentanes respectively.

Clearly such a simple chain mechanism is not, however,
occurring in a two-step alkylation process (6). Neither does
this simple chain mechanism occur when butyl sulfate decomposes,
when TMP's are degraded by sulfuric acid, or when C_4 olefins
react in the presence of sulfuric acid. Furthermore, even during
conventional alkylations, there is extensive evidence that at
least some olefins react with the acid during the initial stages
of the reaction (2,3); in such cases, isobutane reacts to a
greater extent during the final stages of the reaction to form
TMP's. Such evidence strongly suggests that other methods for
production of TMP's are also often important, even during

conventional alkylation.

Differences of opinion still exist when sulfuric acid is used as the catalyst as to the main route for production of DMH^+'s (the precursors for DMH's). Although both Hofmann and Schriesheim (8) and Albright and Li (11) have indicated reasons why the reaction in which 1-butene reacts with a t-butyl cation to form DMH^+ is generally of little or no importance, recent authors still refer to the 1-butene route as an important one but without rebutting the earlier arguments. An even more complete summary of reasons why the 1-butene reaction is of little importance when sulfuric acid is the catalyst is reported below.

First almost identical amounts of DMH's and TMP's are generally produced when either 1-butene or 2-butenes are used as the olefins. A favorite explanation given is that 1-butene and 2-butene rapidly isomerize to form an equilibrium mixture of the two. Although isomerization is without question fast (5), the isomerization explanation alone is not sufficient as next indicated:

(1) The TMP/DMH ratio in alkylates produced by both 1-butene and 2-butene alkylations varies significantly - from about 1 to as high as 15 depending on the operating conditions employed. For example, increased levels of agitation act to increase the ratios significantly. Production of TMP^+'s (from 2-butenes) and of DMH^+'s (from 1-butene) would mean that the ratio of TMP/DMH would depend primarily on the equilibrium ratio of 2-butenes/1-butene. Obviously factors other than the equilibrium ratio are however of importance because this ratio does not change because of operating variables such as agitation.

(2) If isomerization of 1-butene (or 2-butenes) was rapid and if the main routes to produce TMP^+'s and DMH^+'s occurred when 2-butenes and 1-butene reacted with t-butyl cations, then use of a C^{14}-tagged n-butene would result in TMP's and DMH's with similar amounts of C^{14}. The results of Hofmann and Schriesheim (8) however indicate that DMH's (and also LE's and HE's) contain higher amounts of C^{14} than do TMP's.

(3) Purdue results (2,3) have shown that DMH's (and also LE's and HE's) are produced to a larger extent by initial reactions involving predominantly olefins; later reactions involving predominantly isobutane lead primarily however to TMP's. The 1-butene method for producing DMH^+'s would however result in simultaneous production of TMP's and DMH's.

Second, there is the indirect evidence: DMH's are often formed by routes in which 1-butene does not exist in significant amounts. One such reaction is alkylation of isobutane with isobutylene; more DMH's are produced than in comparable reactions

with either 1-butene or 2-butenes. It seems quite unlikely that
isobutylene isomerizes to 1-butene. A second reaction is the
degradation of TMP's or of isobutane when contacted with sulfuric
acid at 0°-30°C (7).
The above discussion was devoted to alkylations with sulfuric
acid. In the case of HF alkylations, considerable amounts of
DMH's are produced however by means of the 1-butene reaction as
will be considered in detail later.
Although the composition of sulfuric acid phase in conventional alkylation is known to have an important effect on
alkylate quality (1,4,8,12), the exact role of acid-soluble hydrocarbons during alkylation has as yet not been well defined.
During the start-up of an alkylation process that employs fresh
sulfuric acid as the catalyst, an apparent induction period
occurs (13). During this period, considerable acid-soluble
hydrocarbons are produced, but relatively little alkylate is
produced. A somewhat similar induction period also sometimes
occurs when TMP's are degraded in the presence of sulfuric acid
(7). Definitely there is a need to better describe the roles
of acid-soluble hydrocarbons during alkylation.
Increased levels of agitation have been found to be a
major factor in obtaining improved quality alkylates (1,3).
Agitation clearly has an important effect on the physical steps
of the overall process. The interfacial surface is one such
variable that would be increased by increased agitation. Careful consideration needs to be given in any proposed mechanism
as to the role of agitation.

Proposed Mechanism

The proposed mechanism is divided into six broad types of
reactions as follows:
1) Formation of t-butyl cations. Six methods have been
identified.
2) Reactions of C_4 olefins. Specific reactions have been
identified for both n-butenes and isobutylene when they
react with t-butyl and heavier cations.
3) Reactions of butyl sulfates.
4) Reactions of acid-soluble hydrocarbons.
5) Production of TMP's. TMP^+'s are formed by at least
three different methods.
6) Production of undesired (or at least less desired)
products such as LE's, DMH's, HE's, and acid-soluble
hydrocarbons.

Production of t-Butyl Cations. Table I outlines six methods
for production of t-butyl cations. The first two listed
(Reactions A and B) are considered to be of major importance in
most commercial reactors.
Reaction A is always of importance whenever isobutylene is

TABLE I

PRODUCTION OF t-BUTYL CATIONS

(A) Formation from isobutylene

Isobutylene + H^+ → t-$C_4H_9^+$

(B) Initiation when isobutane is contacted with acids containing acid-soluble hydrocarbons

R^+(polymer cations) + i-C_4H_{10} → RH(acid-soluble polymer)
$$+ \ t\text{-}C_4H_9^+$$

(C) Initiation when mixtures of isobutane and n-butenes are contacted with sulfuric acid

n-C_4H_8 + H^+ → sec-$C_4H_9^+$

sec-$C_4H_9^+$ + i-C_4H_{10} → n-butane + t-$C_4H_9^+$

(D) Reactions between isoalkyl cations and isobutane

C_5 and higher cations + i-C_4H_{10} → C_5 and higher + t-$C_4H_9^+$
isoparaffins

(E) Isomerization of 2-butene and production of t-$C_4H_9^+$

(1) Isomerization via TMP^+

t-$C_4H_9^+$ + 2-C_4H_8 ⇌ 2,2,3-TMP^+

2,2,3-TMP^+ ⇌ 2,2,4-TMP^+

2,2,4-TMP^+ ⇌ t-$C_4H_9^+$ + i-C_4H_8

i-C_4H_8 + H^+ → t-$C_4H_9^+$

Some t-$C_4H_9^+$'s are also formed by related reactions (fragmentation of heavier isoalkyl cations, as shown in Table III).

(2) Via conjunct polymers (11); R^+ is an acid-soluble cation.

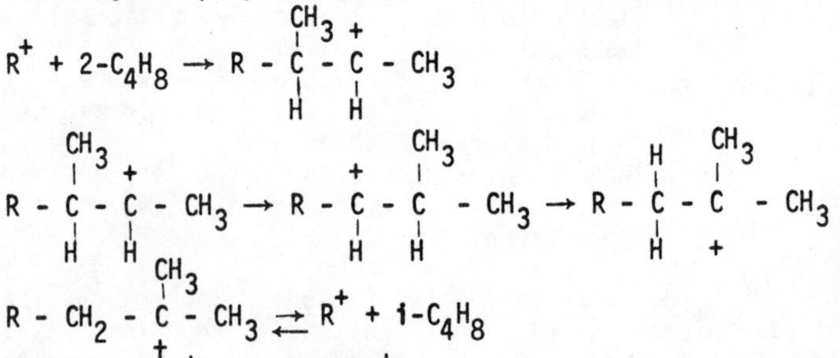

$$R^+ + 2\text{-}C_4H_8 \rightarrow R - \overset{\overset{\textstyle CH_3}{|}}{C} - \overset{\overset{\textstyle +}{|}}{C} - CH_3$$
$$\overset{|}{H} \quad \overset{|}{H}$$

$$R - \overset{\overset{\textstyle CH_3}{|}}{\underset{\underset{\textstyle H}{|}}{C}} - \overset{\overset{\textstyle +}{|}}{\underset{\underset{\textstyle H}{|}}{C}} - CH_3 \rightarrow R - \overset{\overset{\textstyle +}{|}}{\underset{\underset{\textstyle H}{|}}{C}} - \overset{\overset{\textstyle CH_3}{|}}{\underset{\underset{\textstyle H}{|}}{C}} - CH_3 \rightarrow R - \overset{\overset{\textstyle H}{|}}{\underset{\underset{\textstyle +}{|}}{C}} - \overset{\overset{\textstyle CH_3}{|}}{C} - CH_3$$

$$R - CH_2 - \overset{\overset{\textstyle CH_3}{|}}{\underset{\underset{\textstyle +}{|}}{C}} - CH_3 \rightleftharpoons R^+ + i\text{-}C_4H_8$$

i-C_4H_8 + H^+ ⇌ t-$C_4H_9^+$

Table I (continued)

(F) Oxidation of isobutane with sulfuric acid (7).

$$i\text{-}C_4H_{10} + 4H_2SO_4 \rightarrow t\text{-}C_4H_9^+ + 2H_3O^+ + 3HSO_4^- + SO_2$$

Other isoparaffins including both TMP's and DMH's are also
oxidized in a comparable manner (7,17,18).

Notes:

1) Reactions A, B, and D are predominant routes for most alkylations using sulfuric acid as the catalyst.
2) Reaction C is of more importance when HF is the catalyst.
3) Reactions E-1 and F are of importance for degradation of
isoparaffins with sulfuric acid (7).
4) Reactions E-2 are probable when isobutane reacts with acid-
soluble hydrocarbons to form alkylate (6).

TABLE II

REACTIONS OF C_4 OLEFINS, EXCLUDING $t\text{-}C_4H_9^+$ FORMATION

(G) Reaction with $t\text{-}C_4H_9^+$ to form C_8^+'s

1) 2-butene or $+$ $t\text{-}C_4H_9^+$ $\rightleftharpoons$ TMP$^+$'s
    isobutylene

2) 1-butene $+$ $t\text{-}C_4H_9^+$ $\rightleftharpoons$ DMH$^+$'s
    Reaction G-2 that produces DMH$^+$'s (dimethylhexyl
    cations) is of minor importance (11) when sulfuric
    acid is used as the catalyst, but is of major importance
    when HF is the catalyst (19).

(H) Isomerization of n-Butenes

1) 1-butene $\overset{H^+}{\rightleftharpoons}$ trans-2-butene $\overset{H^+}{\rightleftharpoons}$ cis-2-butene
    Isomerization occurs readily in presence of sulfuric
    acid (5), but less so in presence of HF (19).

2) 2-butenes $\rightarrow$ isobutylene (see Reactions E-1 and E-2 of
    Table I)

Table II (continued)

(I) Formation of Butyl Sulfates

1) n-butenes + H_2SO_4 $\overset{H^+}{\underset{\leftarrow}{\rightarrow}}$ sec-butyl sulfate

Butyl sulfate formation is the primary intermediate to alkylate when n-butenes are used with sulfuric acid at temperatures up to at least 0°C (6). Butyl fluorides are also produced at low temperatures with HF (23).

2) isobutylene + H_2SO_4 $\overset{H^+}{\underset{\leftarrow}{\rightarrow}}$ t-butyl sulfate

The importance of Reaction I-2 has not yet been proved; however, some as yet unidentified acid-soluble hydrocarbons (formed from isobutylene) react with isobutane to form alkylate (6).

(J) Complexing or reactions with acid-soluble ions (R^+)

R^+ + $\begin{array}{c} \text{2-butene} \\ \text{or} \\ \text{isobutylene} \end{array}$ $\overset{H^+}{\underset{\leftarrow}{\rightarrow}}$ $R\text{-}C_4H_8^+$ (see Reaction E-2)

(K) Polymerization of C_4 olefins

C_8^+ (TMP's or DMH's) + $\begin{array}{c} \text{2-butene} \\ \text{or} \\ \text{isobutylene} \end{array}$ $\rightleftarrows$ $i\text{-}C_{12}^+$, $i\text{-}C_{16}^+$, etc.

$i\text{-}C_8^+ \rightleftarrows i\text{-}C_8^= + H^+$

$i\text{-}C_{12}^+ \rightleftarrows i\text{-}C_{12}^= + H^+$

$i\text{-}C_{16}^+ \rightleftarrows i\text{-}C_{16}^= + H^+$

Polymers (olefins) were formed in small amounts when alkylate leaving a reactor was quickly quenched with a caustic solution (1). When, however, the acid and hydrocarbon phases were allowed to separate by decanting, few if any olefins were detected.

Isobutylene is particularly susceptible to formation of C_8 and heavier cations (5,6); such cations are probably important intermediates whenever isobutylene is used as olefin for alkylation.

Note: Reactions I and/or J are of major importance in conventional alkylation since olefins react in some manner with sulfuric acid to form acid-soluble hydrocarbons that later react with isobutane to form alkylate (2,3).

present in the feed steeam. It is well known that t-butyl cations form rapidly whenever isobutylene contacts strong acid. The reaction may not stop however at the t-butyl cation; as will be discussed later, polymerization reactions also occur forming i-C_8^+, i-C_{12}^+, etc., that eventually results in a wide variety of isoparaffins and conjunction polymers.

Although Reactions C have been widely accepted as the major method of initiating the formation of t-$C_4H_9^+$ whenever n-butenes are employed as feed olefins, Reaction B involving polymer cations (R^+) dissolved in the acid phase is considered to be more important especially when sulfuric acid is employed as the catalyst. Albright and Li (11) have previously discussed reasons for this choice, and recent experimental information further substantiates this choice. Two reasons are as follows:

1) No n-butane is produced when sulfuric acid is employed as the catalyst. When HF is used, however, some propane and n-butane are produced when propylene and n-butenes are used as olefins. Reactions C are of more importance when HF is employed probably because in part at least isobutane is more soluble in HF and the isobutane is hence more available to react.

2) When fresh sulfuric acid is employed, there is a start-up or induction period. During this period, little alkylation occurs, the quality of the initial alkylate is poor, and considerable amounts of acid-soluble hydrocarbons are produced. Some polymers are also formed. When, however, an acid containing dissolved hydrocarbons is employed, no induction period is noted. These acid-soluble hydrocarbons are highly ionized (14); they could readily accept hydride ions from isobutane, as shown in Reaction B.

The relative importance of Reactions B and C depend to at least some extent on the concentrations of R^+ and of sec-butyl cations at the reaction site, thought to be the interface between the two liquid phases (7,15,16). In general, there will always be a fairly high concentration of R^+ after the induction period, but the concentration of sec-butyl ions will presumably always be low. It also seems quite probable that R^+'s are more reactive than sec-butyl cations relative to isobutane.

Reaction D includes reactions of isoalkyl cations (C_5 and higher cations) with isobutane to form isoparaffins (that become part of the alkylate) and t-$C_4H_9^+$. This reaction is thought to be of some but not major importance for sulfuric acid alkylations, as will be discussed in more detail later. Most isoparaffins are, however, probably formed by the reaction of isoalkyl cations with RH (acid-soluble polymers).

Reactions E and especially F are probably of minor importance in commercial alkylators. Reactions E-1 and F were, however, of major importance whenever TMP's, DMH's, and isobutane degraded (or reacted) in the presence of sulfuric acid (7,17,18). Reactions E-2

were first suggested by Albright and Li (11); although no conclusive evidence has yet been obtained to prove these reaction
steps, the reactions appear most probable based on experimental
findings relative to the role of acid-soluble hydrocarbons during
alkylations, as described later. A seventh method for producing
t-butyl cations is described in Table III (to be presented later).

Main Olefin Reactions. Although some olefins react to form
t-butyl cations, i.e. to initiate the alkylation steps as shown
in Table I, most olefins react by Reactions G through K, as
summarized in Table II. Reaction G-1 is the reaction in which
either 2-butene or isobutylene reacts with a $t-C_4H_9^+$ to form a
TMP^+. This reaction is of major importance, and it is widely
accepted as being a key step in alkylation (9,10). It should be
emphasized, however, that the reaction as postulated here can be
but is not necessarily part of a chain sequence of reactions.
Instead much of the 2-butene or isobutylene is likely regenerated
as a result of the decomposition of butyl sulfates or of acid-
soluble hydrocarbons, as will be discussed later.

When sulfuric acid is employed as the catalyst, relatively
few dimethylhexyl cations (DMH^+) are produced by reactions of
1-butene with t-butyl cations (11), see Reaction G-2; the reasons
for this conclusion have already been discussed in detail earlier
in this paper. Yet when HF is the catalyst, considerably more
DMH's and less TMP's are produced with 1-butene than with 2-butene
(19). Isomerization (Reaction H-1) of 1-butene to 2-butenes is
considerably slower with HF as compared to sulfuric acid. Hence
Reaction G-2 is of importance in the presence of HF.

As indicated earlier (2,3) for sulfuric acid alkylations,
the rates of reaction for C_4 olefins are often higher during the
initial stages of alkylation than the rates of reaction for isobutane. In the case of n-butenes, isomerization of the n-butenes
occurs readily in the presence of sulfuric acid, see Reactions H-1
and H-2. Considerable information on these isomerizations were
reported earlier (5); the rates of isomerization increase as the
amounts of excess acid or as the acidity of the acid phase increase.
Certainly at the conditions employed in commercial alkylators,
isomerization would be very rapid.

Recent Purdue results (5) have now shown at -10°C or less
and in the presence of relatively little excess sulfuric acid
that n-butenes both isomerize and form sec-butyl sulfate, see
Reaction I-1. It seems safe to conclude that significant amounts
of sulfate formation also occur in commercial alkylation reactors
operated at about 5 to 15°C. The evidence supporting the formation
of sulfates is as follows:

    (a) Butyl sulfates are present in small quantities in
         sulfuric acid leaving commercial alkylation units (20).
         Butyl sulfates react readily with isobutane; the rates
         of reaction increase rapidly with increased temperatures
         in the range of -30° to 0°C (5). Even faster rates

presumably occur at 10° and 15°C. Consequently it is
not surprising that only small amounts of butyl sulfates
are detected in the exit acid.

(b) Experiments conducted in a four-staged reactor indicated
that acid-soluble hydrocarbons that in part could have
been butyl sulfates reacted in significant amounts with
isobutane at 10°C to form TMP's (2).

Albright and Li (11) have also proposed that at least some
olefins react with cations present in the acid-soluble hydrocarbons (Reaction E-2 of Table I and Reaction J of Table II).
No direct evidence has yet been obtained to support this hypothesis,
but on the other hand no evidence is known that is contrary to the
above postulate. Definitive evidence is clearly needed.

First effect results with isobutylene (5) clearly indicate
that it forms some acid-insoluble low polymers, mostly in the C_8
to C_{16} range (see Reactions G-1 and K of Table II), when it is
contacted with sulfuric acid. Small amounts of polymers (that
were olefins) were also noted when the hydrocarbon effluent of a
laboratory alkylation reaction was quenched with a caustic solution (1); 1-butene was employed as olefin in this system. It can
be concluded that during alkylation some C_8-C_{16} olefins are
present, but these olefins are ionized in the presence of the
acids to form C_8-C_{16} cations (that are highly reactive as will be
discussed later). Of the C_4 olefins, isobutylene is without
question the easiest to polymerize.

Isobutylene was also found to react in the presence of sulfuric acid to form acid-soluble hydrocarbons that reacted with
isobutane to form alkylate (5). Although the exact nature of
these acid-soluble hydrocarbons is not known, it is thought that
they are in part at least t-butyl sulfates (see Reaction I-2) or
that they complex (or react) with the conjunct polymer cations
(R^+), as shown in Reaction J. In both cases, isobutylene would
be liberated by reverse reactions, and the isobutylene would then
alkylate isobutane.

When propylene is in the feed, it will react with t-$C_4H_9^+$ to
form dimethylpentyl cations; this reaction is very similar to
Reaction G shown in Table II. In addition, propylene reacts with
sulfuric acid to form sec-propyl sulfates similar to Reaction I-1.
Goldsby (21,22) has described the production of sec-propyl sulfates; in some cases, di-sec-propyl sulfate is formed. These
propyl sulfates are more thermally stable than butyl sulfates.

Reactions of Butyl Sulfates. At 0°C or higher, sec-butyl
sulfate is unstable (5). When it decomposes in the absence of
isobutane, both a low quality (i.e. low octane number) hydrocarbon mixture and acid-soluble hydrocarbons are produced. The
products obtained by such decompositions are very similar to
those produced when pure C_4 olefins are contacted with sulfuric
acid at 10°C (1,3). It seems safe to conclude that the initial
reaction is the reverse of Reaction I-1 as shown next:

$$(I-1) \quad \text{sec-butyl sulfate} \rightarrow \begin{array}{c} \text{trans-2-butene} \\ \text{and} \\ \text{cis-2-butene} \end{array} + H_2SO_4$$

Some 1-butene may also be formed, but based on equilibrium considerations, 2-butenes are formed in much larger amounts. The rate of decomposition certainly increases with increased temperatures (5).

Since sec-butyl sulfate is soluble in the acid phase, presumably 2-butenes will be released primarily in the acid phase. Two points can be made based on the hypothesis that C_4 olefins are regenerated as a result of butyl sulfate decomposition in the alkylation reactors.

(a) The regeneration of C_4 olefins is relatively slow even at 10-15°C. The ratio of isobutane to regenerated C_4 olefins is hence high at the acid-hydrocarbon interface where the main alkylation reactions occur (7,15,16). High ratios for these two reactants promote high ratios of TMP's to LE's. This is because production of i-C_{12}^+ and heavier isoalkyl cations (Reactions K) are minimized. Of interest, there was a high TMP's/LE's ratio in the alkylate of two-step alkylations (6).

(b) In the acid phase itself, however, the ratio of isobutane to regenerated C_4 olefin may be low since isobutane is only slightly soluble in sulfuric acid. In the acid phase, some released C_4 olefins polymerize with the production of heavy ends and acid-soluble hydrocarbons (or conjunct polymers). Sufficient agitation should be provided then to transfer the olefins to the acid-hydrocarbon interface. This conclusion is supported by the experimental results for two-step alkylations in which relatively large amounts of heavy ends were produced.

Based on the above reasoning, rather significant differences likely occur when HF is used for two-step alkylations as compared to two-step processes using sulfuric acid. For HF alkylations, butyl fluorides have been reported to be intermediates (23). Since isobutane is much more soluble in HF as compared to sulfuric acid, there is a strong possibility that less heavy ends or conjunct polymers would be produced in two-step alkylations using HF.

When propyl sulfates are present, they will decompose relatively slowly to produce propylene and free sulfuric acid. The resulting propylene will then react primarily with t-butyl cations and heavier cations forming dimethylpentyl and still heavier cations.

When aluminum chloride catalysts are use, esters undoubtedly also are produced as they are with sulfuric acid and HF. The results of Roebuck and Evering (24) strongly imply that the olefins initially reacted forming esters that then slowly reacted with isobutane to form alkylate. The low yields noted for low temperature runs resulted because insufficient time was allowed for more complete reactions of the esters.

<u>Decomposition of Acid-Soluble Hydrocarbon Complexes</u>. As reported earlier (11), it has been postulated that 2-butenes may react with the acid-soluble hydrocarbon cations (see Reaction E-2 of Table I). Isobutylene may also react similarly. The resulting acid-soluble ion could later decompose via β-scission to release isobutylene:

$$R-CH_2-\overset{\overset{\displaystyle CH_3}{|}}{\underset{\underset{\displaystyle CH_3}{|}}{C^+}} \rightarrow R^+ + i\text{-}C_4H_8$$

The isobutylene released would be either at the acid-hydrocarbon interface or in the acid phase itself. The exact location at which the isobutylene is released would depend to a considerable extent on the location of the initial ion in the acid phase. Although no direct information is available, the acid-soluble hydrocarbon cations (14) appear to have surfactant-type characteristics and may be located primarily at the acid-hydrocarbon interface.

<u>Production of TMP$^+$ and TMP</u>. TMP's are the preferred hydrocarbons produced by alkylation, and they are produced in all cases by transfer of a hydride ion to TMP$^+$'s. There are at least three methods by which TMP$^+$'s are formed during alkylation:

(1) Reaction G-1 that was shown in Table II is always the predominant route for alkylations employing C_4 olefins. In this case, either 2-butene or isobutylene reacts with a t-$C_4H_9^+$. This reaction has been considered in detail by Schmerling (9,10).

(2) i-C_{12}^+, i-C_{16}^+, or other heavy isoalkyl cations fragment via β-scission. A wide variety of isoalkyl cations and iso-olefins are produced including TMP$^+$'s and trimethylpentene (see fragmentation of heavy isoalkyl cations as described in Table III). This method is an expansion of the one described by Hofmann and Schriesheim (8); it will be considered in detail later when the production of LE's and DMH's are discussed.

(3) Trimethylpentenes protonate rapidly to form TMP's. As indicated above, trimethylpentenes can be formed by fragmentation of heavy isoalkyl cations. Furthermore the protonations of trimethylpentenes and other olefins are to some extent at lease reversible. Hence, small concentrations of trimethylpentenes are present in alkylation reactors during especially the initial stages of alkylation; Mosby and Albright (1) for example found small amounts of olefins when the alkylation product was quenched and neutralized.

TABLE III
FRAGMENTATION OF HEAVY ISOALKYL CATIONS

(L) Fragmentation of Heavy Isoalkyl Cations

1) Isoalkyl cations in the C_8 and higher range fragment. β-scission of 2,2,4-TMP$^+$ as shown in Reaction E is an example. Isoalkyl cations that frequently fragment are thought to include i-C_{12}^+ and i-C_{16}^+. Examples of how they might fragment are as follows:

$$i\text{-}C_{12}^+ \rightarrow i\text{-}C_5^+ + i\text{-}C_7^= \text{ or } i\text{-}C_6^+ + i\text{-}C_6^=$$
$$i\text{-}C_{16}^+ \rightarrow i\text{-}C_5^+ + i\text{-}C_5^= + i\text{-}C_6^=$$

2) Protonation of olefins

Olefins rapidly protonate to form isoalkyl cations, e.g.

$$i\text{-}C_5^= + H^+ \rightleftharpoons i\text{-}C_5^+$$

Note: Isoalkyl cations formed by fragmentation and protonation of olefins generally contain 4 to 10 carbon atoms. Some t-$C_4H_9^+$'s are formed in this manner, as are both DMH$^+$'s and TMP's. This method is thought to be the major method for production of the precursors for LE's and for DMH's (when sulfuric acid is used as the catalyst).

TABLE IV
HYDRIDE ION TRANSFER STEPS

(M) Hydride Ion Transfer Steps

1) From Acid-Soluble Polymers (RH)

Isoalkyl cation + RH $\rightarrow$ Isoalkane + R$^+$ (polymer cations)

2) From Isobutane and Other Isoalkanes

Isoalkyl cation + i-C_4H_{10} $\rightleftharpoons$ Isoalkane + t-$C_4H_9^+$

Isoalkyl cation + Isoalkane $\rightleftharpoons$ Another + Another
                                        Isoalkyl   Isoalkane
                                        cation

This reaction is an expansion of Reaction D shown on Table I; the solubilities of isobutane and isoalkanes in the acid may be key factor affecting importance of this reaction. Higher isoalkanes are less soluble in acid phase.

3) From Isobutylene

Isoalkyl cation + i-C_4H_8 $\rightarrow$ Isoalkane + $CH_2\text{-}\overset{\overset{\displaystyle CH_3}{|}}{\underset{+}{C}}\text{-}CH_2$

Note: Hydride transfer leads to production of TMP's, DMH's, LE's, HE's, and isobutane.

Transfer of a hydride ion to an isoalkyl cation is often the rate controlling step in the overall reaction scheme (16). Isoalkyl cations including C_5 and higher ones react forming the various isoparaffin molecules found in alkylate, e.g., TMP^+'s react to form TMP's.

Schmerling (9,10) had originally postulated that a hydride ion transfer from isobutane was both the most important method of hydride transfer and also part of the chain set of reactions (see Reaction C of Table I and Reaction M-2 of Table IV). Other hydride transfer steps that have now been suggested include transfer with the acid-soluble hydrocarbons (RH), see Reaction M-1 of Table IV (8,11), and with isobutylene, see Reaction M-3 (8). Reaction M-3 is however considered to be of minor importance since only trace amounts of free isobutylene are likely ever present at the acid-hydrocarbon interface (the probable location of alkylation reactions); isobutylene quickly protonates to form $t\text{-}C_4H_9^+$. Reaction M-1 is considered to be more important than Reaction M-2 especially when sulfuric acid is used as the catalyst for the reasons listed as follows:

(a) Alkylate (or isoparaffins) are produced preferentially as compared to polymers (or olefins) only after acid-soluble hydrocarbons form in the acid phase. Hence, hydride transfer becomes important only after acid-soluble hydrocarbons are produced.

(b) When several percent of acid-soluble hydrocarbons are present in the acid, improved alkylate occurs as compared to fresh acids (that contain little or no acid-soluble hydrocarbons). Obviously such acid-soluble hydrocarbons would be present in significant amounts at or close to the acid-hydrocarbon interface and are hence available for hydride transfer. Isobutane is however only slightly soluble in sulfuric acid, and there is probably only limited availability of it for the reaction. When HF is used however, isobutane is much more soluble and reaction M-2 is likely of greater importance.

(c) If Reaction M-2 was of major importance when sulfuric acid is used, it would be expected that hydride transfer with other isoalkanes such as TMP's and LE's would also be of some importance; there is no evidence that such isoparaffins react to any appreciable extent.

Additional discussions on the role of acid-soluble hydrocarbons as a reactant during alkylation has been discussed earlier (4,8,11). The acid-soluble hydrocarbons clearly act as a reservoir of hydride ions. Hydride ions are furnished to the acid-soluble hydrocarbons primarily from isobutane (Reaction B of Table I) and they are withdrawn from the hydrocarbons principally by Reaction M-1.

The acid-soluble hydrocarbons furthermore also tend to increase the solubility of isobutane and other isoparaffins in the

acid phase. In addition to the viscosity, the interfacial surface
tension is changed. Such factors are important since transfer
of reactants to the interface and/or of products away are obviously rate-controlling steps. Such a conclusion is obvious since
agitation was found to be an important, if not most important,
operating variable relative to the quality of alkylate produced
(3).

Production of Light Ends and Dimethylhexanes. Hofmann and
Schriesheim (8) proposed two mechanisms and evidence to support
each. The mechanism based on fragmentation of large isoalkyl
cations (see Table III) is preferred as the major route for
production of both LE's and DMH's; the mechanism proposed here
has been modified somewhat as compared to the original one that
suggested i-C_{12}^+ cations fragmented. Evidence supporting the
fragmentation route is presented as follows:

(1) LE's and relatively small isoparaffins are produced
     when heavy olefins are used as feedstock for alkylation.
     The products obtained when isobutylene or 2,2,4-tri-
     methylpentene-1 is used as the olefin for alkylation
     resulted in almost identical products including LE's
     and DMH's (6,19). Even heavier olefins produce LE's,
     and fragmentation is obviously of major importance.
(2) Degradation of TMP's or DMH's in the presence of sulfur-
     ic acid leads to the formation of significant, and even
     major amounts of isobutane and LE's (7). Fragmentation
     reactions obviously must be occurring.
(3) The ratio of LE's to DMH's produced is generally relative-
     ly constant regardless of the operating conditions or
     the olefin used. Similar constant ratios are also
     obtained in the products of conventional alkylations,
     two-step alkylations (in which butyl sulfates react),
     and degradation reactions between TMP's and sulfuric
     acid. This finding suggests that a common intermediate
     is involved in the production of most LE's and DMH's.
     1-Butene cannot be considered this intermediate since
     it is sometimes present only in at most minute quanti-
     ties. 1-Butene is mentioned since it has been suggested
     as the C_4 olefin that leads to production of DMH's (9,
     10). Although 1-butene is probably of minor importance
     for production of DMH's when sulfuric acid is used as a
     catalyst (11), considerable DMH's are likely formed
     via Reaction G-2 (with 1-butene) when HF is the catalyst.
(4) There is a question whether i-C_{12}^+ cations are the
     most important cations that fragment, as was originally
     proposed by Hofmann and Schriesheim (8). If i-C_{12}
     cations were the most important intermediates, then the
     amounts of C_5 and C_7 isoparaffins produced would likely
     be essentially identical (see Reaction L). Yet iso-
     pentane (the only C_5 isoparaffin produced) is generally

formed in much larger amounts than the C_7 isoparaffins. Possibly part of the C_5 or C_7 cations or iso-olefins react, but C_5's and not C_7's would probably be the more reactive. The experimental results can, however, be explained if i-C_{16} cations were the key intermediates. Two isopentane molecules and one C_6 isoparaffin could be produced from each i-C_{16} cation (as shown by Reaction L); such a ratio is approximated in many alkylates.

(5)   i-C_{12} and heavier isoalkyl cations are produced during alkylation primarily because of polymerization reactions involving olefins. The C^{14} results of Hofmann and Schriesheim (8) indicating that higher levels of olefin reacted during production of DMH's and LE's as compared to TMP's is consistent with this postulate.

(6)   Production of heavy isoalkyl cations, including i-C_{12} and i-C_{16} cations, would logically occur during conventional alkylations, two-step alkylations, degradation of TMP's and DMH's, and reactions between olefins and sulfuric acid at 10°C or higher. In the case of alkylations, higher ratios of isobutane to C_4 olefin at the reaction site result in relatively smaller amounts of heavy isoalkyl cations and hence lower amounts of DMH's and LE's. This experimental finding is consistent with the fragmentation mechanism.

Hofmann and Schriesheim (8) also postulated that DMH's were formed when isobutylene reacted with an allylic cation (formed via Reaction M-3) as follows:

(N)   CH$_2$=C=CH$_2$ + CH$_2$=C(CH$_3$)CH$_3$ → CH$_2$=C(CH$_3$)—CH$_2$-CH$_2$-C(CH$_3$)-CH$_3$

Transfer of two hydride ions and one proton would result in DMH. Since the methyl groups could migrate on the chain, DMH's other than 2,5-DMH could be produced. Some t-butyl cations dissociate into isobutylene and protons; hence this method could occur during alkylation with olefins other than isobutylene. Reaction N is probably only of minor importance in most cases, however, since only small concentrations of free isobutylene are thought to occur at the acid-hydrocarbon interface; most isobutylene quickly protonates to form t-butyl cations.

Production of Heavy Ends and Acid-Soluble Hydrocarbons. Heavy ends (HE's) are produced in large amounts by reactions involving primarily olefins; these reactions are primarily polymerization-type reactions. The olefins are quite soluble in the acid phase, but isobutane is not. High ratios of isobutane to olefins and high levels of agitation are necessary to minimize formation of heavy ends (3). The large isoalkyl cations formed by polymerization-type reactions obviously fragment to some extent to produce some C_9 and C_{10} isoparaffins (found in appreciable quantities in

HE's). It is thought however, that significant fractions of the
HE's result from acid phase reactions as compared to reactions
at or close to the acid-hydrocarbon interface where most alkylation reactions occur; this latter conclusion is based on the
results for two-step alkylations (6).

Both acid-insoluble and acid-soluble hydrocarbons are also
produced to a small, and probably as a rule insignificant, extent
from isobutane and other isoparaffins when sulfuric acid is used
as the catalyst (7). The sulfuric acid acts as a mild oxidizing
agent producing sulfur dioxide even at temperatures as low as $0°C$.

The acid-soluble hydrocarbons apparently have in all cases
similar structures to the acid-soluble hydrocarbons removed from
used alkylation acids by Miron and Lee (25). These hydrocarbons
contained C_5 cyclic rings; were highly unsaturated and frequently
had conjugated double bonds; had fairly high molecular weights;
and had carbon-to-hydrogen ratios higher than those of the
reacting hydrocarbons. Clearly polymerization and dehydrogenation
steps occur in the production of these acid-soluble hydrocarbons.
Deno and associates (26,27) have shown that some t-butyl cations
react forming cyclopentenyl cations. These latter ions likely
are precursors to acid-soluble hydrocarbons formed during alkylation.

Purdue findings have shown that the viscosity and color of
the acid-soluble hydrocarbons removed from the acid increased
as the acid was used and as the acid was aged in drums stored at
room temperature. Sulfur dioxide was also produced. Undoubtedly
some olefins added to or polymerized with the acid-soluble cations
during alkylation. The net result of these reactions was that
the molecular weight of these acid-soluble hydrocarbons increased.
The sulfur dioxide produced indicates that some sulfuric acid
had acted as an oxidizing agent (and apparently also as a dehydrogenation agent causing double bond formation in the acid-soluble
hydrocarbons). Recently Doshi and Albright (7) presented evidence
that suggests significant fractions of the acid-soluble hydrocarbons are often produced while the acid phase is outside of
the alkylation reactor, i.e., while the acid is being separated
from the organic phase and while it is being recycled. Clearly
more information is, however, needed to determine the complete
role of acid-soluble hydrocarbons during alkylation.

Physical Steps During Alkylation. Physical steps that occur
during alkylation of isobutane have key roles in affecting the
overall process and in the composition (and quality) of the
alkylate produced (3). The transfer of isobutane to the reaction
site (which is at or close to the interface between the two phases)
is in general the controlling physical step. It is affected by
several operating variables. Agitation is, of course, an obvious
variable since it affects not only the isobutane transfer step
but also the interfacial area. Other variables that affect the
isobutane transfer step include the isobutane-to-olefin feed ratio,

the residence time in the reactor, the concentration of inerts in hydrocarbon phase, the volumetric ratio of acid to hydrocarbon phases, and which phase is the continuous phase in the emulsion. Variables affecting both the isobutane transfer step and also the chemical kinetics include temperature, composition of acid, and the specific olefin used (4).

Although the main alkylation reactions are thought to occur at or close to the interface between the acid and hydrocarbon phases (7,15,16), the acid boundary layer at the interface seems more probable. This conclusion is based on the fact that other isoparaffins (including LE's, TMP's, DMH's, and HE's) appear to be much less reactive relative to hydride transfer steps as compared to isobutane; yet many of these isoparaffins contain one or more tertiary carbon-hydrogen bonds. These heavy isoparaffins are, however, even less soluble in the acid phase than isobutane.

Conclusions

Alkylation of isobutane with light olefins in the presence of sulfuric acid involves considerable more chemical steps than have generally been considered in the past. Olefins tend to react to a higher extent during the initial stages of the alkylation reactions and isobutane during the latter stages. Most alkylation reactions occur at the acid-hydrocarbon interface, but reactions resulting primarily in the production of heavy ends and acid-soluble hydrocarbons may occur primarily in the acid phase. Alkyl sulfates and acid-soluble hydrocarbons have key roles relating to the overall chemistry of alkylation. Although experimental data for alkylations using sulfuric acid and primarily C_4 olefins were employed in development of the proposed mechanism, the mechanism has been expanded to include alkylations with HF, other acids, and other olefins.

Literature Cited

1. Mosby, J. F. and Albright, L. F., Ind. Eng. Chem. Prod. Research Dev. 5, 183 (1966).
2. Shlegeris, R. J. and Albright, L. F., Ind. Eng. Chem. Process Des. Dev., 8, 92 (1969).
3. Li, K. W., Eckert, R. E. and Albright, L. F., Ind. Eng. Chem. Process Des. Dev. 9, 434 (1970).
4. Li, K. W., Eckert, R. E. and Albright, L. F., Ind. Eng. Chem. Process Des. Dev. 9, 441 (1970).
5. Albright, L. F., Doshi, B. M., Ferman, M. A., and Ewo, A., "Two-Step Alkylation of Isobutane with C_4 Olefins: Reactions of C_4 Olefins with Sulfuric Acid", This Book, Chapter 5, 1977.
6. Albright, L. F., Doshi, B. M., Ferman, M. A., "Two-Step Alkylation of Isobutane with C_4 Olefins: Reaction of Isobutane with Initial Reaction Products", This Book, Chapter 7, 1977.

7. Doshi, B. M. and Albright, L. F., Ind. Eng. Chem. Process Des. Dev. <u>15</u>, 53 (1976).
8. Hofmann, J. E. and Schriescheim, A., J. Am. Chem. Soc., <u>84</u>, 953-961 (1962).
9. Schmerling, L., Ind. Eng. Chem., <u>45</u>, 1447 (1953).
10. Schmerling, L., in "Friedel Crafts and Related Reactions in Alkylation and Related Reactions". Olah, G. A. Ed., Interscience Publishers (1964).
11. Albright, L. F. and Li, K. W., Ind. Eng. Chem. Process Des. Dev. <u>9</u>, 451 (1970).
12. Albright, L. F., Houle, L., Sumutka, A. M., and Eckert, L. E., Ind. Eng. Chem. Process Des. Dev. <u>11</u>, 446 (1972).
13. Stewart, T. D. and Calkins, W. H., J. Amer. Chem. Soc. <u>70</u>, 1006 (1948).
14. Leftin, H. P. and Hobson, M. C., Advan. Catalysis <u>14</u>, 189, (1963).
15. Kramer, G. M., This Book, Chapter 1 (1977).
16. Kramer, G. M., J. Org. Chem. <u>30</u>, 2671 (1965); U.S. Pat. 3,231,633 (Jan. 25, 1966).
17. Hofmann, J. E., J. Org. Chem. <u>29</u>, 3627 (1964).
18. Kramer, G. M., J. Org. Chem. <u>32</u>, 920 (1967).
19. Innes, R. A., This Book, Chapter 3, 1977.
20. Goldsby, A. R. and Gross, H. H., U.S. Pat. 3,083,247 (March 26, 1963).
21. Goldsby, A. R., U.S. Pat. 3,234,301 (Feb. 6, 1966).
22. Goldsby, A. R., U.S. Pat. 3,422,164 (Jan. 14, 1969).
23. McCaulay, D. A., U.S. Pat. 3,280,211 (Oct. 18, 1966).
24. Roebuck, A. K. and Evering, B. L., Ind. Eng. Chem. Product Res. Develop. <u>9</u>, 76 (1970).
25. Miron, S. and Lee, R. J., J. Chem. Engr. Data <u>8</u>, 150 (1963).
26. Deno, N. C., Chem. Engr. News, Oct. 5, 1964, pp 88-100.
27. Deno, N. C., Boyd, D. B., Hodge, J. D., Pittman, C. U., and Turner, J. O., J. Amer. Chem. Soc. <u>86</u>, 1745 (1964).

Free Radical-Induced Monoethylation with Ethylene

LOUIS SCHMERLING

UOP Inc., Des Plaines, IL 60016

The free radical-induced reaction of a saturated hydrocarbon with ethylene produces chiefly greasy telomers formed by the addition of a hydrocarbyl radical at one end of a polymethylene chain and a hydrogen atom at the other end ($\underline{1}$). For example, when a solution of a catalytic amount of di-$\underline{t}$-butyl peroxide in cyclohexane is heated at 130-140°C (the temperature range at which the rate of decomposition of the peroxide is appreciable) under ethylene pressure, there is formed viscous, high-boiling, grease-like product, some of which is solid; it is a mixture of ethylcyclohexane and telomers (i.e., butyl-, hexyl-, octyl-, and higher molecular weight alkylcyclohexanes in which the alkyl groups contain an even number of carbon atoms, ranging from 2 to 40 or more). The reaction apparently involves the following free radical chain mechanism:

$$(CH_3)_3COOC(CH_3)_3 \longrightarrow 2(CH_3)_3CO \cdot \qquad (1)$$
$$(CH_3)_3CO \cdot + \underline{c}\text{-}C_6H_{12} \longrightarrow (CH_3)_3COH + \underline{c}\text{-}C_6H_{11} \cdot \qquad (2)$$
$$\underline{c}\text{-}C_6H_{11} \cdot + CH_2=CH_2 \longrightarrow \underline{c}\text{-}C_6H_{11}CH_2CH_2 \cdot \qquad (3)$$
$$\underline{c}\text{-}C_6H_{11}CH_2CH_2 \cdot + nCH_2=CH_2 \longrightarrow \qquad (4)$$
$$\underline{c}\text{-}C_6H_{11}(CH_2CH_2)_nCH_2CH_2 \cdot$$
$$\underline{c}\text{-}C_6H_{11}(CH_2CH_2)_nCH_2CH_2 \cdot + \underline{c}\text{-}C_6H_{12} \longrightarrow \qquad (5)$$
$$\underline{c}\text{-}C_6H_{11}(CH_2CH_2)_nCH_2CH_3 + \underline{c}\text{-}C_6H_{11} \cdot$$

The cyclohexyl radical formed in Equation 5 starts a new cycle as in Equation 3.

Inhibition of Telomerization

Procedure. A glass liner containing the quantities of saturated hydrocarbon, alkyl chloride or ether, of hydrochloric acid (aqueous or anhydrous) and di-$\underline{t}$-

butyl peroxide tabulated in Tables I thru V was sealed
into an Ipatieff-type rotating autoclave of 850 ml
capacity. Ethylene was charged to an initial pressure
of from 15 to 40 atm. (about 0.5 to 1.3 mols ethylene)
and the autoclave was rotated while being heated during
four hours from 130 to 140°C, a range at which the
peroxide decomposition proceeds at a practical rate.
The final pressure was usually about 5-25 atm. at room
temperature. By charging ethylene to a lower initial
pressure, lesser amounts of ethylene were employed.
The pressure drop showed that the reaction was usually
completed in less than two hours, but heating was
continued for four hours to ensure optimum conversions.
The autoclave was allowed to stand overnight, the
gaseous product was discharged, the autoclave was
opened and the liquid product (usually chiefly in the
liner) was recovered. The organic product was washed
with water, dried and distilled and/or inspected by
gas chromatography.

Experiments with gaseous reactants (propane and
isobutane, Table III) were carried out by charging the
liquefied gas from a weighed stainless-steel sampling
cylinder into the sealed autoclave containing hydrochloric acid and, when used, a liquid alkane after
which ethylene was charged and the autoclave was
rotated and heated.

Cyclohexane. It is the purpose of the present
paper to discuss a unique modifier for the reaction
which very markedly inhibits the telomerization and
results in the formation of ethylcyclohexane as the
principal single reaction product together with some
butylcyclohexane and diethylcyclohexanes.

The reaction modifier consists of hydrogen
chloride which may be added as the anhydrous gas or as
an aqueous solution. Its effectiveness in the monoethylation of saturated hydrocarbons, alkyl halides
and ethers will be described. Thus, when 47 g (0.47
mol HCl) of 38% hydrochloric acid and 96 g (1.14 mols)
of cyclohexane containing 6 g of dissolved di-t-butyl
peroxide were heated at 130-140° under 40 atm. (1.3
mols) initial ethylene pressure (51 atm. at 130°C) in a
glass liner in a rotating autoclave of 850 ml capacity
for four hours, the final pressure at room temperature
was 22 atm. The product included 24 g (18 mol % yield)
of ethylcyclohexane and 18 g of higher-boiling product,
consisting chiefly of n-butylcyclohexane, diethylcyclohexanes, and n-hexylcyclohexane (and dialkyl isomers)
plus a smaller amount of higher-boiling compounds
(Expt. 2, Table I). On the other hand, when the same

Table I

Reaction of Cycloalkanes with Ethylene

Expt.[a]	c-RH	mols	C_2H_4	Kind	mols	Kind	g	%
	<-Reactants, mols-->			<--HCl--->		<---Chief Products--->		
1	C_6H_{12}	1.07	1.3[ℓ]	H_2O	0	c-C_6H_{11}Et	1	1
						Higher mol wt	29[b]	
2	C_6H_{12}	1.14	1.3	38%	0.5	c-C_6H_{11}Et	29	18
						Higher mol wt	18[c]	
3	C_6H_{12}	1.15	1.3	100%	0.1	c-C_6H_{11}Et	20	16
						Higher mol wt	26[c]	
4	C_6H_{12}	1.37	1.0	38%	0.2	c-C_6H_{11}Et	36	32
						Higher mol wt	19[c]	
5	C_6H_{12}	1.72	0.5	100%	0.1	c-C_6H_{11}Et	19	34
						Higher mol wt	3[c]	
6	C_6H_{12}	1.15	0.6	38%[d]	0.1	c-C_6H_{11}Et	24[e]	36
						$C_{10}H_{20}$	7[e]	17
						Higher mol wt	6	
7	C_6H_{11}Me	1.28	1.3	38%	0.1	MeC_6H_{10}Et	22[f]	13
						c-$C_6H_x C_5 H_y$	7[g]	7
						Higher mol wt	12	
8	C_5H_{10}	1.57	1.3	38%	0.2	c-C_5H_9Et	31	24
						C_9H_{18}	16[h]	20
						$C_{11}H_{22}$	7[i]	10
						Higher mol wt	21	
9[j]	C_5H_9Me	1.34	1.3	19%	0.5	MeC_5H_8Et	38[k]	26
						Higher mol wt	28	

(a) The cycloalkane solution of 0.04 mol di-t-butyl peroxide and hydrochloric acid (or anhydrous hydrogen chloride) was heated at 130-140° during four hours in a glass liner in an Ipatieff-type rotating autoclave under ethylene pressure.
(b) Chiefly greasy liquid and wax.
(c) Includes much butylcyclohexane and diethylcyclohexanes.
(d) Deuterochloric acid in 99% D_2O (38% by weight).
(e) Includes deuterated product; cf. text.
(f) Consists of 1-methyl-1-ethylcyclohexane mixed with smaller amounts of cis-1-methyl-3-ethylcyclohexane, cis- and trans-1-methyl-4-ethylcyclohexane and other isomers.
(g) Mixture of methylbutylcyclohexanes, and methyldiethylcyclohexanes.
(h) B.p. 150-155°, gc shows 3 major peaks.
(i) B.P. 192-200°; gc shows 2 major peaks.
(j) Heated at 120-128° during 16 hours.
(k) Chiefly 1-methyl-1-ethylcyclopentane.
(ℓ) 40 atm pressure.

amounts of cyclohexane and the peroxide were heated
with ethylene under the same conditions, but in the
presence of water instead of hydrochloric acid, the
product contained only 1 g of ethylcyclohexane and 29 g
of higher-molecular weight liquid, greasy and waxy
telomers (Expt. 1). It was shown by gas chromatography
(gc) followed by mass spectroscopy (ms) that ethyl-,
n-butyl-, n-hexyl- and n-octylcyclohexane made up a
large portion of the product, the remainder consisting
of decreasing amounts of higher n-alkylcyclohexanes
having an even number of carbon atoms up to at least
n-tetracontylcyclohexane. Minor amounts of tri- and
tetraethylcyclohexane were also detected.

The higher-boiling products obtained in the hydrogen chloride-promoted reaction consisted chiefly of
$C_{10}H_{20}$, $C_{12}H_{24}$, and $C_{14}H_{28}$ hydrocarbons. The $C_{10}H_{20}$
fraction was a mixture of n-butylcyclohexane and
diethylcyclohexanes. The $C_{12}H_{24}$ product contained n-
hexylcyclohexane and more ethylbutylcyclohexanes than
triethylcyclohexanes. n-Octylcyclohexane was
definitely identified in the $C_{14}H_{28}$ mixture.

Another byproduct of the reaction was a mixture of
alkyl chlorides in which the alkyl groups had an even
number of carbon atoms. These were obviously formed by
telomerization of ethylene and hydrogen chloride (2).

$$Cl\cdot + CH_2=CH_2 \longrightarrow ClCH_2CH_2\cdot \xrightarrow{n-C_2H_4}$$
$$ClCH_2CH_2(CH_2-CH_2)\cdot n \xrightarrow{HCl}$$
$$Cl(CH_2CH_2)_nCH_2CH_3 + Cl\cdot \qquad (6)$$

A higher consumption of the hydrogen chloride in
the aqueous acid, accompanied by a higher yield of
alkyl chloride, occurred when concentrated hydrochloric
acid rather than diluted acid (15-20%) was used.

As is apparent from the results summarized in
Table 1, both hydrochloric acid (dilute or concentrated)
and anhydrous hydrogen chloride were effective in
greatly increasing the ratios of ethylcyclohexane to
higher-boiling product.

Mechanism. The hydrogen chloride apparently
behaves as a chain transfer agent, furnishing a
hydrogen atom to the cyclohexylethyl radical (1) and
forming a chlorine atom which starts a new cycle by
abstracting hydrogen from cyclohexane, producing a new
cyclohexyl radical and regenerating hydrogen chloride.

$$\underset{\underset{\underset{1}{\sim}}{}}{\text{S}}\text{-CH}_2\overset{\bullet}{\text{C}}\text{H}_2 + \text{HCl} \longrightarrow \underset{}{\text{S}}\text{-CH}_2\text{CH}_3 + \text{Cl}\cdot \quad (7)$$

$$\text{Cl}\cdot + \underset{}{\text{S}} \longrightarrow \text{HCl} + \underset{}{\text{S}}^{\bullet} \quad (8)$$

The cyclohexylethyl radical (1) appears to abstract hydrogen more rapidly from hydrogen chloride than from cyclohexane and more rapidly than it adds to ethylene to yield cyclohexylbutyl or higher molecular weight radicals. Chain termination presumably occurs by condensation or disproportionation of a pair of free radicals, or by any other of the chain-terminating reactions which normally occur. No attempt was made to isolate or identify the so-formed byproducts.

The pathway by which the hydrogen chloride inhibits the telomerization and thus produces monoethylcyclohexane was supported by use of a solution of deuterium chloride (38%) in heavy water (99% deuterium oxide) as promoter under the standard conditions. The reaction mixture included 1.15 mols of cyclohexane and 0.1 mol of deuterium chloride as the acid. Ethylcyclohexane was obtained in 36% yield; diethylcyclohexane plus n-butylcyclohexane, in 17% yield. Mass spectrometric analysis of the monoethylcyclohexane showed that despite the high ratio (138:1) of hydrogen in cyclohexane to deuterium in deuterium chloride in the reaction mixture, the monoethylated compound consisted of only about 60% undeuterated compound; there was present about 32% monodeuterated, about 8% dideuterated, and about 1% trideuterated hydrocarbon. About 60% of the deuterium was in the ethyl group and the remaining 40% was in the cyclohexane ring. About 19% of the recovered cyclohexane was monodeuterated, 3% was dideuterated and less than 1% was trideuterated. Similarly, the polyethylated product contained mono- and polydeuterated compounds. The recovered acid was chiefly HCl dissolved in H_2O plus HOD.

It may be concluded that the role of the deuterochloric acid in the ethylation reaction is that suggested by Equations 7 and 8. The deuterium chloride acts as a chain transfer agent by terminating the cycle and yielding a chlorine atom which initiates a new cycle by abstracting hydrogen from cyclohexane:

$$\underline{c}\text{-C}_6\text{H}_{11}\text{CH}_2\overset{\bullet}{\text{C}}\text{H}_2 + \text{DCl} \longrightarrow \underline{c}\text{-C}_6\text{H}_{11}\text{CH}_2\text{CH}_2\text{D} + \text{Cl}\cdot \quad (9)$$
$$\text{Cl}\cdot + \underline{c}\text{-C}_6\text{H}_{12} \longrightarrow \text{HCl} + \underline{c}\text{-C}_6\text{H}_{11}\cdot \quad (8)$$

Formation of ethylcyclohexane which is deuterated
in the ring suggests that the cyclohexyl radical formed,
for example as in Equation 8, underwent intermediate
exchange with the deuterium chloride.

$$\text{S}^{\cdot} + DCl \longrightarrow \text{S}^{D} + Cl^{\cdot} \tag{10}$$

$$\text{S}^{D} + Cl^{\cdot} \longrightarrow \text{S}\!\!-\!\!D + HCl \tag{11}$$

$$\text{S}\!\!-\!\!D + C=C \longrightarrow \text{S}\!\!-\!\!D \underset{\underset{\sim}{2}}{\overset{}{}} C\!-\!C^{\cdot} \tag{12}$$

$$\underset{\underset{\sim}{2}}{\overset{\dfrac{HCl}{\text{or } \underline{c}\text{-}C_6H_{12}}}{\longrightarrow}} \text{S}\!\!-\!\!D\,_{C-CH} + Cl^{\cdot} \text{ (or } \underline{c}\text{-}C_6H_{11}{}^{\cdot}) \tag{13}$$

$$\xrightarrow{DCl} \text{S}\!\!-\!\!D\,_{C-CD} + Cl^{\cdot} \tag{14}$$

Equation 10 explains the recovery of deuterated
cyclohexane. Equation 14 illustrates a pathway for the
formation of a polydeuterated compound.
Calculations showed that the amount of deuterium
in the cyclohexane, ethylcyclohexane and other hydrocarbons was actually more than that present in the
deuterium chloride charged. Hence, it was concluded
that the deuterium present in the deuterium oxide also
underwent exchange. This was confirmed by analysis of
the recovered acid solution. Its mass spectrum showed
that most of the D_2O was converted into DOH and H_2O.
Exchange of the proton of hydrogen chloride with
deuterium of the "heavy water" occurred via formation
of a hydrated proton, $(H \cdot OD_2)^+$ in which the hydrogen
and deuteriums are indistinguishable.
Decreasing the ratio of ethylene to cyclohexane
from 1.14 to 0.73 resulted in an increase in the yield
of monoethylcyclohexane (from 18 to 32%) and a marked
decrease in the yield of higher molecular weight compounds (cf. Expts. 2 and 4). Further decrease in the
ratio to 0.29 resulted in a smaller increase in the
yield of monoethylcyclohexane (34%, Expt. 5).

Other Modifiers. No other substance seems to
exhibit the chain-transfer activity shown by hydrogen
chloride. Even hydrobromic acid was ineffective.

Other reagents which did not inhibit telomerization
induced by di-t-butyl peroxide at 130-140° included
hydriodic acid, acetic acid, trichloroacetic acid, tri-
fluoroacetic acid, phosphoric acid, sulfuric acid,
hydrogen sulfide, hydrogen, and ammonium chloride. The
chief condensation product obtained when any of these
reagents replaced hydrogen chloride or hydrochloric
acid were high-molecular weight telomers of greasy
("Vaseline-like") or waxy composition.

Other Cycloalkanes. Hydrogen chloride promoted
monoethylation of other cycloparaffins (e.g., cyclopentane, methylcyclopentane and methylcyclohexane) and
many paraffins (e.g., propane, isobutane, n-pentane,
isopentane, 2,2-dimethylbutane, 2,3-dimethylbutane,
n-heptane and 2,2,4-trimethylpentane).
The reaction of methylcyclohexane with an equimolar quantity of ethylene in the presence of di-t-
butyl peroxide and hydrochloric acid resulted in
ethylation both at the tertiary carbon atom and at
secondary carbon atoms (Expt. 7). The methylethylcyclo-
hexane which was obtained in 13% yield consisted
(according to infrared (ir) comparison with authentic
samples) chiefly of 1-methyl-1-ethylcyclohexane mixed
with smaller amounts of 1-methyl-cis-3-ethylcyclohexane
and 1-methyl-cis- (and trans-)4-ethylcyclohexane, and
other isomers. The compounds produced by the reaction
of 2 mols of ethylene per mol of cyclohexane (7% yield)
consisted of a mixture of methylbutylcyclohexanes and
methyldiethylcyclohexanes.
Products formed by the reaction of cyclopentane
with 1, 2 or 3 molecular proportions of ethylene were
the chief products of the reaction of approximately
equimolar amounts of the cycloalkane and the olefin
under the standard conditions (Expt. 8).
1-Methyl-1-ethylcyclopentane mixed with less than
5% of 1-methyl-cis-3-ethylcyclopentane was obtained in
26% yield by the ethylation of methylcyclopentane
(Expt. 9). The presence of methyldiethylcyclopentane
and methylbutylcyclopentane in the higher molecular
weight by-product was suggested by gc coupled with ms.

Liquid Alkanes. The peroxide-induced reaction of
ethylene with a molar excess of n-pentane in the
presence of hydrochloric acid produced heptanes in 27%
yield together with only a relatively small amount of
higher-boiling product (Expt. 10, Table II). The chief
heptane was 3-methylhexane (3) which was obtained in
more than three times the quantity of 3-ethylpentane (4).

Table II

Reaction of Liquid Alkanes with Ethylene

Expt.[a]	<-Reactants, mols-> RH	mols	C_2H_4	<--HCl---> Kind	mols	<-----Chief Products-----> Kind	g	%
10	$\underline{n}$-C_5H_{12}	1.78	0.8	35%	0.3	C_7H_{16}	22^b	27
						Higher mol wt	6	
11	$\underline{i}$-C_5H_{12}	1.76	0.8	38%	0.3	C_7H_{16}	25^c	31
						Higher mol wt	3	
12	Me_3CEt	1.46	0.8	20%	0.4	Octanes	6^d	6
						Higher mol wt	3	
13	BIP[e]	1.34	1.3	18%	0.5	$2,3,3$-$Me_3C_5H_9$	10^f	7
						$2,3,3$-$Me_3C_7H_{13}$	6	7
						Higher mol wt	17^g	
14	$\underline{n}$-C_7H_{16}	1.58	0.6	100	0.1	C_9H_{20}	8^h	10
						Higher mol wt	8	
15	TMP[i]	0.63	0.6	19	0.3	$2,2,4,4$-$Me_4C_6H_{10}$	7	8
						$2,2,4,4,5$-$Me_5C_7H_{11}$[j]	1	2
						Higher mol wt	2	

(a) The cycloalkane solution of 0.04 mol di-$\underline{t}$-butyl peroxide and hydrochloric acid (or anhydrous hydrogen chloride) was heated at 130–140° during four hours in a glass liner in an Ipatieff-type rotating autoclave under ethylene pressure.

(b) Includes 69% 3-methylhexane and 21% 3-ethylpentane (by gc + ir).

(c) Includes 86% 3,3-dimethylpentane and 13% 2,3-dimethylpentane.

(d) Consists of 75% 2,2,3-trimethylpentane, 19% 2,2- and 6% 3,3-dimethylhexane.

(e) 2,3-Dimethylbutane (i.e., biisopropyl).

(f) Also about 2 g acetone and $\underline{t}$-butyl alcohol and 3 g 1-chloro-hexane.

(g) Includes 1-chloroalkanes with even-numbers of carbon atoms and branched-chain paraffins, most of which had the 2,3,3-trimethyl-alkane structure.

(h) 75% 3-Methyloctane and small amounts of 3- and 4-ethylheptane.

(i) 2,2,4-Trimethylpentane.

(j) Or 2,2,4,4-tetramethyloctane.

Based only on the presence of four hydrogen atoms
attached to the second carbon atom from either end of
the chain and the presence of two hydrogen atoms
attached to the middle carbon atom, it would be expected that there would be formed twice as much 3 as 4.

```
C-C-C-C-C        C-C-C-C-C
    |  .             .
    ↓  C=C           ↓C=C
C-C-C-C-C        C-C-C-C-C
    |  C-C·            C-C·
HCl |-Cl ·       HCl |-Cl·
    ↓                ↓
C-C-C-C-C        C-C-C-C-C
      C-C             C-C
     3                4
```

More than three times as much 3 as 4 was obtained
because free radicals tend to abstract a hydrogen atom
attached to a penultimate carbon atom of an n-alkane
more readily than one attached to a more internally-
located carbon atom (see, for example, the results of
the ethylation of n-heptane, Expt. 14). There are in
pentane twice as many penultimate carbon atoms as the
internally located carbon atom; however, the more ready
reaction at the penultimate carbon atom is confirmed by
the products obtained with n-heptane.

The reaction of isopentane with ethylene under the
same conditions as those used for n-pentane resulted in
about the same yield of heptanes (31%) and a very small
amount of higher-boiling product (Expt. 11). The
heptanes consisted of more than 6.5 times as much 3,3-
dimethylpentane as 2,3-dimethylpentane. It may be
concluded that abstraction of the single hydrogen atom
attached to the tertiary carbon atom takes place much
more readily than does abstraction of one of the two
hydrogen atoms attached to the secondary (penultimate)
carbon atoms.

A low yield of octanes was formed when ethylene
was heated with 2,2-dimethylbutane (i.e., neohexane) in
the presence of di-t-butyl peroxide and 20% hydrochloric acid (Expt. 12). The low conversion was probably due to the difficulty in abstracting a hydrogen
atom attached to a neopentyl carbon atom (i.e., a
secondary carbon atom attached to the tertiary carbon
atom of a t-butyl group). The principal octane (about
75% of the octane product) was 2,2,3-trimethylpentane
formed by ethylation at the secondary carbon atom; 2,2-
dimethylhexane formed by condensation at a primary
carbon atom (the neohexyl carbon atom) was obtained in

19% yield while 3,3-dimethylhexane formed by reaction at one of the three primary neopentyl ions was obtained in only 6% yield.

On the other hand there was a fair amount of conversion when equimolar quantities of 2,3-dimethyl-butane and ethylene were reacted (Expt. 13). 2,3,3-Trimethylpentane was produced by ethylation at a tertiary carbon atom and more highly end-chain ethylated compounds (up to 2,3,3-trimethylpentadecane) were formed by telomerization as were chloroalkane by-products.

As indicated in the discussion of the reaction of n-pentane, ethylation of n-heptane took place chiefly at a penultimate carbon atom, yielding 3-methyloctane as the principal product; only small amounts of 3- and 4-ethylheptane were produced (Expt. 14).

Rather low yields of ethylation products were obtained by reaction of 2,2,4-trimethylpentane with ethylene. It seems probable that the abstraction of hydrogen even from the tertiary carbon atom was diffi-cult because it was a neohexyl carbon atom. It was suggested by gc and ms analysis that the major reaction did occur at the tertiary carbon atom, yielding 2,2,4,4-tetramethylhexane (5). It was mixed with about 10% of its weight of what appears to be either 2,2,4,4,5-pentamethylheptane (6) formed by ethylation at the external secondary carbon atom of 5 or 2,2,4,4-tetramethyloctane (7) formed by end-chain ethylation.

$$
\underset{\overset{\displaystyle C}{|}}{C}-\underset{\overset{\displaystyle C}{|}}{C}-\overset{\displaystyle \cdot}{C}-C \;+\; C{=}C \;\xrightarrow[-Cl\cdot]{HCl}\; C-\underset{\overset{\displaystyle C}{|}}{C}-C-\underset{\overset{\displaystyle C}{|}}{C}-C-C \tag{15}
$$

$$
\underset{5}{}
$$

$$
C-\underset{\overset{\displaystyle C}{|}}{C}-C-\underset{\overset{\displaystyle C}{|}}{C}-\overset{\displaystyle \cdot}{C}-C \;+\; C{=}C \;\xrightarrow[-Cl\cdot]{HCl}\; C-\underset{\overset{\displaystyle C}{|}}{C}-C-\underset{\overset{\displaystyle C}{|}}{C}-C-C-C \tag{16}
$$

$$
\underset{6}{}
$$

$$
C-\underset{\overset{\displaystyle C}{|}}{C}-C-\underset{\overset{\displaystyle C}{|}}{C}-C-\overset{\displaystyle \cdot}{C}. \;+\; C{=}C \;\xrightarrow[-Cl\cdot]{HCl}\; C-\underset{\overset{\displaystyle C}{|}}{C}-C-\underset{\overset{\displaystyle C}{|}}{C}-C-C-C-C \tag{17}
$$

$$
\underset{7}{}
$$

Gaseous Alkanes. Normally gaseous alkanes con-taining a secondary or tertiary carbon atom could be monoethylated by the peroxide-induced reaction with ethylene in the presence of hydrogen chloride (Table III). The yields were markedly increased by adding a normally liquid saturated hydrocarbon as

Table III

Reaction of Gaseous Alkanes with Ethylene

Expt.[a]	<-Reactants, mols--> RH	mols	C_2H_4	<--HCl---> Kind	mols	<---Chief Products---> Kind	g	%
16	C_3H_8	1.70	0.8	19%	0.3	Liquid product	< 1	
17	C_3H_8[b]	1.70	0.8	18%	0.3	C_5H_{12}	3[c]	5
						C_7H_{16}	2[d]	5
						Higher mol wt	15[e]	
18	$\underline{i}-C_4H_{10}$	0.96	1.0	38%	0.4	Alkanes	10[f]	
						Alkenes	25	
						Alkyl chlorides	5	
19[g]	$\underline{i}-C_4H_{10}$[h]	1.72	0.8	19%	0.3	Me_3CEt	5	7
						Higher mol wt	39[i]	
20	$\underline{i}-C_4H_{10}$[j]	1.72	0.8	19%	0.3	Me_3CEt	9	13
						Higher mol wt	28[k]	

(a) The hydrocarbon reactant, hydrochloric acid, and 0.04 mol di-
    $\underline{t}$-butyl peroxide was heated at 130-140° during four hours in a
    glass liner in an Ipatieff-type rotating autoclave under
    ethylene pressure.

(b) Also 0.17 mol $\underline{n}$-heptane.

(c) 81% Isopentane (by gc).

(d) Other than $\underline{n}-C_7H_{16}$. Gc suggests 75% 2,3-dimethylpentane
    and/or 2-methylhexane, 18% 3,3-dimethylpentane and 7% 3-
    methylhexane.

(e) Includes about 6 g nonanes (28% yield based on $\underline{n}$-heptane).

(f) Gc showed the presence of 28.6 wt % 2,2-dimethylbutane, 4.9%
    2-methylpentane, 14.9% 2,2-dimethylhexane, 5.5% other dimethyl
    hexanes, 12.0% decanes, 7.6% dodecanes and 4.9% tetradecanes.

(g) Heated at 125-135° during 16 hours.

(h) Also 0.37 mol cyclohexane.

(i) Contains at least 0.08 mol (22% yield) ethylcyclohexane.

(j) Also 0.51 mol $\underline{n}-C_7H_{16}$.

(k) Other than $\underline{n}-C_7H_{16}$, about 0.25 mol of which was recovered.

solvent for the gaseous alkane. For example, very
little reaction occurred when propane was heated at
130-140° with ethylene, di-t-butyl peroxide, and hydrochloric acid (Expt. 14). Repetition of the reaction
under the same conditions but in the added presence of
n-heptane solvent resulted in a 5 mol-% yield of
pentane (chiefly isopentane) and a 5 mol-% yield of
branched-chain heptanes (chiefly 2,3- and 3,3-dimethyl-
pentanes (Expt. 17). A comparatively large amount of
nonanes (largely 3-methyloctane together with smaller
amounts of 3- and 4-ethylheptane) were formed, suggesting that about 28 mol-% of the heptane solvent (used in
smaller amount) was monoethylated.

The low yield of pentane by the reaction of propane
in the absence of solvent was probably due to the fact
that most of the propane was in the vapor phase whereas
the ethylation reaction proceeds principally in the
liquid phase.

Similarly, in the absence of added liquid saturated
hydrocarbon solvent, ethylation of isobutane even in
the presence of hydrochloric acid yielded a mixture of
2,2-dimethylbutane, 2-methylpentane, 2,2- and other dimethylhexanes, and uncharacterized decanes, dodecanes
and tetradecanes (Expt. 18). The paraffins, however,
comprised only 30 vol-% of the reaction product which
contained 58 vol-% liquid olefins and 12 vol-% alkyl
chlorides. Alkylation occurred in better yield when
the reaction was carried out in the presence of cyclohexane solvent which caused more of the reaction to
occur in the liquid phase (Expt. 19). 2,2-Dimethyl-
butane was obtained in about 7-mol % yield and ethylcyclohexane in about 22 mol-% yield; a minor amount of
2-methylpentane was formed by condensation as a primary
carbon atom. Use of n-heptane as solvent resulted in a
13 mol-% yield of 2,2-dimethylbutane (Expt. 20).
Higher-molecular weight product consisted largely of
alkanes having an even number of carbon atoms. The
secondary carbon atoms in heptane were apparently less
readily ethylated than the tertiary hydrogen atom in
isobutane.

Alkyl Chlorides. Hydrochloric acid also promoted
the peroxide-induced monoethylation of chloroalkanes.
Results with n-butyl chloride, isobutyl chloride and
isopentyl chloride are described briefly in this paper.

Very little reaction occurred when n-butyl
chloride w a s heated at 130-140° with ethylene in the
presence of di-t-butyl peroxide (Expt. 21, Table IV).
On the other hand, a fair yield of reaction products
was obtained when the experiment was repeated in the

Table IV

Reaction of Alkyl Chlorides with Ethylene

Expt.[a]	<-Reactants, mols-> RCl	mols	C_2H_4	<--HCl---> Kind	mols	<---Chief Products-----> Kind	g	%
21	n-BuCl	0.79	0.8	None	0.0	Little reaction		
22	n-BuCl	0.79	0.6	38%	0.2	EtCHMeCH$_2$CH$_2$Cl	7	10
						PrCHClEt	4	6
						n-C$_6$H$_{13}$Cl	1	4
						C$_8$H$_{11}$Cl's	6	14
						Miscellaneous	1	
23	i-BuCl	0.50	0.6	19%	0.2	EtCMe$_2$CH$_2$Cl	5	8
						PrCHMeCH$_2$Cl	1	2
						n-HexCl	1	4
						i-PrCHClEt	0.5	1
						BuCMe$_2$CH$_2$Cl	0.2	5
24	i-PenCl	0.31	0.6	None	0.0	Telomer[b]	12	
25	i-PenCl	0.49	0.6	19%	0.2	EtCMe$_2$CH$_2$CH$_2$Cl	12	18
						i-BuCClEt		
						n-HexCl }	2	
						C$_9$H$_{19}$Cl's		

(a) The alkyl chloride containing 0.04 mol dissolved di-t-butyl peroxide was heated at 130–140° with the hydrochloric acid during four hours in a glass liner in an Ipatieff-type rotating autoclave under ethylene pressure.

(b) Boiled chiefly from 200° to above 400°.

added presence of concentrated hydrochloric acid (Expt.
22). The major product was 1-chloro-3-methylpentane
which was formed by ethylation at the penultimate
carbon atom of the butyl chloride via abstraction of
the hydrogen atom attached to the penultimate carbon
atom chiefly by the chlorine atom formed from the
hydrogen chloride. Product which was formed in about
one-half the amount of this chlorohexane was an isomer,
3-chlorohexane, formation of which involved alkylation
at the carbon atom holding the chlorine atom. The
presence of a very minor quantity of 1-chlorohexane was
also noted. This was formed either by alkylation at
the methyl carbon atom or by telomerization of ethylene
and hydrogen chloride.

More isomeric octyl chlorides than hexyl chlorides
were formed by a telomerization reaction involving a
second molecule of ethylene. Nuclear magnetic resonance (nmr) suggested that the product was a mixture of
about equal weight of primary and secondary chlorides.
Since many octyl chloride isomers exist, no conclusion
was reached as to the probable structures.

It may be concluded that ethylation of n-butyl
chloride occurs most readily at the penultimate carbon
atom and next most readily at the carbon atom holding
the chlorine atom. It probably also occurs at the
other primary carbon atom.

The peroxide-induced ethylation of isobutyl
chloride in the presence of 19% hydrochloric acid
involved monoethylation at all of the carbon atoms in
the molecule (Expt. 23). As might be expected, the
chief product was 1-chloro-2,2-dimethylbutane, produced
via abstraction of the hydrogen atom attached to the
tertiary carbon atom. Also formed were 1-chloro-2-
methylpentane (ethylation at a methyl group) and 3-
chloro-2-methylpentane (ethylation at the carbon atom
holding the chlorine atom). Some 1-chlorohexane was
also obtained; in this case, its formation was undoubtedly due to telomerization of the ethylene with
hydrogen chloride rather than by a reaction involving
the isobutyl chloride.

The presence of small amounts (a) of 1-chloro-2,2-
dimethylhexane (9) formed by further reaction with
ethylene of the radical (8) responsible for the formation of the major product (1-chloro-2,2-dimethylbutane)
and (b) of 3-chloro-2-methylheptane (11) by further
reaction with ethylene of the radical (10) responsible
for the production of 3-chloro-2-methylpentane was also
observed.

$$Cl-C-\underset{\underset{C}{|}}{C}-C \xrightarrow[-HCl]{Cl\cdot} Cl-C-\underset{\underset{C}{|}}{\overset{\cdot}{C}}-C \xrightarrow{C_2H_4} Cl-C-\underset{\underset{C}{|}}{\overset{\overset{C}{|}}{C}}-C-C\cdot \qquad (18)$$

$$\underset{\underset{\sim}{8}}{}$$

$$8- \left[\begin{array}{l} \xrightarrow{HCl} Cl-C-\underset{\underset{C}{|}}{\overset{\overset{C}{|}}{C}}-C-C + Cl\cdot \qquad\qquad (19) \\[2em] \xrightarrow{C_2H_4} Cl-C-\underset{\underset{C}{|}}{\overset{\overset{C}{|}}{C}}-C-C-C-C\cdot \xrightarrow[-Cl\cdot]{HCl} Cl-C-\underset{\underset{C}{|}}{\overset{\overset{C}{|}}{C}}-C-C-C-C \quad (20) \end{array} \right.$$

$$Cl-C-\underset{\underset{C}{|}}{\overset{\cdot}{C}}-C \xrightarrow[-HCl]{Cl\cdot} Cl-\overset{\cdot}{C}-\underset{\underset{C}{|}}{C}-C \xrightarrow{C_2H_4} \cdot C-C-\overset{\overset{Cl}{|}}{C}\!-\!\underset{\underset{C}{|}}{C}-C \qquad (21)$$

$$\underset{\underset{\sim}{10}}{}$$

$$10- \left[\begin{array}{l} \xrightarrow{HCl} C-C-\overset{\overset{Cl}{|}}{C}\!-\!\underset{\underset{C}{|}}{C}-C + Cl\cdot \qquad\qquad (22) \\[2em] \xrightarrow{C_2H_4} \cdot C-C-C-C-\overset{\overset{Cl}{|}}{C}\!-\!\underset{\underset{C}{|}}{C}-C \xrightarrow[-Cl\cdot]{HCl} C-C-C-C-\overset{\overset{Cl}{|}}{C}\!-\!\underset{\underset{C}{|}}{C}-C \; (23) \end{array} \right.$$

$$\underset{\underset{\sim}{11}}{}$$

The product formed in largest amount by the hydrochloric acid-promoted and peroxide-induced reaction of isopentyl chloride with ethylene was also that formed by alkylation at the tertiary carbon atom, namely 1-chloro-3,3-dimethylpentane (12) (Expt. 25). The remaining constituents of the reaction product were all obtained in very minor amount and were all alkyl chlorides. Among these were 4-chloro-2-methylhexane (14), 1-chlorohexane (formed by telomerization) and some chlorononanes including 5-chloro-3,3-dimethylheptane (16) formed by ethylation of 12, 4-chloro-2-methyloctane (15) and 1-chloro-3-methyloctane (17).

$$ClC-C-\underset{\underset{C}{|}}{C}-C \xrightarrow{Cl\cdot} Cl-C-C-\underset{\underset{C}{|}}{\overset{\cdot}{C}}-C \xrightarrow{C_2H_4} ClC-C-\underset{\underset{C}{|}}{\overset{\overset{C}{|}}{C}}-C-C\cdot$$

$$\text{ClC-C-}\overset{\overset{\text{C}}{|}}{\underset{\underset{\text{C}}{|}}{\text{C}}}\text{-C-C·} \quad\longrightarrow\quad \begin{cases} \xrightarrow[-Cl·]{HCl} \quad \text{ClC-C-}\overset{\overset{\text{C}}{|}}{\underset{\underset{\text{C}}{|}}{\text{C}}}\text{-C-C} \\ \qquad\qquad\qquad \underset{\widetilde{}}{12} \\ \\ \xrightarrow{C_2H_4} \quad \text{ClC-C-C-}\overset{\overset{\text{C}}{|}}{\underset{\underset{\text{C}}{|}}{\text{C}}}\text{-C-C-C-C·} \end{cases}$$

$$HCl \Big|{-Cl·}$$

$$\overset{\downarrow}{\text{Cl-C-C-}}\overset{\overset{\text{C}}{|}}{\underset{\underset{\text{C}}{|}}{\text{C}}}\text{-C-C-C-C} \qquad (24)$$

$$\underset{\widetilde{}}{13}$$

Also, $\text{ClG-C-}\underset{\underset{\text{C}}{|}}{\text{C}}\text{-C} \xrightarrow{Cl·} \text{Cl}\overset{·}{\text{C}}\text{-C-}\underset{\underset{\text{C}}{|}}{\text{C}}\text{-C}$

$$\text{Cl}\overset{·}{\text{C}}\text{-C-}\underset{\underset{\text{C}}{|}}{\text{C}}\text{-C} \xrightarrow{C_2H_4} \text{·C-C-}\overset{\overset{\text{Cl}}{|}}{\text{C}}\text{—C-}\underset{\underset{\text{C}}{|}}{\text{C}}\text{-C} \xrightarrow[-Cl·]{HCl} \text{C-C-}\overset{\overset{\text{Cl}}{|}}{\text{C}}\text{-C-}\underset{\underset{\text{C}}{|}}{\text{C}}\text{-C}$$

$$\underset{\widetilde{}}{A} \qquad\qquad\qquad \underset{\widetilde{}}{14}$$

$$\underset{\widetilde{}}{A} \xrightarrow{C_2H_4} \text{·C-C-C-C-}\overset{\overset{\text{Cl}}{|}}{\text{C}}\text{—C-}\underset{\underset{\text{C}}{|}}{\text{C}}\text{-C} \xrightarrow[-Cl·]{-HCl} \text{C-C-C-C-}\overset{\overset{\text{Cl}}{|}}{\text{C}}\text{—C-}\underset{\underset{\text{C}}{|}}{\text{C}}\text{-C} \quad (25)$$

$$\underset{\widetilde{}}{15}$$

$$12 \xrightarrow[-HCl]{Cl·} \text{Cl-}\overset{·}{\text{C}}\text{-C-}\underset{\underset{\text{C}}{|}}{\text{C}}\text{-C-C} \xrightarrow[-HCl]{C_2H_4} \text{C-C-}\overset{\overset{\text{Cl}}{|}}{\text{C}}\text{—C-}\underset{\underset{\text{C}}{|}}{\text{C}}\text{-C-C} \qquad (26)$$

$$\underset{\widetilde{}}{16}$$

$$\text{Cl-C-C-}\underset{\underset{\text{C}}{|}}{\text{C}}\text{-C} \xrightarrow[-HCl]{Cl·} \text{Cl-C-C-}\underset{\underset{\text{C}}{|}}{\text{C}}\text{-C·} \xrightarrow{C_2H_4} \text{Cl-C-C-}\underset{\underset{\text{C}}{|}}{\text{C}}\text{-C-C-C·}$$

$$\underset{\widetilde{}}{B}$$

$$\underset{\widetilde{}}{B} \xrightarrow{C_2H_4} \text{Cl-C-C-}\underset{\underset{\text{C}}{|}}{\text{C}}\text{-C-C-C-C·} \xrightarrow[-Cl·]{HCl} \text{Cl-C-C-}\underset{\underset{\text{C}}{|}}{\text{C}}\text{-C-C-C-C}$$

$$\qquad\qquad\qquad\qquad\qquad\qquad\qquad\qquad\qquad (27)$$

$$\underset{\widetilde{}}{17}$$

Under the same conditions but in the absence of
hydrochloric acid, telomerization occurred (Expt. 24).
A trace amount of monoethylation product may have been
produced, but the reaction product boiled chiefly from
about 200° to higher than 400°.
   It may be concluded that primary alkyl chlorides
undergo peroxide-induced, hydrogen chloride-promoted,
alkylation with ethylene to yield products formed by
alkylation at a tertiary carbon atom, at a penultimate
secondary carbon atom, or at a primary carbon atom
holding a chlorine atom. In the absence of hydrochloric acid, n-butyl chloride underwent little
peroxide-induced reaction with ethylene presumably
because hydrogen chloride is necessary for propagating
the reaction chain via abstraction of hydrogen from the
hydrogen chloride to produce the ethylated product and
a chlorine atom which maintains the chain by abstraction from the alkyl chloride.

   Ethers. Low yields of several compounds were
obtained when ethyl ether was heated at 130-140° under
ethylene pressure in the presence of di-t-butyl
peroxide and hydrochloric acid (Expt. 26, Table V).
Ethylation took place in the normal fashion to yield
ethyl sec-butyl ether by monoethylation together with
at least three ethers having eight carbon atoms; di-
sec-butyl ether, (ethylation at both secondary carbon
atoms of the ethyl ether), ethyl 1-methyl-1-ethylpropyl
ether (ethylation at the tertiary carbon atom of the
primary product) and ethyl-1-methylpentyl ether formed
by telomerization of the primary radical with two
molecules of ethylene). Some C_{10} ethers were also
formed.
   n-Butyl chloride was produced in about the same
amount as ethyl sec-butyl ether. Its formation presumably involved telomerization of ethylene with
hydrogen chloride.
   Ethylation of tetrahydrofuran took place in good
yield under the standard conditions (Expt. 17). The
principal product was 2-ethyltetrahydrofuran mixed with
a smaller amount of 3-ethyltetrahydrofuran. Diethyl-
tetrahydrofurans and butyltetrahydrofurans were also
formed as was much high-boiling product.

Table V

Reaction of Ethers with Ethylene

Expt.[a]	Ether	mols	C_2H_4	Kind	mols	Kind	g	%
	<----Reactants, mols->			<--HCl--->		<----Chief Products---->		
26	Et_2O	1.49	1.0	38	0.2	EtOBu-s	4	4
						n-BuCl	4	9
						$C_8H_{16}O$[b]	5	8
						Miscellaneous[c]	1	
27	THF[d]	1.97	1.0	38	0.2	EtC_4H_7O[e]	22	22
						Higher-boiling	25	
28	p-D[f]	1.05	1.3	None		p-D[f]	82	89
						EtDioxane	8	7
						Telomer[g]	46	
29	p-D[f]	0.92	1.3	19	0.2	2-EtDioxane	3	3
						$(ClCH_2CH_2)_2O$	13	10
						$ClC_2H_4OC_2H_4OH$	3	3
30	p-D[f]	0.78	0.6	38	1.2	2-EtDioxane	8	11
						$(ClCH_2CH_2)_2O_2$	20	18
						$ClC_2H_4OC_2H_4OH$	3	3

(a) The ether, hydrochloric acid (when used) and 0.04 mol di-t-butyl peroxide were heated at 130-140°C in a glass liner in an Ipatieff-type rotating autoclave under ethylene pressure.

(b) Includes Et-O-C(Me)Et$_2$, Et-O-CH(Me)Bu, and (s-Bu)$_2$O.

(c) Includes C_{10} ethers and n-$C_6H_{13}Cl$ and n-$C_8H_{11}Cl$.

(d) Tetrahydrofuran.

(e) Mixture of 2- and 3-ethyltetrahydrofuran.

(f) p-Dioxane.

(g) Product of higher molecular weight than the ethyldioxane. Most (44 g) boiled above 195° and was a very viscous yellow oil at room temperature.

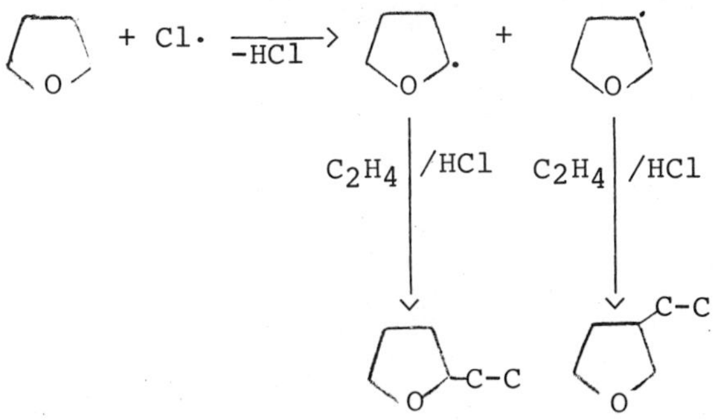

(28)

2-Ethyldioxane (18) was a product of the reaction of p-dioxane with ethylene in the presence of hydrochloric acid and di-t-butyl peroxide at 130-140°C (Expts. 29 and 30). However, the major product was bis-(2-chloroethyl) ether (20) and a smaller amount of 2-chloroethyl 2-hydroxyethyl ether (19). These were formed by hydrolysis of the p-dioxane by reaction with the hydrogen chloride.

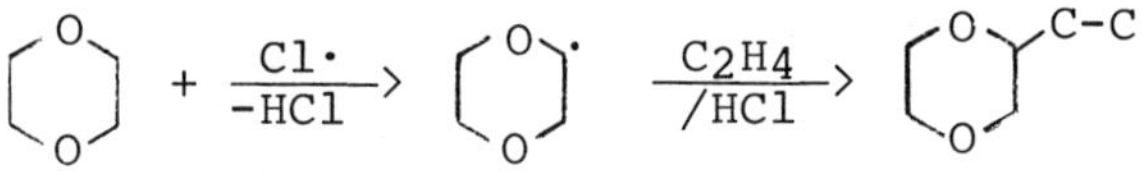

18   (29)

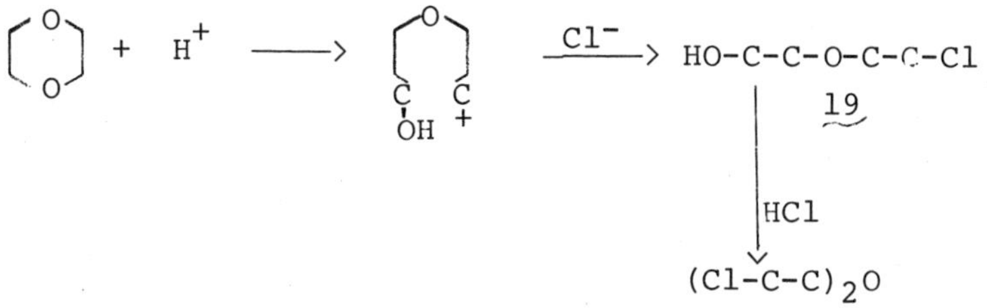

$$HO-C-C-O-C-C-Cl$$
19

$$(Cl-C-C)_2O$$ (30)

20

More hydrolysis occurred in the presence of a larger proportion of more concentrated hydrochloric acid (cf. Expts. 29 and 30).

In the absence of hydrochloric acid, telomerization occurred, yielding very high molecular weight telomer (Expt. 28). Hanford and Roland (3) found that the benzyl peroxide-induced reaction of dioxane with ethylene at 80°C resulted in products formed by reaction of 54 mols of ethylene per mol of p-dioxane. In Expt. 28 using di-t-butyl peroxide at 130-140° 14 mols of ethyl-

ene was consumed for each mol of p-dioxane which was
not recovered. After distilling off the unreacted
dioxane, there remained a reaction product containing
some ethyldioxane (7 mol % yield), the remainder consisting of product, over 95 wt % of which was a very
viscous oil boiling above 195°.

It may be concluded that hydrochloric acid
promotes the peroxide-induced monoethylation of p-
dioxane. However, this selective alkylation reaction
is accompanied by hydrolysis of the dioxane.

Literature Cited

1. Schmerling, Louis, U.S. Patent 2,769,849
(Nov. 6, 1956, originally filed May 31, 1946).
2. Hanford, W. E., and Harmon, J., U.S. Patent
2,418,832 (April 15, 1947, filed June 17, 1942).
3. Hanford, W. E., and Roland, J. R., U.S.
Patent 2,402,137 (June 18, 1946, filed January 1, 1943).
It is interesting to note that the word "telomer" seems
to have been introduced in this patent.

Coupling of Alkyl Groups Using Transition Metal Catalysts

JAY K. KOCHI

Department of Chemistry, Indiana University, Bloomington, IN 47401

The formation of carbon-carbon bonds is one of the most important operations in organic synthesis, and it can be represented by the coupling of organometallic reagents (including alkyllithium and Grignard reagents) with organic derivatives, such as alkyl halides among others.

$$RMgX + R'\text{-}X \xrightarrow{\text{[M]}} R\text{-}R' + MgX_2 \qquad [1]$$

The organic moieties, R and R' in Equation 1, can either be saturated alkyl, aryl, vinyl or acetylenic groups leading to a wide variety of hydrocarbon structures. The most effective catalysts represented in Equation 1 as [M] are derived from transition metal complexes.

The role of the metal catalysts is varied in these reactions, but they are most commonly involved in the formation of organometallic intermediates RM which subsequently reductively eliminate to the coupled product. Reactions leading to the formation of the key intermediate R-M and the elucidation of the pathways for its decomposition are thus central to the understanding of these catalytic processes. We will first summarize briefly the processes involved in the formation and destruction of RM, which will be followed by our studies of various catalytic systems leading to the coupling of alkyl groups.

Formation of Alkylmetal Complexes

Organometals R-M are commonly prepared by metathesis of a transition metal complex with substitution-labile carbanionoid reagents such as Grignard and lithium derivatives.

$$\underset{\underset{\text{I}}{|}}{\overset{\overset{\text{CH}_3}{|}}{CH_3\text{-}Au^{\text{III}}\text{-}PPh_3}} + CD_3Li \longrightarrow \underset{\underset{\text{CD}_3}{|}}{\overset{\overset{\text{CH}_3}{|}}{CH_3\text{-}Au^{\text{III}}\text{-}PPh_3}} + LiI \qquad [2](\underline{1})$$

167

The formal oxidation state of the metal under these circumstances does not change. Similarly, no change in the formal
oxidation state of the metal results from the insertion of an
olefin into a ligand-metal bond.

$$L_2Pt\overset{\text{II}}{\underset{Cl}{\overset{H}{<}}} + CH_2 = CH_2 \longrightarrow L_2Pt\overset{\text{II}}{\underset{Cl}{\overset{CH_2CH_3}{<}}} \qquad [3](\underline{2})$$

On the other hand, when either alkyl radicals or alkyl
carbonium ions (or their precursors) are used as alkylating
agents, the metal center undergoes a change in the formal oxidation state of either one or two, respectively.

$$CH_3\cdot + Cr^{\text{II}}en_2^{+2} \longrightarrow CH_3-Cr^{\text{III}}en_2^{+2} \qquad [4](\underline{3})$$

$$CD_3I + (CH_3)_2Au^{\text{I}}Li \xrightarrow{PPh_3} CD_3-\underset{\underset{CH_3}{|}}{\overset{\overset{CH_3}{|}}{Au^{\text{III}}}}-PPh_3 + LiI \qquad [5](\underline{4})$$

Alkylations of metal centers under these circumstances are considered as oxidative additions.

Decomposition of Alkylmetal Complexes

There are a number of modes by which carbon-metal bonds
can be cleaved. Conceptually, they can be represented by the
microscopic reverse of each of the processes in Equations 2-5
which lead to the alkylation of the metal center. Thus, the
reverse of Equation 2 is represented by the well-known electrophilic cleavage of organometals.(\underline{5})

$$(CH_3)_3Au^{\text{III}}PPh_3 + HOAc \longrightarrow (CH_3)_2Au^{\text{III}}(OAc)PPh_3 + CH_3-H \qquad [6]$$

Similarly, β-elimination of hydrogen is probably the most
common route by which alkylmetals decompose.

$$L_2Pt\overset{CH_2CH_2CH_2CH_3}{\underset{CH_2CH_2CH_2CH_3}{<}} \longrightarrow \begin{array}{c} CH_2 = CHCH_2CH_3 \\ + \\ L_2Pt\overset{H}{\underset{CH_2CH_2CH_2CH_3}{<}} \end{array} \quad etc. \qquad [7](\underline{6})$$

The homolytic cleavage of alkylmetal bonds, particularly
those of Main Group metals, is known from Paneth's classic
experiments to occur at high temperatures. The reverse of
Equation 4, however, does not usually represent the energeti-

cally most favored pathway in the decomposition of R-M. Alkyl
coupling as a result of homolysis to free radicals, followed by
dimerization

$$R\text{-}M \longrightarrow [M\cdot + R\cdot] \longrightarrow R\text{-}R, \quad etc. \qquad [8]$$

is not ubiquitous, although such views were widely held due to
the misguided belief that carbon-metal bonds, particularly those
involving transition metals, are extremely weak. If alkyl radicals are intermediates in the catalyzed coupling of alkyl groups,
they should undergo disproportionation in addition to dimerization.

$$2\ CH_3CH_2\cdot \begin{cases} \xrightarrow{k_d} CH_2=CH_2 + CH_3CH_3 & [9a] \\ \xrightarrow{k_c} CH_3CH_2CH_2CH_3 & [9b] \end{cases}$$

Either bimolecular process is not possible without the other,
since the ratio k_d/k_c of the rate constants is fixed by the structure of the radical.(7) However, a number of metal-catalyzed
couplings are known to proceed without the formation of any disproportionation products, certainly in the amounts dictated by
the values of k_d/k_c. Further, the coupling represented in Equation 1 can occur without the scrambling of R and R', as would be
expected of free radical intermediates.(8)
    The microscopic reverse of oxidative addition in Equation
5 is represented by reductive elimination which can be intermolecular (Equation 10) or intramolecular (Equation 11).

$$dmgCo^{IV}CH_3 + Br^- \longrightarrow dmgCo^{II} + CH_3Br \qquad [10](\underline{9})$$

$$(CH_3)_3Au^{III}PPh_3 \longrightarrow CH_3Au^{I}PPh_3 + CH_3CH_3 \qquad [11](\underline{10})$$

Indeed, the combination of oxidative addition in Equation 5 and
reductive elimination in Equation 11 is the basis for a catalytic
mechanism for alkyl coupling.

Alkyl Transfers from Organometallic Intermediates
in Catalytic Processes

    The oxidation-reduction reactions of organometallic inter-
mediates presented in the foregoing description can be applied,
in combination, to a variety of catalytic processes, such as the
metal-catalyzed alkyl transfer reactions of Grignard reagents
originally investigated by Kharasch and coworkers.(11) We have
found that the catalytic reactions between labile organometals and
alkyl halides can be generally classified into two categories,
coupling in Equation 12 and disproportionation in Equation 13,

depending on the catalyst. For example, silver(I) and copper(I) are effective catalysts in the coupling of alkyl groups, whereas

$$R\text{-}m + R\text{-}X \longrightarrow \begin{cases} R\text{-}R + mX & [12] \\ RH + R(\text{-}H) + mX & [13] \end{cases}$$

iron effects only disproportionation except when aryl and vinylic halides are employed. Each catalyst shows unique features which are best described within the following mechanistic context.

A. Nickel Catalysis in the Cross Coupling of Aryl Halides with Alkylmetals. The Role of Arylalkylnickel(II) Species as Intermediates. For the study of nickel catalysis in the formation of aralkanes, we employed the system consisting of aryl bromides and methyllithium or methylmagnesium bromide.

$$Ar\text{-}Br + CH_3\text{-}m \xrightarrow{(Et_3P)_2NiBr_2} Ar\text{-}CH_3 + m\text{-}Br \qquad [14]$$

$$where\ m = Li\ or\ MgX$$

Although the triethylphosphine complexes of nickel(II) may not necessarily represent optimum examples of catalysts,(12)(13) they allowed us access to the key intermediate, viz., the aryl-methyl-nickel complexes such as Ia and b. Any catalytic cycle

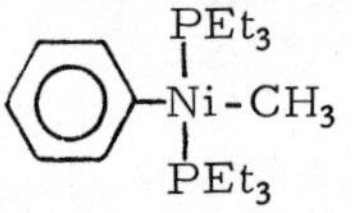

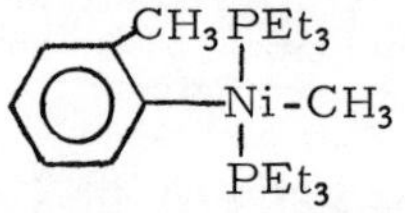

which is developed around this intermediate must depend on the availability of pathways for its rapid formation and facile decomposition.(14) The most straightforward formulation simply consists of a sequence of steps involving (a) reductive elimination followed by (b) oxidative addition and (c) metathesis as shown in Scheme 1 (L = triethylphosphine).

Scheme 1

$$L_2Ni\diagdown\substack{Ar \\ CH_3} \xrightarrow{\ a\ } ArCH_3 + L_2Ni \qquad [15]$$

$$ArX + L_2Ni \xrightarrow[fast]{\ b\ } L_2Ni\diagdown\substack{Ar \\ X} \qquad [16]$$

$$L_2Ni\diagdown\substack{Ar \\ X} + CH_3m \xrightarrow[fast]{\ c\ } Xm + L_2Ni\diagdown\substack{Ar \\ CH_3} \quad etc. \qquad [17]$$

Indeed, each step in this catalytic cycle can be documented

separately. We have shown that reductive elimination in step (a) proceeds to a nickel(0) species which can undergo rapid oxidative addition in step (b) as does the metathesis in step (c). Similar observations have been made qualitatively by Parshall and others.(15)(16) There are, however, two serious points to be raised on this simple mechanism. Most importantly, we find that the rate of reductive elimination in step (a) is too slow to allow for a catalytic process. Secondly, aryl-methyl-bis(triethylphosphine)nickel(II) complexes Ia and b exist exclusively in the trans configuration. Oxidative addition to nickel(0) species in step (b) also affords only the trans adducts. Our demonstration of the intramolecular route for reductive elimination of Ia and b demands that it proceed in a concerted fashion via a transition state in which the aryl and methyl groups are juxtaposed. Thus, a cis square planar or tetrahedral geometry for reductive elimination would be similar to that previously shown for the somewhat analogous example of the trialkylgold(III) complexes in Equation 18.(17) Since we could find no evidence of the isomerization of the trans to either a cis isomer or a tetrahedral

$$R-\overset{\overset{\displaystyle CH_3}{|}}{\underset{\underset{\displaystyle CH_3}{|}}{Au}}^{III}PPh_3 \longrightarrow R-CH_3 + CH_3\overset{I}{Au}PPh_3 \qquad [18]$$

species even at 80°, we tentatively conclude that reductive elimination of Ia and b does not proceed via these stereoisomeric forms. In such an event, reductive elimination can proceed directly from the trans complex via a tetrahedral-like transition state. On the other hand, retardation of reductive elimination by added triethylphosphine suggests that it proceeds by a dissociative mechanism, since no five-coordinate species can be detected in the ^{31}P NMR spectrum(18) in sufficient concentrations to account quantitatively for the rate reduction.

$$(Et_3P)_2Ni{\overset{\displaystyle Ar}{\underset{\displaystyle CH_3}{<}}} \rightleftharpoons \dot{E}t_3P + Et_3PNi{\overset{\displaystyle Ar}{\underset{\displaystyle CH_3}{<}}} \quad etc. \qquad [19]$$

The three-coordinate intermediate in Equation 19 would then allow for greater mobility of the groups from which reductive elimination could proceed.(19) Rearrangement followed by a very rapid reductive elimination of the cis isomer is also a possibility. However, our unsuccessful attempts to synthesize cis-aryl-methyl-nickel(II) complexes still leave open the question of reductive elimination from such stereoisomers. The results we obtained in studies with the bidentate diphosphine ligand (dppe), though qualitatively in this direction, unfortunately lack definitiveness as yet.

The foregoing discussion indicates that other considerations must be invoked in order to allow for the facile expulsion of

aryl and methyl groups from Ia and b necessary for a catalytic
cycle. Indeed, the observation that oxygen induces the rapid
reductive elimination of Ia and b provides the key to this understanding. In order to develop this thesis, we present first the
autoxidation of organometals within the general context of electron transfer processes.

The rapid reaction of a variety of organometallic complexes with molecular oxygen proceeds from a prior electron
transfer step such as Equation 20,

$$O_2 + [R\text{-}M] \longrightarrow O_2^{\overline{\cdot}} + [R\text{-}M]^{\overline{\cdot}+} \text{ , etc.} \qquad [20]$$

in which oxygen acts as an electron acceptor. This process is
important independently of whether the subsequent steps in
autoxidation involve a stoichiometric or radical chain process.(20)
Similarly, organic halides such as aryl bromides are also known
to be acceptors in electron transfer processes.(8)(21) Furthermore, organometals being generally electron-rich species are
capable of acting as donors in a number of electron transfer
reactions.(22) The enhanced lability of the resultant cation-
radical toward reductive elimination has been discussed recently
with regard to some main group as well as transition metal-
organic complexes.(23)

As structurally different as they are, nonetheless oxygen
and aryl bromides bear interesting resemblances, in that their
interaction with aryl-methyl-nickel compounds Ia and b both
lead to the elimination of aryl and methyl groups. For purposes
of further discussion, we propose a charge transfer interaction
common to both, in which reactions are promoted by a prior
electron transfer from I to oxygen or aryl bromide acting as
electron acceptors (**A**).

$$\mathbf{A} + (Et_3P)_2Ni{\overset{\displaystyle Ar}{\underset{\displaystyle CH_3}{\big\langle}}} \longrightarrow \mathbf{A}^{\overline{\cdot}}(Et_3P)_2\overset{+}{Ni}{\overset{\displaystyle Ar}{\underset{\displaystyle CH_3}{\big\langle}}} \text{ , etc.} \qquad [21]$$

A concerted process for reductive elimination from the nickel(III)
species(24) in Equation 21 is indicated by the reasonable yields
of ArCH$_3$ obtainable even when **A** is oxygen. Moreover, the
partial scrambling observed when **A** is aryl bromide (and the
aryl groups on the nickel compound I and the aryl bromide are
different) suggests that the ion pair in Equation 21 may be intimately associated. According to this formulation, the scrambling of the aryl groups in the cross coupled product must occur
after electron transfer, since the recovered aryl bromide is not
exchanged. Several possibilities can be envisaged, including
irreversible collapse to a five-coordinate intermediate(25) such
as:

$$(Et_3P)_2\overset{+}{Ni}{\overset{\displaystyle Ar}{\underset{\displaystyle CH_3}{\big\langle}}} ArBr^{\overline{\cdot}} \longrightarrow (Et_3P)_2\overset{\displaystyle Ar}{\underset{\displaystyle CH_3}{Ni{\overset{+Ar}{\big\langle}}}}Br^- \longrightarrow (Et_3P)_2Ni{\overset{\displaystyle Ar}{\underset{\displaystyle Br}{\big\langle}}} + ArCH_3 \quad [22]$$

followed by the elimination of the aryl and methyl groups in a
subsequent step.

We have also observed that the sensitivity of aryl-methyl-
nickel compounds Ia and b to oxygen is greatly enhanced by the
addition of methyllithium. Under these conditions, the presence
of Ia and b as nickelate complexes III is indicated by isotopic
exchange studies. These anionic nickel complexes should be
even better donors than their neutral counterparts I,(26) and they
are thus expected also to show enhanced reactivity to aryl bromides in those interactions proceeding by electron transfer.
Reductive eliminations similar to those presented for I can be
formulated as:

$$(Et_3P)_2Ni{\overset{Ar}{\underset{(CH_3)_2Li}{<}}} + ArX \longrightarrow (Et_3P)_2Ni{\overset{Ar}{\underset{CH_3}{<}}} + ArCH_3 + LiX \quad [23]$$

The various stoichiometric reactions of the aryl-methyl-
nickel compounds of Ia and b leading to reductive elimination can
be assembled as components to a catalytic cycle in the cross
coupling process. Scheme 2 is consistent with the facts in hand,
including the products, stoichiometry, kinetics, exchange and
scrambling experiments.

Scheme 2

$$L_2Ni(Ar)CH_3 + ArX \xrightarrow{\text{slow}} [L_2Ni(Ar)CH_3^+ ArX^-] \quad [24]$$

$$L_2Ni(Ar)CH_3^+ \xrightarrow{\text{fast}} L_2Ni^+ + ArCH_3 \quad [25]$$

$$L_2Ni^+ + ArX^- \xrightarrow{\text{fast}} L_2NiArX \quad [26]$$

Scheme 2 is presented stepwise only to emphasize the mechanistic factors brought out by the study of Ia and b. Thus, it is probably that Equations 24-26 occur in one to two steps such as
Equation 22, or in rapid succession before any of the intermediates diffuse apart. Furthermore, Scheme 2 is obviously incomplete without considering the nickelate species III as participating in reactions such as Equation 23 which are mechanistically
equivalent to the rate-limiting step shown by I in Equation 24 of
the catalytic cycle.

Finally, there are several mechanistic points yet to be
clarified. For example, the origin of side products such as
reduced arene is unclear. Preliminary results suggest that the
arene arises from an aryl-methyl exchange process such as,

$$(Et_3P)_2Ni{\overset{Ar}{\underset{CH_3}{<}}} + CH_3Li \rightleftharpoons (Et_3P)_2Ni{\overset{Ar}{\underset{(CH_3)_2Li}{<}}} \rightleftharpoons (Et_3P)_2Ni{\overset{CH_3}{\underset{CH_3}{<}}} + ArLi \quad [27]$$

since deuterolysis of the reaction mixture leads partly to deuterated arene. It does not appear, however, that all of the hydrogen is derived from protonolysis, and it is possible that some other source such as that from the methyl group may be involved.

B. Iron Catalysis in the Cross Coupling of Alkenyl Halides and Grignard Reagents. Grignard reagents are cross coupled stereospecifically with alkenyl halides such as 1-bromopropene in the presence of catalytic amounts of iron complexes.(27)

$$RMgBr + {>}C=C{<}^{Br} \xrightarrow{\text{(Fe)}} {>}C=C{<}^{R} + MgBr_2 \qquad [28]$$

Iron(III) complexes are employed, but they are rapidly reduced by Grignard reagent in situ to generate a catalytically active reduced iron species, presumably iron(I). Among various iron complexes examined, tris-dibenzoylmethidoiron(III), Fe(DBM)₃, was found to be the most effective, particularly with respect to deactivation of the catalyst.(28) The yields of olefins obtainable by this catalytic process vary according to the structure of the alkyl moiety in the Grignard reagent as shown in Table I. Thus, high yields of cross coupled products are obtainable with methylmagnesium bromide. Under the same conditions, ethylmagnesium bromide afforded ethane and ethylene as side products in addition to the expected cross coupled product. The difference can be attributed to the availability of β-hydrogens in the latter, a factor which is also important in a variety of other organometallic reactions.(29)

    In this study we have carried out a thorough analysis of the products formed during the reaction of ethylmagnesium bromide with (Z)- and (E)-1-bromopropene in the presence of tris-dibenzoylmethido-iron(III). A complete accounting of the material balance as well as the electron balance has been achieved. Together with stereochemical and isotopic labelling studies, they provide substantial mechanistic information about this interesting catalytic process.

    Five major types of side products are produced during the catalytic process, in greater or lower yields depending on the relative concentrations of the reactants, the temperature of the reaction and the structure of the alkylmagnesium bromide. Thus, alkene R(-H) and alkane RH from the Grignard component as well as propylene, propenylmagnesium bromide and 2,4-hexadiene from 1-bromopropene are always formed. No simple relationship could be found for the formation of these side products in relationship to the predominant cross coupled product. The latter suggests that the side products are intimately connected with the principal reaction, and that both processes involve common reactive intermediates. Alternatively, the side products could arise via concurrent but largely independent reactions from the cross coupling process. The rigorous delineation between these basic mechanistic categories is extremely difficult to make in a catalytic system in which the isolation of interme-

Table I.   Cross Coupling of Primary, Secondary and Tertiary Alkyl Grignard Reagents with 1-Bromopropene[a]

RMgBr (mmol)	Products (mmol)						Mat. Bal.[d] (%)
	R(-H)[b]	RH	C3H6	C3H5R(%)[c]	C6H10	R2	
1.02 CH3CH2-	0.32	0.12	0.19	0.41Z(49) 0.05E	nd[e]	tr[f]	96
1.04 CH3CH2CH2-	0.45	0.03		0.65Z(73) 0.02E	0.02 (Z,Z)	tr	101
1.13 CH3(CH2)4-	0.38	nd	0.29	0.64Z(60)	0.08 (Z,Z)	nd	96
1.09 (CH3)2CHCH2-	0.12	0.01	0.11	0.79Z(79)	nd	tr	97
1.04 (CH3)2CH-	0.07	0.02		0.79(82)[g]	0.05[h]	0.05	98
1.01 CH3CH2(CH3)CH-	0.04	0.03	0.02	0.5Z(77)[j] 0.2E	0.05[h]	0.01	94[j]
1.00 CH3CH2CH2(CH3)CH-	0.13	nd	0.02	0.5Z(75)[j] 0.2E	nd	nd	102[j]
1.01 cyclo-C6H11-	0.04	nd	tr	0.56Z(76) 0.15E	nd	nd	85
0.95 (CH3)3C-	0.1	nd	0.08	0.37Z(60) 0.15E	nd	0	86

[a]In reactions containing 2.96 mmol 1-bromopropene (95% Z and 5% E) and 3.6 x 10^{-3} mmol Fe(DBM)_3 in 9 ml THF at 25° for 1 hr. [b]Alkene by loss of β-hydrogen. [c]Cross coupled product [Z is cis and E is trans isomer]; yields based on RMgX consumed including 0.08 mmol in catalyst preparation. [d]Based on RMgX consumed (determined by hydrolysis). [e]Not determined, nd. [f]< 0.01 mmol detected. [g]Mixture of (E) and (Z) isomers. [h]0.04 (Z,Z) and 0.01 (Z,E). [j]Value approximate due to unavailability of authentic product (see experimental section).

diates is impractical.

Any mechanistic formulation of the catalytic process must take into account the diversity of side products, as well as the isotopic labelling and stereochemical results. In the following discussion we wish to present a reaction scheme which is consistent with the available data, while at the same time keeping the number of intermediates to a minimum.

The Catalyst is best described as an iron(I) species formed by the facile reduction of the iron(III) precursor by the Grignard reagent.(30) It is a metastable species subject to deactivation on standing, probably by aggregation. Formally, iron(I) species consist of a d^7 electron configuration, isoelectronic with manganese(0) and cobalt(II). Only a few complexes of iron(I) have been isolated, but a particularly relevant one is the paramagnetic hydrido complex, HFe(dppe)$_2$, which is stabilized by the bisphosphine ligand, dppe[Ph$_2$PCH$_2$CH$_2$PPh$_2$]. A toluene solution shows a strong esr signal centered at $<g> = 2.085$ with poorly resolved fine structure.(31) A broad intense esr spectrum ($<g> = 2.08$) is also obtained if $\overline{Fe}$(DBM)$_3$ is treated with excess ethylmagnesium bromide in THF solutions at $-40°$ to $0°C$. A broad resonance is also observed when FeCl$_3$ [$<g> = 2.15$] or Fe(acac)$_3$ [$<g> = 2.07$] are employed. The spectrum centered at $<g> = 2.00$ retains the same general features when n-pentyl- or sec-butyl-magnesium bromides are employed as reducing agents.(12) It is destroyed immediately by molecular oxygen, and is the dianion-radical of the ligand, PhCOCHCOPh$\cdot^{-2}$.

The Mechanism of the cross coupling reaction can be accommodated by an oxidative addition of 1-bromopropene to iron(I) followed by exchange with ethylmagnesium bromide and reductive elimination. Scheme 3 is intended to form a basis for discussion and further study of the catalytic mechanism. In order to maintain the stereospecificity, the oxidative addition of bromo-propene in step a should occur with retention. Similar stereochemistry has been observed in oxidative additions of platinum(0) and nickel(0) complexes.(32)(33) The metathesis of the iron(III) intermediate in step b is expected to be rapid in analogy with other alkylations.(34) The formation of a new carbon-carbon bond by the reductive elimination of a pair of carbon-centered ligands in step c has been demonstrated to occur

Scheme 3

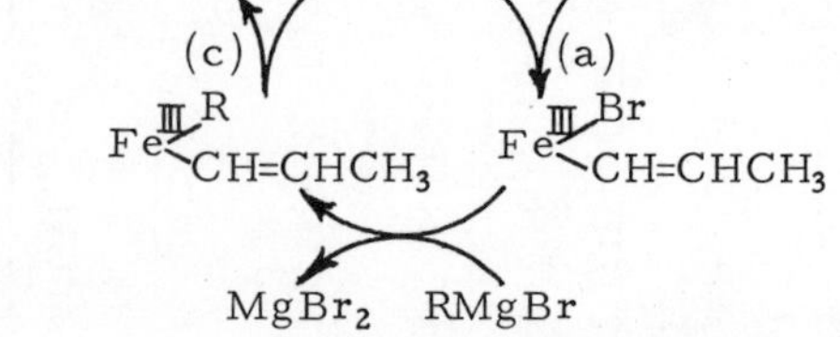

[29]

with organogold(III), organonickel(II), organoplatinum(IV) and organorhodium(III) complexes.(17)(35)

The iron(III) intermediates in Scheme 3 serve as focal points for the formation of the side products. For example, metathetical exchange of propenyl-iron(III) with Grignard reagent would afford propenylmagnesium bromide in Equation 30.(36)

$$Fe^{III}-CH=CHCH_3 + R-MgBr \longrightarrow Fe^{III}-R + CH_3CH=CHMgBr \quad [30]$$

Further exchange in Equations 31 and 32 would produce bis-(1-propenyl)iron(III) species which reductively eliminate to produce the 2,4-hexadienes stereospecifically.(37)

$$Fe^{III}\begin{smallmatrix}Br\\CH=CHCH_3\end{smallmatrix} + CH_3CH=CHMgBr \longrightarrow Fe^{III}\begin{smallmatrix}CH=CHCH_3\\CH=CHCH_3\end{smallmatrix} + MgBr_2 \quad [31]$$

$$Fe^{III}\begin{smallmatrix}CH=CHCH_3\\CH=CHCH_3\end{smallmatrix} \longrightarrow Fe^{I} + CH_3CH=CH-CH=CHCH_3 \quad [32]$$

Disproportionation products are postulated to arise from alkyliron(III) and/or dialkyliron(III) species formed by an analogous metathesis between iron(III) species and alkylmagnesium bromide. Thus, the disproportionation processes may proceed as follows (vide infra):

$$Fe^{III}\begin{smallmatrix}CH_2CD_3\\CH_2CD_3\end{smallmatrix} \longrightarrow Fe^{I} + CH_2=CD_2 + DCH_2CD_3 \quad [33]$$

$$Fe^{III}\begin{smallmatrix}CH_2CD_3\\CH=CHCH_3\end{smallmatrix} \longrightarrow Fe^{I} + CH_2=CD_2 + CH_3CH=CHD \quad [34]$$

These disproportionations could proceed directly or by a 2-step mechanism involving prior transfer of a β-hydrogen to iron followed by reductive elimination. Similar disproportionation processes have been described with organocopper(I), organomanganese(II), and organoplatinum(II) complexes.(38)

The mechanism in Scheme 3 accommodates much of the extant data on the iron-catalyzed cross coupling reaction of Grignard reagents and alkenyl halides. The side products derive naturally from organoiron(III) intermediates by reasonably well-established pathways. However, there are a number of interesting observations which merit further scrutiny in the light of this mechanism. For example, it is commonly held that organometallic compounds such as the alkyl- and propenyl-iron-(III) species in Scheme 3 undergo elimination of β-hydrogens in the order: tert.-R > sec.-R > prim.R. However, the results presented in Table I run counter to this expectation. Furthermore, if the oxidative addition of 1-bromopropene to iron(I) is rate-limiting, the reactivities of the (Z) and (E) isomers should be

relatively independent of the Grignard reagent. It is found, however, the (Z)-bromopropene is more reactive than the (E)isomer
with primary alkylmagnesium bromide, but the converse is true
of methyl, secondary and tertiary alkylmagnesium bromides.
The degree of association and complex formation of the latter no
doubt affect a quantitative evaluation, but even a qualitative
rationalization of this result remains obscure. Changes in the
concentration of the reactants as well as the temperature of the
reaction could affect the rates and equilibria of the various reactions outlined in Scheme 3 and Equations 30-34 in a manner to
change the product distribution. Nonetheless, the anomalies
presented above ultimately must be resolved before this mechanistic formulation can be accepted with more confidence.

Finally, the mechanism in Scheme 3 bears a resemblance
to that presented above for the nickel-catalyzed reaction of
methylmagnesium bromide and aryl bromides. However, there
are outstanding differences between iron and nickel in their
abilities to effect cross coupling reactions. Iron is a catalyst
which is effective at lower temperatures and concentrations than
used with nickel. Even more importantly, cross coupling can be
effected completely stereospecifically with an iron catalyst and
no alkyl isomerization of the Grignard component has been observed, in contrast to the nickel-catalyzed reactions.

C. Cross Coupling of Grignard Reagents and Alkyl Halides
with Copper(I) Catalysts. Copper(I) specifically catalyzes the
cross coupling (Equation 35) between Grignard reagents and
alkyl bromides when carried out in THF solutions at 0°C or
lower.(39)

$$RMgX + R'X \xrightarrow{Cu^I} R-R' + MgX_2 \qquad [35]$$

The yield of homodimers, R-R and R'-R', under these conditions
is negligibly small. This cross coupling reaction is most facile
with primary alkyl halides, but unlike silver(I) catalysis, the
secondary and tertiary alkyl halides are generally inert and give
poor yields of coupled products and mainly disproportionation.
The structure of the Grignard reagent is not as important, in
analogy with the cross coupling observed with lithium dialkylcuprates.(40)

The coupling of ethylmagnesium bromide and ethyl bromide
to n-butane follows overall third-order kinetics, being first
order in each component and the copper(I) catalyst. There is no
evidence for alkyl radicals in the copper(I)-catalyzed coupling
process, and we propose the following two-step mechanism:

Scheme 4 $$RMgBr + Cu^IBr \longrightarrow RCu^I + MgBr_2 \qquad [36]$$

$$RCu^I + R'Br \xrightarrow{slow} R-R' + Cu^IBr \qquad [37]$$

The rate-limiting step 37 can be shown independently by
examining the stoichiometric reaction of alkylcopper(I) directly
with organic halides. However, the extent to which decomposition of the alkylcopper(I) intermediate competes with the catalytic coupling reaction introduces disproportionation products.
The latter involves a copper(0)-catalyzed sequence(41) similar
to that observed with iron (vide infra), and it is especially
important with secondary and tertiary alkyl systems. The
effects of structural variation are consistent with a rate-limiting
step involving nucleophilic displacement of halide in Equation
37.(40) The involvement of a nucleophilic copper(I) center, i.e.,
oxidative addition, followed by reductive elimination has direct
analogy to the mechanism which has been established with the
analogous gold(I) catalyst. Organocopper(III) intermediates

$$RCu^{I} + R'X \longrightarrow R(R')Cu^{III}X \qquad [38]$$

$$R(R')Cu^{III}X \longrightarrow R-R' + Cu^{I}X \qquad [39]$$

presented in Equation 38 are formally related to the species
formed in the association of alkyl radicals with copper(II) complexes,(42) with both showing a marked propensity for reductive
elimination. Although the observation of these highly metastable
intermediates is unlikely, the analogous organogold intermediates are more stable and can be isolated or observed directly
by NMR. Oxidative addition of alkyl halides to alkyl(PPh₃)-
gold(I) follows the expected pattern: $CH_3I > EtI > i\text{-}PrI$.(3) The
subsequent reductive elimination of trialkylgold(III) complexes to
coupled dimer was described earlier (Equation 11). The parallel
between copper(I) and gold(I) is further shown in the behavior of
the corresponding cuprate(I) and aurate(I) complexes. Thus,
alkylgold(I) reacts with an equimolar amount of alkyllithium to
afford an isolable lithium dialkylaurate(I). The anionic dimethyl-
aurate(I) species formed in this manner is at least 10^6 times
more reactive to oxidative addition of methyl iodide than the
neutral methyl(PPh₃)gold(I). The same pattern is qualitatively
established with organocopper(I) species in comparing the coupling reaction in Equation 37 with that reported for lithium
dialkylcuprates.(40)

D. Homo-Coupling with Silver(I). Silver is an effective
catalyst for the coupling of Grignard reagents and alkyl halides,
and it is especially useful when both alkyl groups are the
same.(43) When different alkyl groups are employed, a mixture
of three coupled products is obtained. Disproportionation
becomes increasingly important with secondary and tertiary

$$RMgX + RX \xrightarrow{Ag^{I}} R-R + MgX_2 \qquad [40]$$

groups, independently of whether they are derived from the Grignard reagent or the alkyl halide. The rate of production of butane from ethylmagnesium bromide and ethyl bromide is roughly first order in silver and ethyl bromide, but zero order in Grignard reagent.(44) The reactivity of alkyl halides follows the order: t-butyl >i-propyl >n-propyl bromide in the ratio 20:3:1, and structural variations in the Grignard reagent show no apparent systematic trend.

The results can be accommodated by Scheme 5, in which the coupling arises from alkyl silver(I) intermediates generated via two largely independent pathways:

Scheme 5

$$R'MgX + Ag^I X \longrightarrow R'Ag^I + MgX_2 \qquad\qquad [41]$$

$$RAg^I, R'Ag^I \longrightarrow [R\text{-}R, R'\text{-}R, R'\text{-}R'] + 2\,Ag^0 \qquad [42]$$

$$Ag^0 + R\text{-}X \longrightarrow R\cdot + Ag^I X \qquad\qquad [43]$$

$$R\cdot + Ag^0 \longrightarrow RAg^I \quad \text{etc.} \qquad\qquad [44]$$

The rate-limiting step in this mechanism is given by Equation 43 in which the alkyl halide is responsible for the re-oxidation of silver(0) produced in Equation 42. This slow step is closely akin to the production of alkyl radicals by the ligand transfer reduction of alkyl halides with other reducing metal complexes.(45) More direct evidence for the selective formation of alkyl radicals from the alkyl halide is shown by trapping experiments as well as stereochemical studies. Thus, the catalytic reaction of cis-propenylmagnesium bromide with methyl bromide yielded cis-butene-2, in accord with the retention of stereochemistry during the reductive coupling of vinyl-silver(I) complexes.(46) On the other hand the reverse combination, cis-propenyl bromide and methylmagnesium bromide, is catalytically converted to a mixture of cis- and trans-butene-2, consistent with the formation and rapid isomerization of the 1-propenyl radical in Scheme 5.

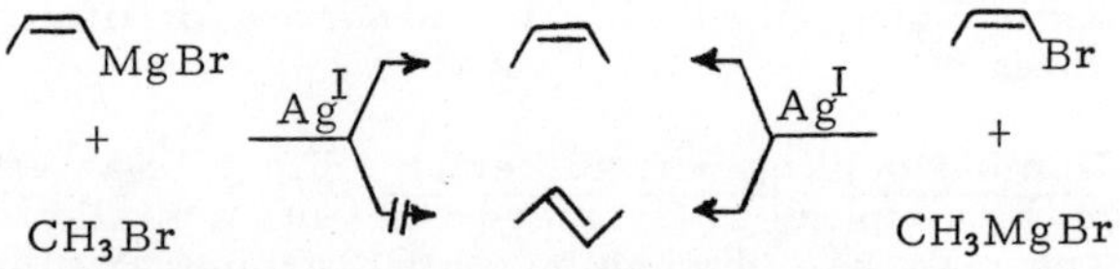

E. Catalysis of Alkyl Disproportionation by Iron. Alkyl disproportionation is the sole reaction observed during the iron-catalyzed reaction of ethylmagnesium bromide and ethyl bromide.(47) The catalyst is a reduced iron species formed in situ by the reaction of iron(II,III) with Grignard reagent, and effective

$$CH_3CH_2MgBr + CH_3CH_2Br \xrightarrow{(Fe)} CH_3CH_3 + CH_2{=}CH_2 + MgBr_2 \quad [46]$$

in concentrations as low as 10^{-5} M. Although the reaction has limited synthetic utility, it merits study since it can provide insight into some of the complications involved with organometallic intermediates discussed above.

The rate of reaction shows first order dependence on the concentration of iron and ethyl bromide, but is independent of the concentration of ethylmagnesium bromide. The rate, however, varies with the structure of the Grignard reagent, and disproportionation usually results except when the alkyl group is methyl, neopentyl or benzyl which possess no β-hydrogens. The reactivities of the alkyl bromides (t-butyl > i-propyl > n-propyl) as well as the kinetics are the same as the silver-catalyzed coupling described above and suggest a similar mechanism:

Scheme 6

$$Fe^{I} + RBr \longrightarrow Fe^{II}Br + R\cdot \qquad [47]$$

$$R\cdot + Fe^{I} \longrightarrow RFe^{II} \qquad [48]$$

$$R'MgBr + Fe^{II}Br \longrightarrow R'Fe^{II} + MgBr_2 \qquad [49]$$

$$RFe^{II}, R'Fe^{II} \longrightarrow [RH, R'H, R{-}H, R'{-}H] + 2\ Fe^{I}, \text{ etc.} \qquad [50]$$

According to this postulate, the difference between coupling with silver and disproportionation with iron rests on the decomposition of the alkylmetal intermediate in Equation 50. Indeed, it has been shown separately in Equation 42 that the decomposition of alkylsilver(I) proceeds by reductive coupling. Unfortunately, the highly unstable alkyliron intermediate in Scheme 6 is not yet accessible to independent study, but the somewhat analogous dialkylmanganese(II) species in Equation 51 undergoes similar reductive disproportionation by a mechanism(48) reminiscent of dialkylplatinum(II) complexes described in Equation 7.

$$R_2Mn^{II} \longrightarrow RH + R{-}H + Mn^{0} \qquad [51]$$

Selective trapping of alkyl radicals from the alkyl halide component during the course of the catalytic disproportionation is the same as the previous observation with silver, and it indicates that the prime source of radicals in the Kharasch reaction lies in the oxidative addition of alkyl halide to reduced iron in Equation 47. Separate pathways for reaction of i-propyl groups derived from the organic halide and the Grignard reagent are also supported by deuterium labelling studies which show that they are not completely equilibrated.(49) Furthermore, the observation of CIDNP (AE multiplet effect) in the labelled propane and propene

derived only from the alkyl halide component can be attributed to
a bimolecular disproportionation of isopropyl radicals arising
from diffusive displacements. However, the latter can only be a
minor fate of the alkyl radicals derived from the alkyl halide,
since the coupled dimer is not formed in amounts required by the
bimolecular reaction of alkyl radicals previously discussed in
Equation 9.$\overline{(48)}\overline{(50)}$

　　Cross coupling of Grignard reagents with 1-alkenyl
halides, in marked contrast to alkyl halides, occurs readily with
the reduced iron catalyst, as described above. The iron-
catalyzed reaction of Grignard reagents with 1-alkenyl halides
can, however, be differentiated from the reaction with alkyl
halides. Thus, a mixture of propenyl bromide and ethyl bromide
on reaction with methylmagnesium bromide afforded butene-2 but
no cross-over products such as pentene-2 or propylene. The
latter certainly would have resulted if a propenyliron species per
se were involved in the catalytic process. Cross coupling under
these circumstances clearly merits further study.

Conclusions

　　The complex catalytic reactions leading to the coupling of
organic substrates induced by metal complexes can be rationally
dissected into a variety of elementary steps involving oxidation-
reduction reactions of organometallic intermediates. Electron
transfer interactions are important considerations in differentia-
ting concerted from stepwise processes, especially with regard
to chain processes. Crucial to the design of new synthetic procedures and the understanding of catalytic processes is the information to be gained from the scrutiny of transient alkylmetal
species, which represent a large potential for a variety of novel
reactions.

Literature Cited

1.　Rice, G. W. and R. S. Tobias, J. Organometal. Chem.
　　(1975), 86, C37.
2.　Clark, H. C., C. Jablonski, J. Halpern, A. Mantovani and
　　T. A. Weil, Inorg. Chem. (1974), 13, 1541; A. J. Deeming,
　　B. F. G. Johnson and J. Lewis, J. Chem. Soc. Dalton
　　(1973), 1848.
3.　Kochi, J. K. and J. W. Powers, J. Am. Chem. Soc. (1970),
　　92, 137.
4.　Tamaki, A. and J. K. Kochi, J. Chem. Soc. Dalton (1973),
　　2620.
5.　Matteson, D. S., "Organometallic Reaction Mechanisms,"
　　Academic Press, New York, 1974; S. Komiya and J. K.
　　Kochi, to be published.
6.　Whitesides, G. M., J. G. Gaasch and E. R. Stedronsky, J.
　　Am. Chem. Soc. (1972), 94, 5258.

7. Gibian, M. J. and C. Corley, Chem. Revs. (1973), 73, 441.
8. The complex mixture of products resulting from free radi-
    cals in Wurtz-type reactions have been described by J. F.
    Garst, "Chemically Induced Magnetic Polarization," A. R.
    Lepley and G. L. Closs, eds., Chapter 6, J. Wiley and
    Sons, New York, 1973. See also Chapter 7, Ibid., H. R.
    Ward, R. G. Lawler and R. A. Cooper.
9. Dodd, D. and M. D. Johnson, J. Organometal. Chem.
    (1973), 52, 1; P. Abley, E. R. Dockal and J. Halpern, J.
    Am. Chem. Soc. (1972), 94, 659.
10. Tamaki, A. and J. K. Kochi, J. Organometal. Chem.
    (1974), 64, 411.
11. Kharasch, M. S. and O. Reinmuth, "Grignard Reagents of
    Nonmetallic Substances," Prentice-Hall, Inc. New York,
    1954.
12. (a) Tamao, K., K. Sumitani and M. Kumada, J. Am. Chem.
    Soc. (1972), 94, 4374. (b) Corriu, R. J. P. and J. P.
    Masse, J. Chem. Soc., Chem. Commun. (1972), 144.
13. Kiso, Y., K. Tamao and M. Kumada, J. Organometal.
    Chem. (1973), 50, C12.
14. Similar species have been invoked in cross couplings
    effected with stoichiometric amounts of organonickel rea-
    gents; see (a) Semmelhack, M. F., Org. Reactions (1972),
    19, 155; (b) Baker, R., Chem. Revs. (1973), 73, 487; (c)
    Nakamura, A. and S. Otsuka, Tetrahedron Letters (1974),
    463.
15. Parshall, G. W., J. Am. Chem. Soc. (1974), 96, 2360.
16. Cf. (a) Hidai, M., T. Kashiwagi, T. Ikeuchi and Y. Uchida,
    J. Organometal. Chem. (1971), 30, 279; (b) Cundy, C. S.,
    Ibid. (1974), 69, 305; (c) Fahey, D. R., J. Am. Chem. Soc.
    (1970), 92, 402; and (d) reference 15.
17. (a) Tamaki, A., S. A. Magennis and J. K. Kochi, J. Am.
    Chem. Soc. (1974), 96, 6140; (b) Cf. also for Pt(IV): M. P.
    Brown, R. J. Puddephatt, C. E. E. Upton, J. Chem. Soc.
    Dalton (1974), 2457.
18. For structural factors in equilibria involved in 4- and 5-
    coordinate Ni(II) complexes with monodentate phosphines:
    L_nNiX_2 (n=2,3) see E. C. Alyea and D. W. Meek, J. Am.
    Chem. Soc. (1969), 91, 5761; Cf. also J. W. Dawson et al.,
    Ibid. (1974), 96, 4428.
19. Interestingly, reductive elimination from alkylgold(III) also
    proceeds from a 3-coordinate species (see ref. 17a).
20. Cf. J. F. Garst in "Free Radicals," J. K. Kochi, ed.,
    Chapter 9, Wiley-Interscience, Inc., New York, 1973; G. A.
    Russell, E. G. Janzen, A. G. Bemis, E. J. Geels, A. J.
    Moye, S. Mak, E. T. Strom, Adv. Chem. Ser. (1965), 51,
    112.
21. (a) Bank, S. and D. A. Juckett, J. Am. Chem. Soc. (1975),
    97, 567; (b) Baizer, M. M., ed., "Organic Electrochemis-
    try," M. Dekker, Inc., New York, 1973; (c) Rogers, R. J.,

H. L. Mitchell, Y. Fujiwara and G. M. Whitesides, J. Org.
Chem. (1974), 39, 857.

22. (a) House, H. O. and M. J. Umen, J. Am. Chem. Soc.
(1972), 94, 5495; (b) Gardner, H. C. and J. K. Kochi, Ibid.
(1975), 97, 1855; (c) Nugent, W. A., F. Bertini and J. K.
Kochi, Ibid. (1974), 96, 4945; (d) Hegedus, L. S. and L. L.
Miller, Ibid. (1975), 97, 459; (e) Ashby, E. C., I. G. Lopp
and J. D. Buhler, Ibid. (1975), 97, 1964.

23. (a) Halpern, J., M. S. Chan, J. Hanson, T. S. Roche and
J. A. Topich, J. Am. Chem. Soc. (1975), 97, 1607; (b)
Anderson, S. N., D. H. Ballard, J. Z. Chrzastowski and
M. D. Johnson, J. Chem. Soc., Chem. Commun. (1972),
685; (c) Kochi, J. K., Acc. Chem. Research. (1974), 7,
351; (d) Costa, G., A. Puxeddu and E. Reisenhofer, Bio-
electrochem. and Bioenergetics. (1974), 1, 29.

24. Oxidation numbers of nickel are included only as a book-
keeping device and are not necessarily intended to denote
actual changes in oxidation states.

25. A stable 5-coordinate σ-phenylnickel(II) species is described
[P. DaPorto and L. Sacconi, Inorg. Chim. Acta (1974), 9,
62].

26. Cf. H. O. House and M. J. Umen, J. Org. Chem. (1973),
38, 3893.

27. Tamura, M. and J. K. Kochi, J. Am. Chem. Soc. (1971),
93, 1487.

28. Neumann, S. M. and J. K. Kochi, J. Org. Chem. (1975),
40, 599.

29. Braterman, P. S. and R. J. Cross, Chem. Soc. Revs.
(1973), 2, 271.

30. Kwan, C. L. and J. K. Kochi, J. Am. Chem. Soc., in
press.

31. Gargano, M., P. Giannocaro, M. Rossi, G. Vasapollo and
A. Sacco, J. Chem. Soc. Dalton (1975), 9.

32. (a) Osborn, J. A. in "Prospects in Organotransition Metal
Chemistry," M. Tsutsui, ed., Plenum Press, New York,
1975; (b) Rajaram, J., R. G. Pearson and J. A. Ibers, J.
Am. Chem. Soc. (1974), 96, 2103.

33. (a) Tamao, K., M. Zembayashi, Y. Kiso and M. Kumada,
J. Organometal. Chem. (1973), 55, C91; (b) Semmelhack,
M. F., P. M. Helquist and J. D. Gorzynski, J. Am. Chem.
Soc. (1972), 94, 9234; but see (c) Zembayashi, M., K.
Tamao and M. Kumada, Tetrahedron Letters (1975), 1719.

34. Chatt, J. and B. L. Shaw, J. Chem. Soc. (1960), 1718;
J. R. Moss and B. L. Shaw, Ibid. (1966), 1793; G. Calvin
and G. E. Coates, Ibid. (1960), 2008.

35. (a) Parshall, G. W., J. Am. Chem. Soc. (1974), 96, 2360;
(b) Morrell, D. M. and J. K. Kochi, Ibid. (1975), 97, 7262.

36. Propenylmagnesium bromide formed in this manner should
retain the stereochemistry of the reactant bromopropene in
contrast to that formed with magnesium metal. Cf. H. M.

Walborsky and M. S. Aronoff, J. Organometal. Chem.
(1973), 51, 53; H. L. Goering and F. H. McCarron, J. Am.
Chem. Soc. (1958), 80, 2287; R. J. Rogers, H. L.
Mitchell, Y. Fujiwara and G. M. Whitesides, J. Org.
Chem. (1974), 39, 857.

37. Cf. G. M. Whitesides, C. P. Casey and J. K. Krieger, J.
Am. Chem. Soc. (1971), 93, 1379.

38. Whitesides, G. M., E. R. Stedronsky, C. P. Casey and J.
San Filippo, Jr., J. Am. Chem. Soc. (1970), 92, 1426;
G. M. Whitesides, E. J. Panek and E. R. Stedronsky,
Ibid. (1972), 94, 232; M. Tamura and J. K. Kochi, Ibid.
(1971), 93, 1483; J. Organometal. Chem. (1972), 42, 205;
Ibid. (1971), 29, 111.

39. Tamao, K., Y. Kiso, K. Sumitani and M. Kumada, J. Am.
Chem. Soc. (1972), 94, 9268; M. Tamura and J. K. Kochi,
J. Organometal. Chem. (1972), 42, 205.

40. Jukes, A. E., Adv. Organometal. Chem. (1974), 12, 215.

41. Tamura, M. and J. K. Kochi, J. Am. Chem. Soc. (1971),
93, 1485.

42. Kochi, J. K., Pure App. Chem. (1971), 4, 3958.

43. Tamura, M. and J. K. Kochi, Synthesis (1971), 303.

44. Tamura, M. and J. K. Kochi, J. Am. Chem. Soc. (1971),
93, 1483.

45. Jenkins, C. L. and J. K. Kochi, J. Am. Chem. Soc. (1972),
94, 843, 856.

46. Whitesides, G. M., C. P. Casey and J. K. Krieger, J. Am.
Chem. Soc. (1971), 93, 1379.

47. Tamura, M. and J. K. Kochi, J. Organometal. Chem.
(1971), 31, 289; Bull. Chem. Soc. Japan (1971), 44, 3063.

48. Tamura, M. and J. K. Kochi, J. Organometal. Chem.
(1971), 29, 111.

49. Allen, R. B., R. G. Lawler and H. R. Ward, J. Am.
Chem. Soc. (1973), 95, 1692.

50. The large enhancement possible in CIDNP may not reflect
its chemical importance until they are quantitatively
related. A small amount of radical combination leading to
CIDNP may have been overlooked in the chemical studies.

11

New Strong Acid Catalyzed Alkylation and Reduction Reactions

M. SISKIN, R. H. SCHLOSBERG, and W. P. KOCSI

Corporate Research Laboratories, Exxon Research and Engineering Company, P. O. Box 45, Linden, NJ 07036

Most of the work described in this paper was carried out using the strong acid system composed of tantalum pentafluoride and hydrogen fluoride. Tantalum pentafluoride (TaF_5) is a white solid which melts at $\sim 97°C$. It is an acidic metal fluoride because of its large metal atom, which is coordinatively unsaturated having only ten electrons around it. The positive nature of the tantalum atom is enhanced by five very electronegative fluorines. This allows the tantalum to accept an anion from a Brönsted acid, such as hydrogen fluoride and generate a proton active enough to protonate the weakly basic hydrogen fluoride solvent (eq. 1).

$$2HF + TaF_5 \rightleftharpoons HF + H^{\delta+}\text{---}F\text{---}TaF_5^{\delta-} \rightleftharpoons H_2F^+ + TaF_6^-\qquad(1)$$

The resultant concentration of $[H_2F^+]$ is responsible for the extraordinary acidity of the system. Furthermore, the TaF_6^- anion is so weakly basic as to be essentially inert in a medium which thus allows the formation of stable, long-lived carbocations in hydrocarbon reactions. Tantalum pentafluoride is also very thermally stable (1) even at 300°C and resistant to reduction reactions, especially, by molecular hydrogen and hydrocarbon ions (2-5). This is in sharp contrast to antimony pentafluoride containing strong acid systems which are readily reduced to the antimony (III) state (6-8). Hydrogen fluoride is a colorless liquid boiling at $\sim 19.5°C$, which when anhydrous is itself a fairly strong acid with a Hammett acidity (Ho) of ~ -11 (9). It is a thermally stable and non-reducible Brönsted acid (10). When TaF_5 and HF are mixed together, a colorless solution is formed which can be classified as a "super acid" having an Ho of -18.85 (11), or $\sim 10^8$ times stronger than anhydrous hydrogen fluoride. In addition to forming a very stable liquid phase strong acid Friedel-Crafts system, it is, because of its high acidity, and therefore its ability to protonate weak organic bases, a good hydrogenation catalyst. We have shown that in the hydrogenation of benzene, in fact, it is protonated benzene, not benzene itself, which undergoes the initial hydrogenation (12).

Introduction

The direct and selective alkylation of benzene by alkanes has long been a desirable goal (13). The ability to ionize the lower alkanes to highly acidic cations in super acids has given further impetus to the challenge (14). Sacrificial species have even been added to the reaction mixture in order to provide a stoichiometric driving force to overcome the very unfavorable thermodynamics (Table I) for the reaction (15, 16). In sharp contrast, less attention has been given to carrying out the direct acid catalyzed alkylation of the lower alkanes with the lower alkenes, (17, 6) although such reactions are very thermodynamically favorable (Table I), especially at low temperatures (25-125°) where antagonistic entropy effects are less important.

TABLE I (18)

THERMODYNAMICS OF ALKANE ALKYLATIONS

	ΔF_f^o (KCAL/MOLE) (°K)		
AROMATICS	300	400	500
$CH_4 + C_6H_6 \longrightarrow C_6H_5CH_3 + H_2$	+10.32	+10.36	+10.31
$CH_3CH_3 + C_6H_6 \longrightarrow C_6H_5CH_2CH_3 + H_2$	+ 8.08	+ 8.18	+ 8.15
ALKENES			
$CH_4 + C_2H_4 \longrightarrow C_3H_8$	- 9.70	- 6.43	- 3.17
$CH_3CH_3 + C_2H_4 \longrightarrow \underline{n}-C_4H_{10}$	-12.46	- 9.14	- 5.86
$CH_4 + C_3H_6 \longrightarrow \underline{i}-C_4H_{10}$	- 7.77	- 3.97	- 0.21

Conventional alkane-alkene alkylation is an acid catalyzed reaction which involves the addition of a tertiary carbenium ion generated from an alkane to an alkene to yield (after hydride addition) a saturated hydrocarbon of higher molecular weight.

Mechanistically, as elucidated by Schmerling (19) and as illustrated for isobutane-ethylene (alkane-alkene) alkylation (Scheme 1), the reaction is initiated by protonation of the alkene (ethylene) to form a very acidic primary ethyl cation (step 1) which rapidly abstracts a hydride ion from an isobutane molecule to generate the chain carrying t-butyl cation (step 2). This can then alkylate another molecule of ethylene to form the secondary-2-methyl-t-butyl carbenium ion (step 3). This cation rapidly undergoes a methyl shift to form the more stable (less acidic) tertiary-dimethylisopropyl carbenium ion which then abstracts a hydride ion from isobutane to form the 2,3-dimethylbutane alkylation product and generate a t-butyl cation which can then react with another ethylene in (step 3), etc., and thus make the reaction catalytic.

SCHEME 1

ISOBUTANE-ETHYLENE ALKYLATION

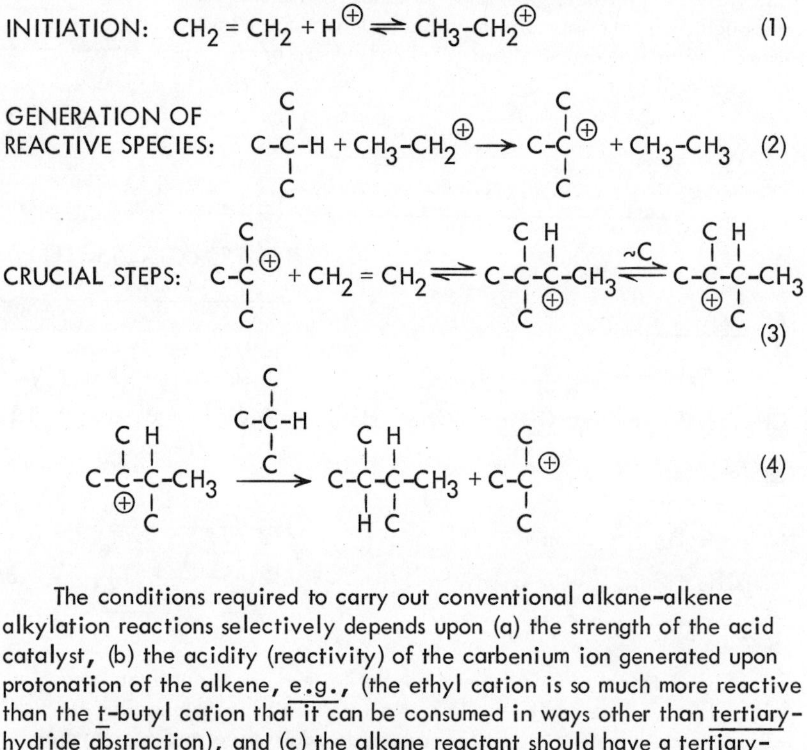

INITIATION: $CH_2 = CH_2 + H^{\oplus} \rightleftharpoons CH_3-CH_2^{\oplus}$ (1)

GENERATION OF
REACTIVE SPECIES:
$$\underset{\underset{C}{|}}{\overset{\overset{C}{|}}{C}}\text{-C-H} + CH_3-CH_2^{\oplus} \longrightarrow C\text{-}C^{\oplus} + CH_3-CH_3 \qquad (2)$$

CRUCIAL STEPS: $C\text{-}C^{\oplus} + CH_2 = CH_2 \rightleftharpoons C\text{-}C\text{-}C\text{-}CH_3 \overset{\sim C}{\rightleftharpoons} C\text{-}C\text{-}C\text{-}CH_3$ (3)

$$C\text{-}C\text{-}C\text{-}CH_3 \longrightarrow C\text{-}C\text{-}C\text{-}CH_3 + C\text{-}C^{\oplus} \qquad (4)$$

The conditions required to carry out conventional alkane-alkene alkylation reactions selectively depends upon (a) the strength of the acid catalyst, (b) the acidity (reactivity) of the carbenium ion generated upon protonation of the alkene, e.g., (the ethyl cation is so much more reactive than the t-butyl cation that it can be consumed in ways other than tertiary-hydride abstraction), and (c) the alkane reactant should have a tertiary-hydrogen because the hydride abstraction steps are easier.

A turning point in the revival of interest in strong acid chemistry was a publication in 1968 in which ionization of the C-H bonds of the extraordinarily unreactive "lower paraffins" methane and ethane in $HSO_3F\text{-}SbF_5$ at 50°C was reported (14). The proposed mechanism (Scheme 2) proposes the existence, in super acid solution, of protonated alkanes or pentacoordinated ions, at least as possible transition states, and attempts quite logically to draw a parallelism between the presence of such species in solution chemistry in super acids and gaseous ion-molecule reactions which occur in the mass spectrometer.

SCHEME 2

METHANE CONDENSATION REACTION

$$\boxed{HSO_3F-SbF_5}$$

$$CH_4 \underset{-H^-}{\overset{+H^+}{\rightleftarrows}} [CH_5]^+ \xrightarrow[\ -H_2\ ]{} CH_3^+ \xrightarrow{CH_4} \left[CH_3 \text{---} \overset{+}{\underset{CH_3}{<}}{}^{H} \right]^+ \xrightarrow{-H_2} CH_3CH_2^+ \xrightarrow{CH_4}$$

$$\left[CH_3 \text{---} \underset{\underset{CH_3}{|}}{\overset{H}{<}}{}^{C-CH_3} \right]^+ \xleftarrow{CH_4} CH_3\overset{+}{C}HCH_3 \xleftarrow{-H_2} \left[CH_3 \text{---} \overset{H}{<}_{CH_2CH_3} \right]^+$$

$$\xrightarrow{-H_2} CH_3\text{-}\overset{\overset{\displaystyle CH_3}{|}}{\underset{\underset{\displaystyle CH_3}{|}}{C}}\text{+} \xrightarrow{H_2O} CH_3\text{-}\overset{\overset{\displaystyle CH_3}{|}}{\underset{\underset{\displaystyle CH_3}{|}}{C}}\text{-OH}$$

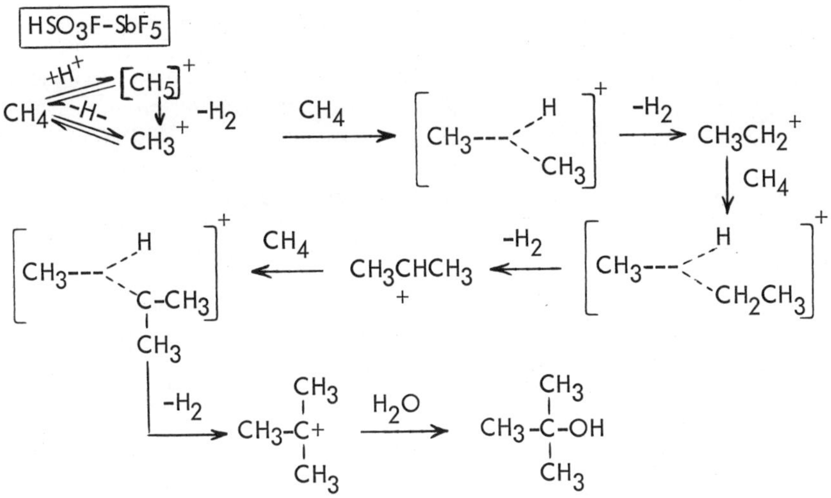

This parallelism is reflected in the proposed mechanism for the ionization of methane which shows that (a) the second step of the scheme involves attack of an ethyl cation on methane, but the reaction cannot stop there, and goes on to (b), the third step, which involves attack of a secondary-isopropyl cation on methane. The primary and secondary alkyl cations are very strongly acidic species and are unstable under the reaction conditions. The condensation reaction essentially terminates with the much more weakly acidic tertiary-butyl ion. Alkane polycondensation and olefin polymerization side reactions producing stable, less acidic, tertiary ions obscured the simple alkylation reactions of the primary and secondary alkyl cations. Implicit in this mechanism, however, is that it is possible to react an acidic energetic primary cation (such as the ethyl cation) with molecules as weakly basic as methane and thus, the door was opened to new chemistry through activation of the heretofore passive, weakly basic, "paraffins" (20-24).

Results and Discussion

Alkene-Alkane Alkylations. A logical approach to achieve the catalytic alkylation reaction of methane was to initiate electrophilic attack of its very unreactive C–H bonds by using a very energetic primary carbenium ion. The simplest way to generate such an ion is to dissolve ethylene in a large excess of a super acid at a moderate temperature (in which it would be as fully protonated as possible) to form the highly acidic and reactive primary ethyl cation. The ion is thus available to react with the strongest base available, i.e., methane, in an alkene-alkane alkylation in contrast with the traditional alkane-alkene alkylations. We have now found that such simple addition reactions can be selectively carried out in the HF–TaF$_5$ catalyst system (25). A methane:ethylene (85.9%: 14.1%) gas mixture was passed through an autoclave containing 50 cc of a 10:1 HF–TaF$_5$ (2.0 mol/0.20 mol) system stirred at 1000 rpm at 40° and maintained at 40 psig. In order to minimize possible competition from ethylene oligomerization reactions a forty-fold excess of acid as well as efficient mixing was maintained and the temperature was not permitted to vary more than ±1°. Gas samples were taken after 1.5 and 2.5 hours and the selectivity to C$_3$ in the product amounted to 58%.

Mechanistically, two pathways are logical (Scheme 3). The ethyl cation can directly alkylate methane via a pentacoordinated carbonium ion (Olah) (path a), or alternatively, although a less favorable pathway (b), the ethyl cation could abstract a hydride ion from methane. The methyl cation thus formed, which is less stable by ~39 kcal/mole (26), could then react directly with ethylene. In the latter case, propylene and/or polymeric material would probably be formed since the hydrogen required for a catalytic reaction has been consumed by the formation of ethane.

SCHEME 3

ETHYLENE-METHANE ALKYLATION AT 40°C

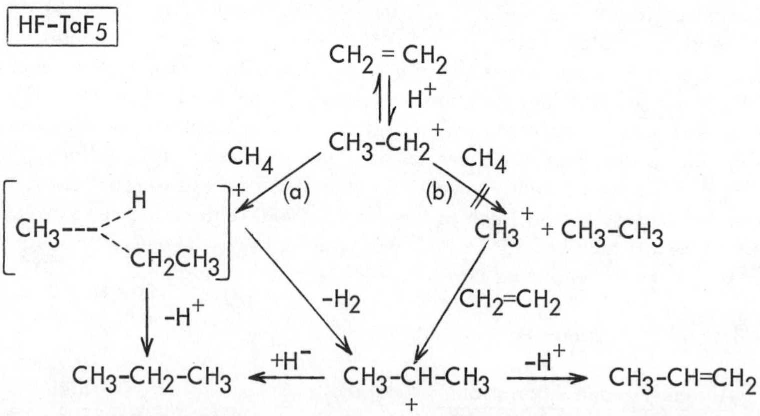

Ethylene alone (15% diluted in helium) does not react in $HF-TaF_5$ at 40°C to form any propane product but rather forms a complex mixture of unsaturated and higher molecular weight products. Methane also does not react at all under these conditions (27). It should be noted that unless a flow system is used, the propane product, which is a substantially better hydride donor than methane, reacts further with the intermediate ethyl cation to ultimately form ethane and propylene (eq. 2).

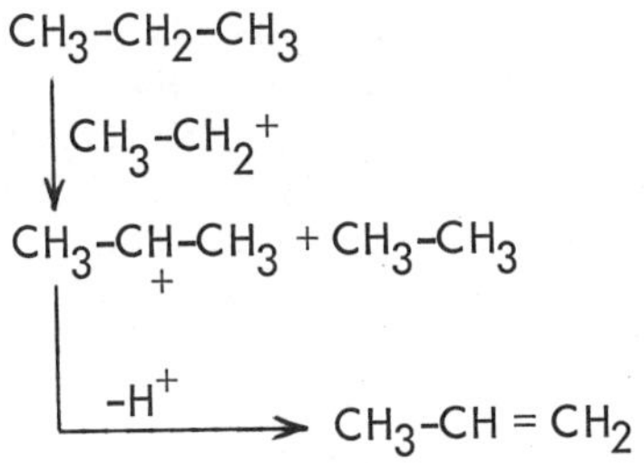

$$CH_3-CH_2-CH_3$$

$$\downarrow CH_3-CH_2^+$$

$$CH_3-\underset{+}{C}H-CH_3 + CH_3-CH_3$$

$$\xrightarrow{-H^+} CH_3-CH = CH_2$$

(2)

In our work, only ~1% of the propylene formed in the flow system reacted with another molecule of methane to form isobutane. Also, based upon the results of acid quenching and analysis of hydrocarbons, only traces of isopentane and isohexanes were present in the acid. No hydrogen or hydrocarbons above C_6 could be detected in the product.

In an attempt to generate primary (trivalent) cations and to simulate the ethylene-methane alkylation, ethyl chloride was reacted with methane (eq. 3) under alkylation reaction conditions (28). When no propane or propylene product was observed, the energetically more favorable reaction of methyl chloride with ethane was carried out (eq. 3a). These two reactions proceeded without any involvement of the alkane and provide evidence that the ethylene-methane alkylation proceeds through a more stabilized species such as a pentacoordinated carbonium ion. The behavior of these alkyl chlorides will be discussed separately after the alkylation chemistry.

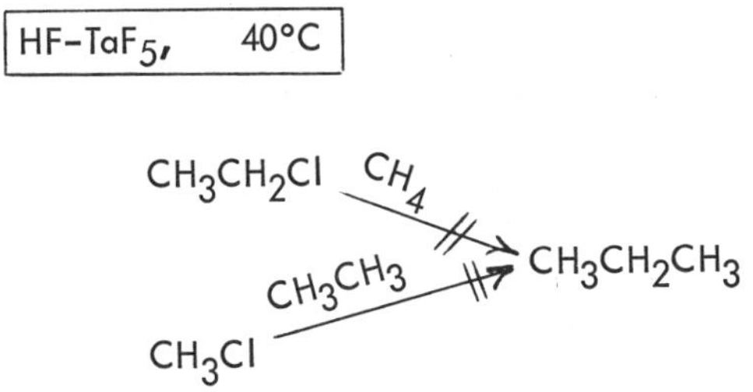

$$\boxed{HF-TaF_5, \quad 40°C}$$

$$CH_3CH_2Cl \xrightarrow{CH_4} CH_3CH_2CH_3$$

$$CH_3CH_3$$

$$CH_3Cl$$

(3)

(3a)

It should also be noted that the propane product is not being formed by degradation of polyethylene, because under similar reaction conditions, but in the presence of hydrogen, we have shown that the polymer ($\overline{MW}$ 800,000) reacts quantitatively to form C_3–C_6 paraffins with isobutanes and isopentanes constituting over 85% of the product. The polymer degradation products can be easily rationalized on the basis of known carbenium ion stabilities in acid media. Olah (29) observed t-butyl cation by reaction of polyethylene in "magic acid." These results further substantiate that the direct alkene–alkane alkylation takes place in the ethylene/methane reaction.

The reaction of ethylene with ethane is of major scientific significance. The only C_4 product, formed in 78% selectivity, in this reaction is normal-butane (Scheme 4) which does not isomerize under these conditions with the HF-TaF$_5$ catalyst (eq. 4) or under similar conditions with the HF-SbF$_5$ catalyst (30). This means that the primary ethyl cation is alkylating a primary ethane position (path a) and that there is no classical free primary normal–butyl cation formed (path b) because such a cation, as would be generated from n–butylchloride, would yield exclusively isobutane upon rearrangement as shown in one experiment carried out where n–butyl chloride with hydrogen in the acid yielded only isobutane (eq. 5).

SCHEME 4

ETHYLENE-ETHANE ALKYLATION AT 40°C

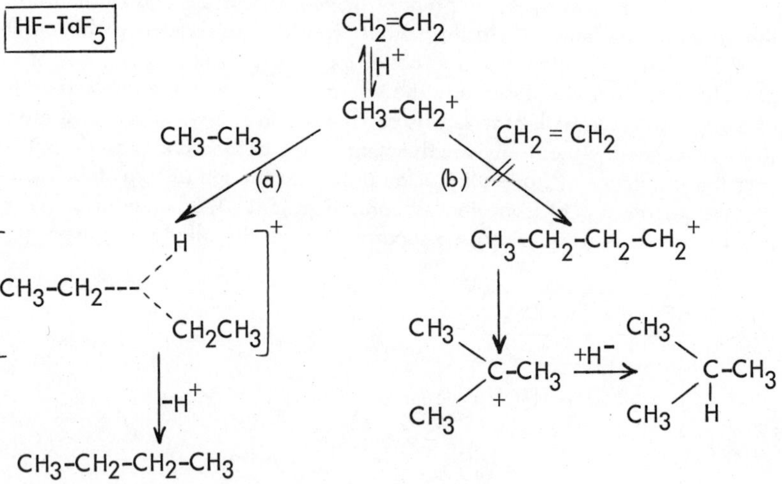

$$CH_3CH_2CH_2CH_3 \xrightarrow[40°C, H_2]{HF-TaF_5} \quad N.R. \qquad (4)$$

$$\underset{(0.05\ mole)}{CH_3CH_2CH_2CH_2Cl} \xrightarrow[H_2\ (0.13\ mole)]{HF-TaF_5,\ 20°C} \quad \underset{CH_3}{\overset{CH_3}{>}} \underset{H}{\overset{|}{C}}-CH_3 \qquad (5)$$

Of further interest is the fact that n-butyl chloride reacts in the presence of excess ethane, also at 40°C, to form butylenes (85%) and some isobutane (15%)(eq. 6a). These products lead to the conclusion that rearrangement of the "free" trivalent carbenium ion is more rapid than hydride abstraction from another n-butyl chloride molecule. The t-butyl carbenium ion thus formed, being too weak an acid to abstract a hydride, deprotonates to form butylene products. No isohexane alkylation products are formed (eq. 6).

$$\underset{(0.05\ mole)}{CH_3CH_2CH_2CH_2Cl} \xrightarrow[CH_3CH_3\ (0.13\ mole)]{HF-TaF_5,\ 40°C} \begin{array}{l} \nearrow\!\!\!\!\!/ \ C_6H_{14}\text{'s} \qquad (6) \\[2ex] \searrow \ \underset{(85\%)\quad (15\%)}{C_4H_8\text{'s} + \underline{i}-C_4H_{10}} \qquad (6a) \end{array}$$

The secondary propyl carbenium ion formed in the reaction of propylene (3.4%) at 40° in 10:1 HF-TaF$_5$, attacks methane (96.6%) to form isobutane (Scheme 5) with 60% selectivity.

Olah (17a) has also reported the alkylation reactions (at -10° with 1:1 HSO$_3$F-SbF$_5$) of n-butane with ethylene to yield 38 weight percent of hexanes and of n-butane with propylene to yield 29 weight percent of heptanes. The former reaction has also been reported by Parker (31) at 60°, but the product in this case more nearly resembles polyethylene degradation products. In our work with 10:1 HF-TaF$_5$ at 40°, in a flow system, ethylene (14.1 wt.%) reacted with n-butane to form 3-methylpentane as the initial product of 94% selectivity (Scheme 6, path a). The alternative, i.e., the direct reaction of ethylene with a secondary-butyl cation (path b), can be ruled out since butane does not ionize under these conditions (vide supra).

The products of the reactions in schemes 4 and 6 suggest strongly that cations or tight ion pairs of the pentacoordinated type proposed by Olah are more stable than the classical type in these reactions.

In pentacoordinated systems three center-two electron bonds help stabilize the system relative to a localized cation. This becomes more

SCHEME 5

ALKYLATION OF METHANE WITH PROPYLENE

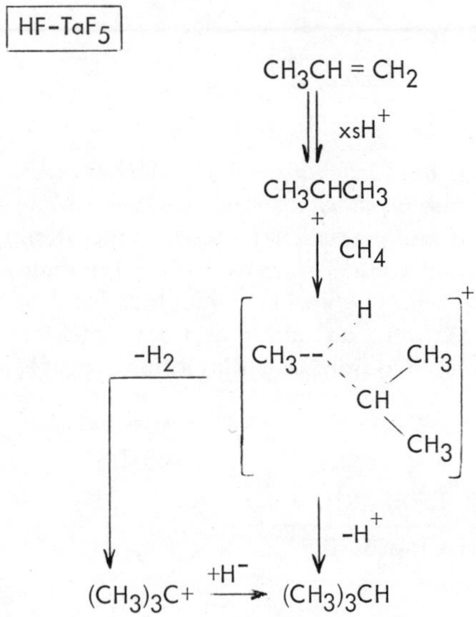

$$\boxed{\text{HF-TaF}_5}$$

$$CH_3CH = CH_2$$

$$\Big\Downarrow \; xsH^+$$

$$\underset{+}{CH_3CHCH_3}$$

$$\Big\downarrow \; CH_4$$

$$-H_2 \quad \left[CH_3 -\!\!\!- \; \overset{H}{\underset{\underset{CH_3}{CH}}{\underset{\diagdown}{\diagup}}} \; CH_3 \right]^+$$

$$\Big\downarrow \; -H^+$$

$$(CH_3)_3C^+ \xrightarrow{+H^-} (CH_3)_3CH$$

important as the carbocation type goes from tertiary to secondary to primary and thus we can rationalize the formation of n-butane rather than i-butane in the ethylene/ethane reaction and of 3-methylpentane rather than 2,3-dimethylbutane in the ethylene/butane reaction.

Direct Alkyl Chloride Reductions. It was previously pointed out that at 40° in HF-TaF$_5$ methyl- and ethyl-chloride do not form the expected alkylation product with alkanes, expected that is, assuming traditional carbenium ion chemistry, but rather underwent an independent reaction. The second part of this paper will describe in more depth what reactions do take place when alkyl chlorides react in the presence or absence of the lower alkanes under our reaction conditions of 40°C.

A century ago, Friedel and Crafts (32) reported that "when a small amount of anhydrous aluminum chloride was added to amyl chloride an immediate vigorous evolution of gas was observed in the cold. The gas was composed of hydrogen chloride accompanied by gaseous hydrocarbons not absorbed by bromine." The precise nature of these saturated hydrocarbons has not been well understood. In the course of our mechanistic studies on alkene-alkane alkylations we observed the direct reduction products of several alkyl chlorides. While there exists an extensive literature on the be-

SCHEME 6

ALKYLATION OF NORMAL-BUTANE WITH ETHYLENE

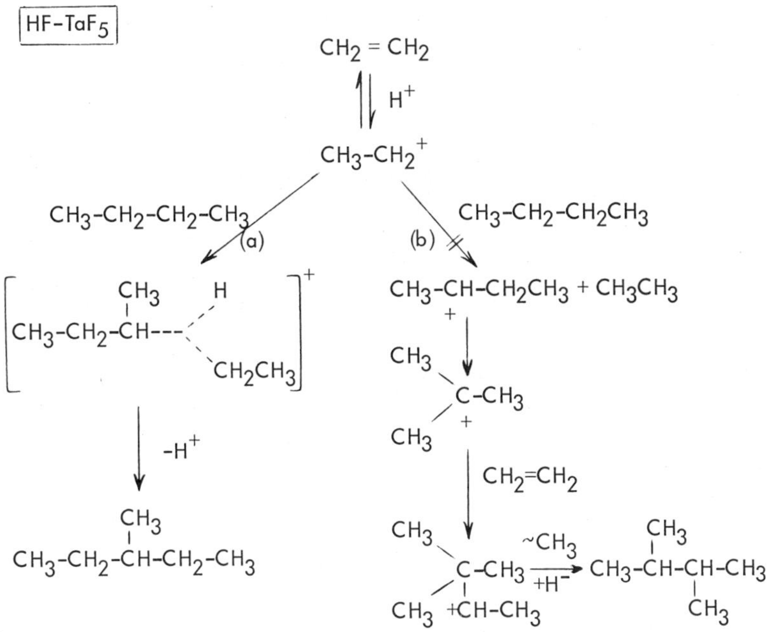

|HF-TaF$_5$|

$$CH_2 = CH_2$$

$$\Updownarrow H^+$$

$$CH_3-CH_2^+$$

$CH_3-CH_2-CH_2-CH_3$ (a)   (b) $CH_3-CH_2-CH_2CH_3$

$$\left[\begin{array}{c} CH_3 \quad H \\ | \\ CH_3-CH_2-CH--- \\ \qquad\qquad CH_2CH_3 \end{array} \right]^+$$

$CH_3-CH-CH_2CH_3 + CH_3CH_3$

$$\begin{array}{c} \overset{+}{C}H_3 \\ \downarrow \\ CH_3 \\ \diagdown \\ \underset{+}{C}-CH_3 \\ \diagup \\ CH_3 \end{array}$$

$\Big\downarrow -H^+$

$\Big\downarrow CH_2=CH_2$

$$\begin{array}{c} CH_3 \\ | \\ CH_3-CH_2-CH-CH_2-CH_3 \end{array}$$

$$\begin{array}{c} CH_3 \\ \diagdown \quad \sim CH_3 \qquad\qquad CH_3 \\ C-CH_3 \xrightarrow{+H^-} CH_3-\overset{|}{C}H-CH-CH_3 \\ \diagup \quad | \qquad\qquad\qquad | \\ CH_3 \; {}^+CH-CH_3 \qquad\qquad CH_3 \end{array}$$

havior or alkyl chlorides in acidic and super-acidic media, we believe that these results represent the first conversion of C_1-C_3 alkyl chlorides as determined by analysis of the primary gas phase products to the corresponding alkane. The reactions proceed in considerable yield (up to 34%), via a direct hydride transfer route.

We have looked at the following systems in HF-TaF$_5$ at 40°C (eq. 7-9): CH$_3$Cl, C$_2$H$_5$Cl, and iso-C$_3$H$_7$Cl. The latter systems and n-C$_3$H$_7$Cl were also studied in HCl:AlCl$_3$ at 140°. In all cases the gas phase was analyzed and, where reaction occurred, the corresponding alkane was found to be the major product.

The first conclusion is that the lower alkane does not participate in the reaction and the alkyl chloride reacts with a second molecule of alkyl chloride to form the direct reduction product which is observed in the gas phase. These gas phase products do not account for all of the alkyl chloride charged to the reactor. Upon quenching the acid, we can recover some alkyl halide and some higher molecular weight condensed-hydrocarbon products.

ANALYSIS OF GAS PHASE

$\boxed{\text{HF–TaF}_5}$

$$CH_3CH_2-Cl \underset{\searrow}{\overset{CH_4 \; \longrightarrow \; CH_3CH_3}{\nearrow}} \qquad (7)$$

$$CH_3CH_3 \qquad (7a)$$

$$CH_3-Cl \underset{\searrow}{\overset{CH_3CH_3 \; \longrightarrow \; CH_4}{\nearrow}} \qquad (8)$$

$$CH_4 \qquad (8a)$$

$$\begin{array}{c} CH_3 \\ \diagdown \\ CH-Cl \longrightarrow CH_3CH_2CH_3 \qquad (9) \\ \diagup \\ CH_3 \end{array}$$

Olah (33) has shown that alkyl fluorides undergo a self-condensation reaction (eq. 10) similar to that previously described for the ionization and condensation of methane so that this additional pathway is not an unexpected one.

SELF CONDENSATION OF ALKYL FLUORIDES IN HF–SbF₅

$$FCH_3 + \text{"CH}_3F-SbF_5\text{"} \rightleftharpoons \left[FCH_2---\overset{H}{\underset{CH_3}{\diagdown}} \right]^+ \qquad (10)$$

$$\overset{-H^+}{\rightleftharpoons} FCH_2CH_3 \xrightarrow{\text{"RF–SbF}_5\text{"}} \longrightarrow \longrightarrow (CH_3)_3C^+, \; \text{Etc.}$$

Using the propyl chlorides as an example (eq. 11, Table II), after five minutes at 140° in a 45 cc Hastelloy C reactor charged with aluminum chloride and 2-chloropropane the pressure was 500 psig of which 71% was propane by mass spectrometric analysis.

$$\left. \begin{array}{c} CH_3CHCH_3 \\ | \\ Cl \\[4pt] CH_2CH_2CH_2Cl \end{array} \right\} \xrightarrow[140°C]{HCl-AlCl_3} CH_3CH_3CH_3 \qquad (11)$$

TABLE II

REACTION OF PROPYL CHLORIDES IN HCl–AlCl$_3$

Compound	Time (Min.)	% C Conv. to C$_3$H$_8$
i-C$_3$H$_7$Cl	5	34
	15	24
n-C$_3$H$_7$Cl	5	18
	15	17

In other words, of the initial charge of isopropyl chloride, 34% of the carbon is converted to propane under these conditions. A comparable sample of 1-chloropropane under identical reaction conditions, of excess alkyl halide vs. Lewis acid, was converted after five minutes to the extent of 18% to propane. After 15 minutes the relative amount of "propane" had decreased as a result of further acid catalyzed polycondensation reactions. Similarly, isopropyl chloride reacts in HBr–AlBr$_3$ at room temperature to give a gas product which is entirely propane after 2 min. In this system we again begin to see a buildup of heavier hydrocarbon species with time. The fact that the reaction proceeds so rapidly here can probably be attributed to a homogeneous hydrocarbon/acid liquid phase. This last result and additional experiments are summarized in Table IV.

Since the ratio of acid to alkyl halide could conceivably affect the acidity of the medium and hence the reaction mechanism, these reactions were also carried out at widely varying ratios (Tables III and IV). The results indicate that in aluminum halide systems the stoichiometry does not alter the primary pathway. In the HF–TaF$_5$ system at the higher acid:alkyl chloride ratio there is a lower selectivity to the alkane with hydrogen the only other major product in the gas phase lending support to the conclusion that condensation is the major competing reaction. 2-Chloropropane undergoes a shift in reaction mechanism at low acid:alkyl chloride ratio to form the direct reduction product propane. The more acidic methyl cation, on the other hand, undergoes only slow polycondensation (~25% in 1 hr.) reaction and no direct reduction at the low acid:alkyl halide ratio.

Mechanistically, alkane formation could be occurring via several pathways.

1. Dialkylchloronium ions (Scheme 7): The existence of halogen substituted cations (34) is well documented in the literature as is the existence of dialkylhalonium ions (35). Support for this possibility in our studies comes from low temperature (-35°C) 60 MHz pmr spectra of the acid layer of the methyl chloride reaction in HF–TaF$_5$ which shows a sharp

TABLE III

REACTIONS OF ALKYL CHLORIDES WITH STRONG ACIDS
- RESULTS -

	% RH on Total HC Gases		
	% CH_4	% C_2H_6	% C_3H_8
TaF_5:RCl(4:1)	57	49	48
TaF_5:RCl(1:4)	0	--	97

$$RCl \quad \frac{HF-TaF_5}{40°C, \ 2 \ hr.}$$

$R=CH_3, C_2H_5, \underline{i}-C_3H_7$

	% C_3H_8 of HC Gases	
	i-PrCl	n-PrCl
$AlCl_3$:RCl(4:1)	97	--
$AlCl_3$:RCl(1:4)	97	97

$$RCl \quad \frac{HCl-AlCl_3}{140°C, \ 5 \ min.}$$

$R=\underline{i}-C_3H_7, \underline{n}-C_3H_7$

TABLE IV

SUMMARY OF RESULTS OF THE REACTIONS
OF ALKYL CHLORIDES WITH STRONG ACIDS

Alkyl Chloride (R–Cl)	MX_n/ RCl	Acid	T (°C)	t (min.)	% RH in Gas Phase	% RH in Total HC Gases
CH_3Cl	4.0	10:1 HF–TaF$_5$[a]	40	30	6.6	21.5[d]
				60	7.5	42.3
				120	8.9	57.0
CH_3Cl	0.2	10:1 HF–TaF$_5$	40	60	0.0	0.0
CH_3CH_2Cl	4.0	10:1 HF–TaF$_5$	40	30	9.7	25.7
				60	10.0	32.2
				120		49.3
$CH_3CH_2CH_2Cl$	0.2	1:1 HCl–AlCl$_3$	140	5	45.0[b]	97.2
				15	47.0[b]	93.0
$(CH_3)_2CHCl$	0.2	1:1 HCl–AlCl$_3$	140	5	70.7[c]	97.1
				15	56.8[c]	92.8
	4.0	1:1 HCl–AlCl$_3$	140	5	> 90.0	96.8
		10:1 HF–TaF$_5$	40	30	30.1	43.0
				60	32.5	46.0
				120	33.4	47.8
	0.2	10:1 HF–TaF$_5$	40	5	55.5	97.0
				60	61.8	96.7
	0.2	1:1 HBr–AlBr$_3$	20	2	~100.0	~100.0[e]
				18	~90.0	~90.0

(a) HF–TaF$_5$ (2.0 mole:0.2 mole) + RCl (0.05 mole) in a 300 cc Hastelloy C Autoclave Engineers autoclave stirred at 1000 rpm.

(b) Yields: 9.5% and 9.0% in 5 and 15 min. respectively.

(c) Yields: 34.2% and 23.6% in 5 and 15 min. respectively.

(d) 21.5% of the hydrocarbons in the gas phase is methane.

(e) Analysis via gas chromatography.

singlet at 4.1 δ in HF solution relative to external TMS (36). This peak
represents about 50% of the unreacted methyl chloride remaining in the acid
layer. This also indicates that the formation of dimethylchloronium ion must
be more rapid than displacement of chloride in methyl chloride by fluoride
ion and that since fluoride is too electronegative to form fluoronium ion
salts the fluoride does not displace chloride from the dimethylchloronium
ion.

SCHEME 7

Via Dialkylhalonium Ions:

$$2R_2CHCl + MXn \rightleftharpoons (R_2CH)_2\overset{+}{Cl}(MXnCl)^-$$

$$(R_2CH)_2\overset{+}{Cl}(MXnCl)^- + R_2CHCl \longrightarrow R_2CH_2 + (R_2CH\overset{+}{Cl}CR_2Cl)(MXnCl)^-$$

2. Chloroalkyl carbenium ions (Scheme 8):   The existence of simple
halomethyl cations has as yet only been the subject of speculation and discussion (37). The stability of such species should become increasingly
favorable as the size of the alkyl group and number of such groups is
increased.

Recent energy calculations by Hehre favor the 1-haloethyl form of the
cation (38). Based upon the temperature dependence of the observed nmr
spectrum, Olah (39) has postulated the formation of the 1-fluoroethyl cation
by the reaction of SbF_5 with 1,1-difluoroethane. He also observed the 1-
chloroethyl and chlorine-bridged ethyl cation in the nmr by reaction of
SbF_5 in sulfuryl chloride and from ^{13}C nmr results predicts a significant
contribution by the $CH_3-CH=Cl^+$ resonance form. That part of these reactions are occurring through this type of species cannot be precluded. We
have not as yet been successful, however, in identifying by nmr or quenching experiments with toluene possible chloroalkyl-halonium ions in the acid
solution.

SCHEME 8

Via Haloalkyl Carbenium Ions:

$$R_2CHCl + MXn \rightleftharpoons R_2\overset{+}{CH}(MXnCl)^-$$

$$R_2\overset{+}{CH}(MXnCl)^- + R_2CHCl \longrightarrow R_2CH_2 + R_2\overset{+}{CCl}(MXnCl)^-$$

For the reaction of the propyl chlorides, the possibility of dehydro-halogenation followed by hydride abstraction from propylene to produce propane, while remote, cannot be discounted (Scheme 9). We believe, however, that propylene formation would produce oligomerization and polymerization products and thus is not consistent with fully one-third of the product going to propane.

SCHEME 9

For Propyl Chlorides:

A. Via Propenyl Cation Formation;

$$CH_3\underset{\underset{Cl}{|}}{C}HCH_3 + MXn \longrightarrow (CH_3\overset{+}{C}HCH_3)(MXnCl)^-$$

$$(CH_3\underset{+}{C}HCH_3)(MXnCl)^- \longrightarrow H^+(MXnCl)^- + CH_3CH=CH_2$$

$$CH_3CH=CH_2 + (CH_3\underset{+}{C}HCH_3)MXnCl^- \longrightarrow$$

$$CH_3CH_2CH_3 + (CH_2\overset{+}{=\!=\!=}CH\overset{}{=\!=\!=}CH_2)MXnCl^-$$

Or,

B. Via Diisopropyl Chloronium Ion;

$$CH_3CH=CH_2 + ([CH_3]_2CH)_2\overset{+}{C}l(MXnCl)^- \longrightarrow$$

$$CH_3CH_2CH_3 + (C_3H_5)\text{-}\overset{+}{C}l\text{-}(C_3H_7)(MXnCl)^-$$

In any case, the conversion of methyl or ethyl chlorides to methane and ethane, respectively, precludes the allylic hydrogen donation pathway for these systems and lends support to a mechanism such as outlined in schemes 7 and 8.

In a related experiment we looked at the chemistry of methanol in HCl-AlCl_3 at 150°C. We were interested in determining whether a reduction reaction of methanol to methane (eq. 12) could be competitive with conversion to methyl chloride (eq. 13). Mass spectral results show copious amounts of methyl chloride in the gas phase after 5 min. and 30 min. of reaction, with little or no methane, thus arguing that, if occurring, the

reduction does not kinetically compete with the conversion to methyl
chloride under our conditions.

COMPETITIVE REDUCTION VS. CHLORINATION OF METHANOL --HCl-AlCl₃ CATALYST--

$$2CH_3OH \xrightleftharpoons{H^+} (CH_3)_2\overset{+}{O}H$$

$$(CH_3)_2\overset{+}{O}H + CH_3OH \longrightarrow CH_4 + HOCH_2-\underset{+}{\overset{H}{\underset{|}{O}}}-CH_3 \qquad (12)$$

$$\boxed{Versus}$$

$$CH_3OH \xrightleftharpoons{H^+} CH_3\overset{+}{O}H_2 \xrightleftharpoons{Cl^-} CH_3Cl \qquad (13)$$

Conclusions

   The complexity of the chemistry involved is apparent from the variety
of plausible pathways which can be invoked to explain that part of the
reaction we can measure. The work to date clearly indicates that differences in reaction products will result and should be expected in comparing
the behavior of carbocations generated from the alkane versus the complex
mixture of cationic species generated from the alkyl chloride in super acid
media.

REACTIONS IN STRONG FRIEDEL-CRAFTS ACIDS PROCEED VIA DIFFERENT STABILIZED SPECIES

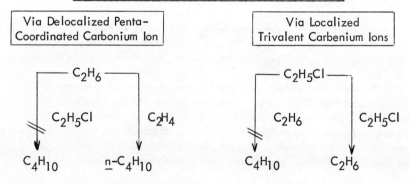

We believe much more work is needed before the true nature of species obtained via the acid catalyzed treatment of haloalkanes is well understood and we are loathe to extrapolate from alkane/acid systems to those involving carbon halogen bonds.

Literature Cited

1. Fairbrother, F., Grundy, K. H. and Thompson, A., *J. Chem. Soc.*, 765 (1965).
2. O'Donnell, T. A., "The Chemistry of Fluorine," Vol. 5, p. 1078, Pergamon Press, Oxford, England, 1973.
3. Canterford, J. H. and O'Donnell, T. A., *J. Inorg. Chem.* 5, 1442 (1966).
4. Muetterties, E. L. and Castle, J. E., *J. Inorg. Nucl. Chem.*, 18, 148 (1961).
5. Emeleus, H. J. and Guttmann, V., *J. Chem. Soc.*, 2115 (1950).
6. Oelderik, J. M., Mackor, E. L., Platteeuw and van der Wiel, A., U.S. Patent 3,201,494, August 17, 1965.
7. Olah, G. A. and Olah, J. A. in "Carbonium Ions," Vol. II, p. 761, Edited by G. A. Olah and P. v. R. Schleyer, Wiley - Interscience, New York, New York, 1970.
8. Olah, G. A., Schilling, P. and Grosse, I. M., *J. Amer. Chem. Soc.*, 96, 876 (1974).
9. Gillespie, R. J. and Peel, T. E., "Advances in Physical Organic Chemistry," Vol. 9, p. 16 and references cited therein, Edited by V. Gold, Academic Press, New York, New York, 1971.
10. Simons, J. H. in "Fluorine Chemistry," Vol. I, p. 225 and references cited therein, Edited by J. H. Simons, Academic Press, New York, New York, 1968.
11. Gillespie, R. J., personal communication.
12. Siskin, M., *J. Amer. Chem. Soc.*, 96, 3641 (1974).
13. Patinkin, S. H. and Friedman, B. S. in "Friedel Crafts and Related Reactions," pp. 254-255, Edited by G. A. Olah, Vol. II, Interscience Publishers, New York, New York 1964.
14. Olah, G. A. and Schlosberg, R. H., *J. Amer. Chem. Soc.*, 90, 2726 (1968).
15. Olah, G. A., Schilling, P., Staral, J. S., Halpern. Y. and Olah, J. A., *ibid.*, 97, 6807 (1975).
16a. Schmerling, L. and Vesely, J. A., *J. Org. Chem.*, 38, 312 (1973).
b. Schmerling, L., *J. Amer. Chem. Soc.*, 97, 6134 (1975).
17a. Olah, G. A., U.S. Patent 3,708,553, January 2, 1973.
b. van Dijk, P., U.S. Patent 3,415,899, December 10, 1968.
c. Pinkerton, R. D., U.S. Patent 2,177,579, October 24, 1939.

18. Calculated from values taken from D. R. Stull, E. F. Westrum, Jr. and G. C. Sinke, "The Chemical Thermodynamics of Organic Compounds," John Wiley and Sons, Inc., New York, New York 1969.
19. Schmerling, L., J. Amer. Chem. Soc., 66, 1422 (1944).
20. Olah, G. A., Klopman, G. and Schlosberg, R. H., J. Amer. Chem. Soc., 91, 3261 (1969).
21. Hogeveen, H., Lukas, J. and Roobeek, C. F., Chem. Commun. 920 (1969).
22. Olah, G. A. and Mo, Y. K., J. Amer. Chem. Soc., 94, 6864 (1972).
23. Roberts, D. T., Jr. and Calihan, L. E., J. Macromol. Sci., Chem., 7, 1629 (1973).
24. Olah, G. A., Yoneda, N. and Parker, D. G., J. Amer. Chem. Soc., 98, 5261 (1976).
25. Siskin, M., ibid, 98, 5413 (1976).
26. Field, F. H. and Franklin, J. L., "Electron Impact Phenomena and Properties of Gaseous Ions,", p. 87, Academic Press, Inc. New York, New York, 1957.
27. Results reported by Z. Vostroknutova and A. A. Shteinman in Kinet. Katal., 13, 324 (1972) indicate that even at 50°C in 1:1 HSO₃F-SbF₅, the ionization of methane is too slow to account for our results in terms of the traditional alkylation mechanism.
28. Schlosberg, R. H., Siskin, M., Kocsi, W. P. and Parker, F. J., J. Amer. Chem. Soc., 98, 7723 (1976).
29. Olah, G. A. and Lukas, J. ibid., 89, 4739 (1967).
30. Brouwer, D. M. and Oelderik, J. M., Rec. Trav. Chim., 87, 721
31. Parker, P. T., U.S. Patent 3,636,129, January 18, 1972.
32. Friedel, C. and Crafts, J. M., Compt. Rend., 84, 1450 (1877).
33. Olah, G. A., DeMember, J. R., Schlosberg, R. H. and Halpern, Y., J. Amer. Chem. Soc., 94, 156 (1972).
34. Olah, G. A., and Comisarow, M. B., J. Amer. Chem. Soc., 91, 2955 (1969).
35. Olah, G. A. and DeMember, J. R., ibid., 2113 (1969).
36. Farcasiu, D.A., personal communication.
37a. Ogata, Y. and Okano, M., J. Amer. Chem. Soc., 78, 5123 (1956).
   b. Olah, G. A. and Yu. S. H., ibid., 97, 2293 (1975).
38. Hehre, W. J. and Hiberty, P. C., ibid., 96, 2665 (1974).
39a. Olah, G. A., Beal, D. A. and Westerman, P. W., ibid., 95, 3387 (1973).
   b. Olah, G. A., Mo, Y. K. and Halpern. Y., J. Org. Chem., 37, 1169 (1972).

Base-Catalyzed Carbon–Carbon Addition of Hydrocarbons and Related Compounds

HERMAN PINES

The Ipatieff Catalytic Laboratory, Department of Chemistry,
Northwestern University, Evanston, IL 60201

The discovery that sodium in the presence of small amounts of organosodium compounds, produced in situ or deposited on alumina, acts as an effective catalyst for double bond isomerization of alkenes and cyclenes (1) triggered much research in this field (2). It was subsequently discovered that base-catalyzed isomerization of olefins may proceed in homogeneous solutions using lithium ethylenediamine (3) or potassium tert-butoxide (t-BuOK) in dimethyl sulfoxide (DMSO) (4).

Base-catalyzed carbon-carbon addition reactions are of synthetic interest because they afford hydrocarbons and related compounds in good yields by a simple one-step procedure. These reactions are made possible by the fact that hydrocarbons and related compounds having a benzylic or allylic hydrogen are carbon acids, having a pK_a of about 35 to 37; they can donate a proton to a base and thus become carbanions. These carbanions can add to olefinic hydrocarbons. The steps involved in the catalytic chain reactions are illustrated by the following set of equations, using toluene and ethylene as reactants, and sodium as catalyst (5).

$$\text{Promoter} + \text{Na} \rightarrow \text{B}^-\text{Na}^+ \qquad (1)$$

Initiation

$$C_6H_5CH_3 + B^-Na^+ \rightleftarrows C_6H_5CH_2^-Na^+ + BH \qquad (2)$$

Addition

$$C_6H_5CH_2^-Na^+ + CH_2{=}CH_2 \rightleftharpoons C_6H_5CH_2CH_2CH_2^-Na^+ \qquad (3)$$

Propagation

$$C_6H_5CH_2CH_2CH_2{}^-Na^+ + C_6H_5CH_3 \rightarrow$$

$$C_6H_5CH_2CH_2CH_3 + C_6H_5CH_2{}^-Na^+ \qquad (4)$$

Step 2, chain initiation, involves the metalation of toluene by an organosodium compound. Since the action of sodium on the reactants does not produce organosodium compounds, a "promoter" is added to accomplish it. o-Chlorotoluene and anthracene were found to be effective promoters since they readily react with sodium to form benzylsodium and disodium anthracene (6), respectively. The promoters also facilitate the dispersion of sodium into a very fine black powder which can form additional organosodium compounds as the reactions progress.

Step 3, the addition reaction, is energetically the least favorable step as it involves the formation of an anion lacking resonance stabilization from a resonance-stabilized benzylic anion. However, once the anion adduct is produced it is instantaneously and irreversibly protonated by the benzylic hydrogen of the toluene present. The facility with which addition occurs depends to a great degree on the olefin used. Addition of benzyl anion to propene requires more drastic conditions than to ethylene, due both to steric and to inductive effects. The addition reaction is greatly facilitated when conjugated alkadienes or styrenes are used as olefins since the anion adducts formed are resonance stabilized by conjugation with a double bond and benzene ring, respectively.

The ease of addition to olefins depends also on the acidity of the alkylarenes used in the reaction. 4-Methylpyridine, having a pK_a of about 29 at $-40°$, adds to isoprene at room temperature in the presence of t-BuOK in DMSO. Toluene, however, having a much higher pK_a does not add to isoprene in the presence of this catalyst.

Of the alkali metals, potassium is a more effective catalyst than sodium, while lithium has only limited applications as catalyst. Unlike sodium, potassium also catalyzes certain cyclialkylation reactions, and thus provides new methods for the synthesis of a variety of cyclic compounds.

Oligomerization of Olefins

Propene. The oligomerization of olefins by bases was

reported in 1956 using simple olefins or olefin pairs in the presence of sodium as catalyst, and anthracene as a chain initiator (7). The olefins used were ethylene, propene, isobutylene, and cyclohexene.

The dimerization of propene at about 210° in the presence of potassium or cesium yielded 4-methyl-1-pentene as the predominant dimer (8). The dimerization proceeds through an initial formation of an organoalkali compound, followed by metalation of the propene.

The dimerization of propene in a flow system over supported potassium or sodium on graphite or potassium carbonate, at 150° and under pressure, gave good yields of dimers, and the copolymerization of ethylene with propene on supported alkali metal catalysts gave 92% pentenes (9).

Isoprene. Using benzene as solvent and sodium as catalyst isoprene undergoes oligomerization at 40° to yield 47% of dimers and 20% of trimers (10). The dimer fraction contained 85% myrcene, 1, and the trimer contained 76% of 2. The formation of these compounds involves an initial metalation of the methyl group in isoprene.

Lithium naphthalene and sodium naphthalene (11) in solvents such as THF, diglyme, or aliphatic amines were found to be effective agents for the hydrodimerization of isoprene. Although this reaction is not strictly catalytic, it contributes to a better understanding of the base-catalyzed oligomerization of unsaturated conjugated hydrocarbons.

In a typical reaction using THF as solvent Li (0.1 g.-at.), naphthalene (0.03 mol), triethylamine (0.001 mol), and isoprene (0.1 mol) at 30° for 8 hrs, a 75% yield of hydrodimers was obtained, consisting of 3 (43%), 4 (52%), and other products (5%).

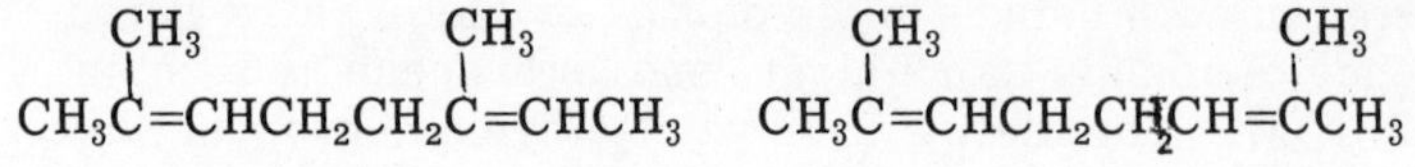

$$\underset{CH_3}{\overset{CH_3}{|}}\underset{}{\overset{}{C}}=CHCH_2CH_2\underset{}{\overset{CH_3}{|}}C=CHCH_3$$

3

$$CH_3\underset{}{\overset{CH_3}{|}}C=CHCH_2CH_2CH=\underset{}{\overset{CH_3}{|}}CCH_3$$

4

Styrenes. Dimerization and cyclidimerization of styrene and of α- and β-methylstyrene have been accomplished in the presence of bases. The type of product depends greatly on the styrene and on the specific catalyst used.

The polymerization of styrene to form macromolecules using alkali metal catalysts has been known and extensively used. The dimerization of styrene is, however, a novel reaction (12). Dimers of styrene, α-methylstyrene and a co-dimer of the above two styrenes are obtained by heating the olefins at about 160° in the presence of catalytic amounts of anhydrous t-BuOK.

α-Methylstyrene yields 93% dimer 5, while styrene forms 11% of 6. When a mixture consisting of 1 mol of styrene and 0.5 mol of α-methylstyrene is reacted, "dimeric" material composed of 7% 5, 48% 6, 32% 7, and 13% 8 results. The initial

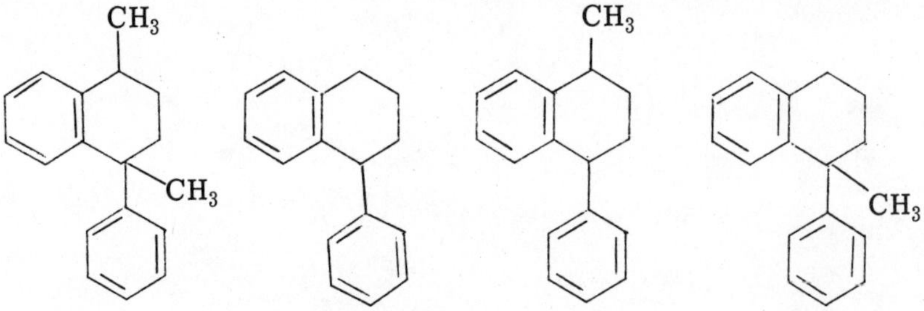

5   6   7   8

step probably involves the formation of an ionized compound through the addition of t-BuOK to the monomer. In the next step, through a concerted addition and elimination reaction an intermediate dimer, 9, is produced which by the usual base-catalyzed isomerization is converted to 5.

The intermediate dimer 9 could also be formed by a Diels-Alder condensation of the monomers (13), followed by t-BuOK catalyzed isomerization of the double bond (14).

α-Methylstyrene, on refluxing in the presence of sodium-benzylsodium, produces a cyclic dimer 10 in a 32% yield based on the reacted olefin (15). The other products are isopropyl-benzene, a trimer, and small amounts of p-terphenyl. The following mechanism was proposed for the formation of the cyclic dimer 10.

(B = benzylsodium or any organosodium compound formed in the reaction)

β -Methylstyrene undergoes dimerization to 1, 5-diphenyl-4-methyl-1-pentene, 11, in 80-90% yields, when heated at 100-155° in the presence of sodium (16). In the presence of the more electropositive potassium, hydrodimerization was the principal reaction, resulting in the formation of compound 12.

$$C_6H_5CH=CHCH_2\overset{\overset{\displaystyle CH_3}{|}}{C}HCH_2C_6H_5 \qquad C_6H_5CH_2\overset{\overset{\displaystyle CH_3}{|}}{C}HCHCH_2C_6H_5$$
$$\underset{|}{}CH_3$$

$$\underline{11} \qquad\qquad\qquad\qquad \underline{12}$$

The two reactions are mechanistically different: dimerization to form 11 is a typical carbanion-catalyzed chain reaction, whereas hydrodimerization leading to compound 12 is a non-catalytic reaction.

Dimer 11 may undergo further reaction leading to the formation of toluene, 1, 3-diphenyl-2-methylpropane, 13, and 3-methyl-4-phenylcyclobutene, 14:

$$11 \underset{RH}{\overset{R^-M^+}{\rightleftharpoons}} [C_6H_5CH=CH\overset{CH_3}{\overset{|}{C}H}-CH-CH_2C_6H_5] M^+ \rightarrow$$

$$C_6H_5CH=CHCH=CHCH_3 + C_6H_5CH_2^-M^+$$

$$\underline{A} \qquad\qquad\qquad \underset{R^-M^+}{\overset{RH}{\rightleftharpoons}} C_6H_5CH_3$$

$$C_6H_5CH_2^-M^+ + C_6H_5CH=CHCH_3 \rightleftharpoons C_6H_5CH_2\overset{\overset{\displaystyle CH_3}{|}}{C}H\bar{C}HC_6H_5 M^+$$

$$\underset{R^-M^+}{\overset{RH}{\rightleftharpoons}} C_6H_5CH_2\overset{\overset{\displaystyle CH_3}{|}}{C}HCH_2C_6H_5$$

$$\underline{13}$$

$$A \xrightleftharpoons[RH]{R^-M^+} [C_6H_5CH\because CH\because CH\because CH\because CH_2]^- M^+$$

$$\left[C_6H_5\bar{C}H \overset{\frown}{C}H \overset{\frown}{=\!\!=} CH_2 \atop CH\!=\!CH \right] M^+ \rightarrow C_6H_5CH\!-\!CH\!-\!CH_2^- M^+ \atop CH\!=\!CH$$

$$\xrightleftharpoons[R^-M^+]{RH} \quad C_6H_5CH\!-\!CHCH_3 \atop CH\!=\!CH$$

14

α, β -Unsaturated Esters and Nitriles

Ethyl crotonate undergoes fast and selective dimerization at 110° in the presence of potassium and a promoter to form the diethyl ester of diacid **15** with a high yield (**16**).

$$\underset{\textbf{15}}{EtOOC\overset{CH_3}{\underset{\|}{C}}C\!-\!\overset{CH_3}{\underset{|}{C}}HCH_2COOEt}$$

Crotononitrile undergoes oligomerization at 110° in the presence of potassium-benzylpotassium catalyst to afford a mixture consisting 20-23% of dimer and 67-75% of a trimer **18**, and 5-15% of high boiling product. The dimer was composed of 67% cis-, **16**, and 33% trans-2-ethylidene-3-methyl-glutaronitrile, **17**.

$$\underset{\textbf{16}}{\overset{H_3C}{\underset{H}{}}C\!=\!C\overset{\overset{CH_3}{|}}{\underset{CN}{CHCH_2CN}}}$$ $$\underset{\textbf{17}}{\overset{H}{\underset{H_3C}{}}C\!=\!C\overset{\overset{CH_3}{|}}{\underset{CN}{CHCH_2CN}}}$$

The trimer consisted of 75-80% of **18**.

$$\underset{CN}{\overset{CN}{\underset{NC}{\overset{H_3C}{\bigcirc}}}} \overset{CH_3}{\underset{CN}{}}$$

18

Reaction of Alkylaromatic Hydrocarbons with Olefins

The base-catalyzed reaction of alkylaromatics with olefins is unique in that it allows the size of the alkyl group of an arylalkane to be increased. Arylalkanes suitable for this reaction are those which contain a benzylic hydrogen. The olefins most useful for this reaction are ethylene, propylene, conjugated alkadienes, and styrene and its derivatives. Sodium and potassium are very effective catalysts. Sodium usually requires the presence of a chain precursor to initiate the reaction.

Ethylation of alkylbenzene proceeds at 150° to 200° with ethylene pressure ranging from 1 to 70 atm. The reaction involves the replacement of benzylic hydrogens by ethyl groups, i. e., toluene forms n-propylbenzene as the primary product of reaction, further ethylation produces 3-phenylpentane and 3-ethyl-3-phenylpropane (5).

This reaction can also be extended to the ethylation of cycloalkylbenzenes, indan, and tetrahydronaphthalene.

Potassium also catalyzes cyclialkylation reaction resulting in the formation of indans. The ratio of indans to monoalkylbenzenes formed depends on the extent of substitution on the α-carbon atom, and in the case of isopropylbenzene the concentration of 1,1-dimethylindan amounted to 49%. Potassium hydride or K-BuLi seem to be more effective catalysts for cycliethylation (20).

Sodium in the presence of o-chlorotoluene promoter is a selective catalyst for the ethylation of alkylnaphthalenes at 175-200° (21):

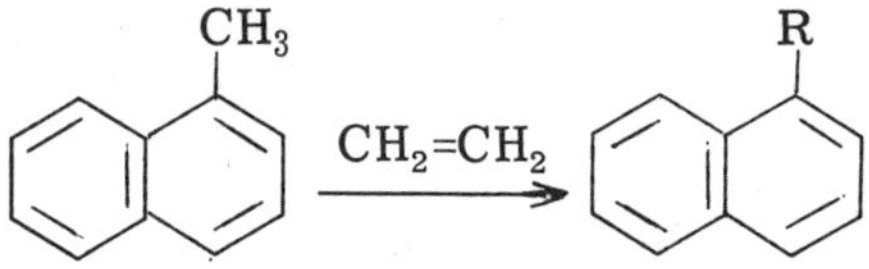

$$R = CH_2CH_2CH_3$$
$$CH(C_2H_5)_2$$
$$C(C_2H_5)_3$$

Only on prolonged heating and stirring a small amount of the heptyltoluene is obtained.

Potassium-catalyzed ethylation of alkylnaphthalenes proceeds at 90-160° and, besides the normal side-chain ethylation, a cyclialkylation also occurs. From 1-methyl-naphthalene compounds 19 and 20 are obtained, and from 2-methylnaphthalene compound 21 is obtained.

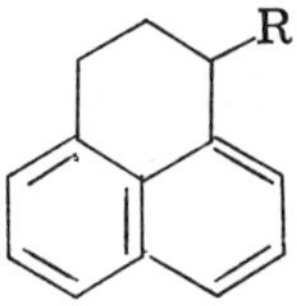

19 R = H

20 R = C$_2$H$_5$

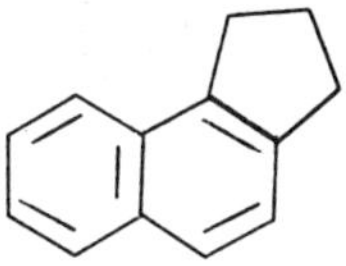

21

Propylation of alkylbenzene with propene requires more drastic conditions than that of ethylation. The main product of side-chain propylation of toluene was isobutylbenzene.

Alkenylation of Alkylbenzenes

The mechanism of alkenylation is similar to that of side-chain alkylation of alkylaromatics and can be illustrated as follows:

$$C_6H_5CH_3 + B^-Na^+ \rightleftharpoons C_6H_5CH_2^-Na^+$$

$$C_6H_5CH_2^-Na^+ + CH_2=CHCH=CH_2$$

$$\rightleftharpoons [C_6H_5CH_2CH_2CH \cdot\cdot\cdot CH \cdot\cdot\cdot CH_2]^-Na^+$$

$$C_6H_5CH_3$$

$$C_6H_5CH_2^-Na^+ + C_6H_5CH_2CH_2CH=CHCH_3$$

The alkenylation reaction proceeds at 100° or below. Yields of 80-91% of monobutenylated alkylbenzene were obtained from toluene, o- and p-xylene, and ethylbenzene using sodium deposited on calcium oxide (23).

The pentenylation of alkylbenzenes with isoprene was carried out at 135° using dispersed sodium or potassium as catalyst (24). Two monoadducts were obtained, A and B, with a ratio of A/B ranging from 2 to 3.

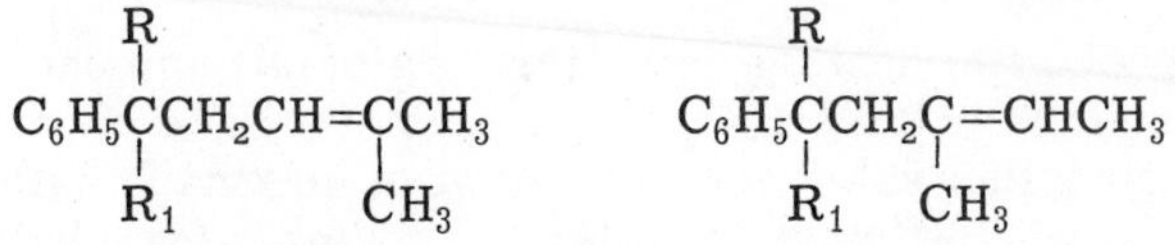

$$C_6H_5\underset{R_1}{\overset{R}{C}}CH_2CH=\underset{CH_3}{C}CH_3 \qquad C_6H_5\underset{R_1}{\overset{R}{C}}CH_2\underset{CH_3}{C}=CHCH_3$$

A   B

$$R = R_1 = H$$
$$R = H, \quad R_1 = CH_3$$
$$R = R_1 = CH_3$$

In the presence of a catalyst composed of sodium naphthalene in THF, the pentenylation occurs at 20° (25).

Aralkylation of Alkylbenzenes

Mono- and diadducts have been produced in relatively good yields in reacting alkylbenzenes with styrene and α- and β-substituted styrene in the presence of either sodium or potassium at 100-120° (26).

Toluene and ethylbenzene react with styrene to form monoadducts 22 and 23 and diadducts 24 and 25

$$C_6H_5CH_2CH_2CH_2C_6H_5 \qquad C_6H_5CH \overset{CH_3}{\underset{CH_2CH_2C_6H_5}{}}$$

22   23

$$C_6H_5CH(CH_2CH_2C_6H_5)_2 \qquad C_6H_5C \overset{CH_3}{\underset{(CH_2CH_2C_6H_5)_2}{}}$$

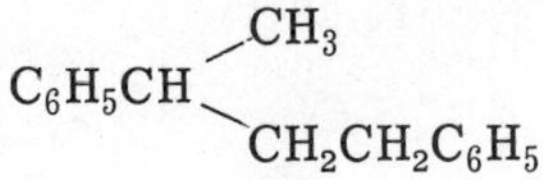

24   25

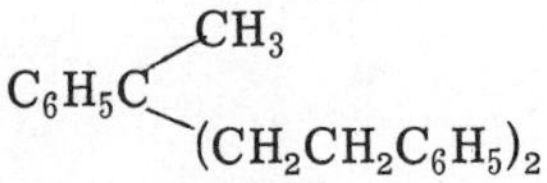

Isopropylbenzene produces with styrene mono- and diadducts 26 and 27

$$C_6H_5C \underset{CH_2CH_2C_6H_5}{\overset{(CH_3)_2}{<}}$$

$$C_6H_5C \underset{\underset{C_6H_5}{\mid}}{\overset{(CH_3)_2}{<}} CH_2CHCH_2CH_2C_6H_5$$

26   27

α-Methylstyrene reacts readily with n-alkylbenzenes in the presence of potassium to form 1,3-diphenylalkanes in yields ranging from 73 to 82%. β-Alkylbenzenes on reaction with toluene form 1,3-diphenyl-2-alkylpropanes in 60-95% yields.

Reaction of Alkylpyridines with Olefins

The picolyl hydrogens of 2- and 4-alkylpyridines are more acidic than the corresponding benzylic hydrogens. For that reason, the reactions of alkylpyridines with olefins occur under mild conditions, and with conjugated dienes, styrenes, and vinylpyridines at below room temperatures. The mechanism of side-chain addition of alkylpyridines to olefins is very similar to that described for alkylbenzenes.

2- And 4-alkylpyridines undergo reaction with ethylene under pressure at 140-150° in the presence of sodium to form the corresponding 2-n-propyl- and (3-pentyl)-2- and 4-pyridines. Yields of 86-94% of monoethylated products are obtained from higher 2- and 4-alkylpyridines (28). The relative ethylation of 4-alkylpyridines is 3.5 to 7.0 times greater than that of 2-alkylpyridines, and is associated with greater acidity of the 4 over the 2 isomers.

3-Alkylpyridines. The reaction between ethylene and 3-alkylpyridines in the presence of alkali metal is much more complicated than similar reactions made with the corresponding 2 and 4 isomers (29). The two primary products from the reaction of 3-ethylpyridine with ethylene in the presence of either sodium or potassium are 3-sec-butylpyridine and cyclic compound 28. Further ethylation occurs with a longer reaction time to form 29.

2- And 2-Alkylpyridine with Diolefins, Styrenes, and Vinylpyridines.

The reaction of alkylpyridines with isoprene, styrenes, and vinylpyridines occurs at 0-25°. The yields of adducts are high, and the compounds formed are those predicted from similar reactions made with alkylbenzenes (30).

The relative rate of alkenylation of 4-alkylpyridines with either butadiene or isoprene is about 8 to 20 times greater than that of 4-methylpyridine, and therefore the alkenylation of 4-methylpyridine is always accompanied by dialkenylation (31).

Homogeneous Carbon-Carbon Addition Reactions

The use of aprotic solvents promotes anionic reactions of very weak organic acids under mild conditions. A variety of alkylaromatic compounds undergo alkenylation and aralkylation in a homogeneous aprotic solvent-t-BuOK system (32). The reactions can be represented by the following equations.

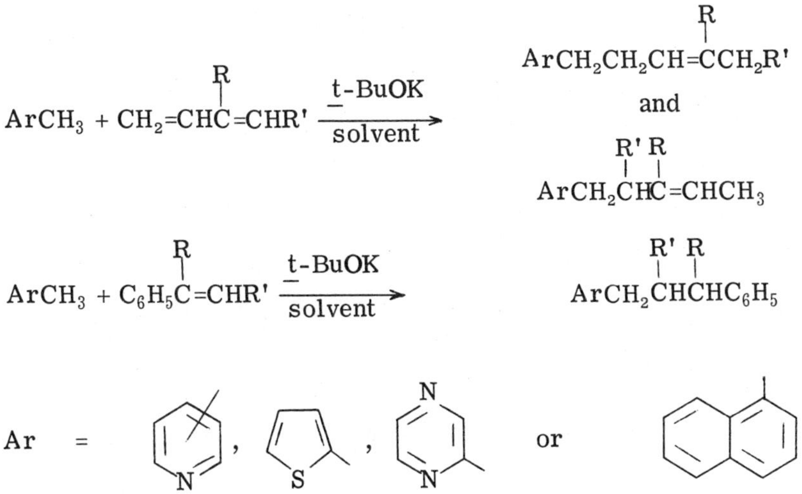

R, R' = H or CH_3

The rate of reaction is greatly influenced by the type of solvents used.

Aprotic solvents such as NM-2-P, 30a, and NM-2-Pi, 30b, may react almost quantitatively with olefins containing an activated double bond, at room temperature and in the presence of *t*-BuOK (33). The general scope of the reaction is presented by

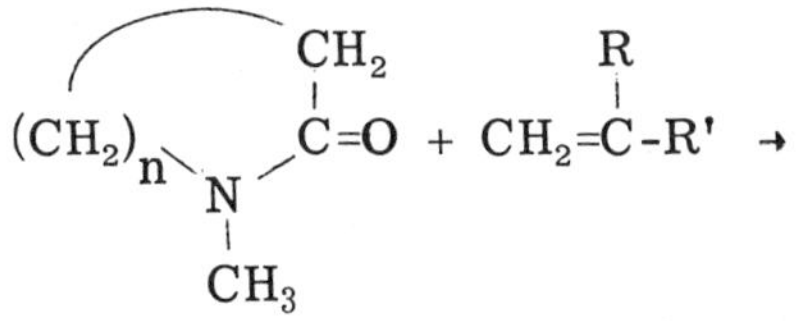

n = 2 = NM-2-P = 30a

n = 3 = NM-2-Pi = 30b

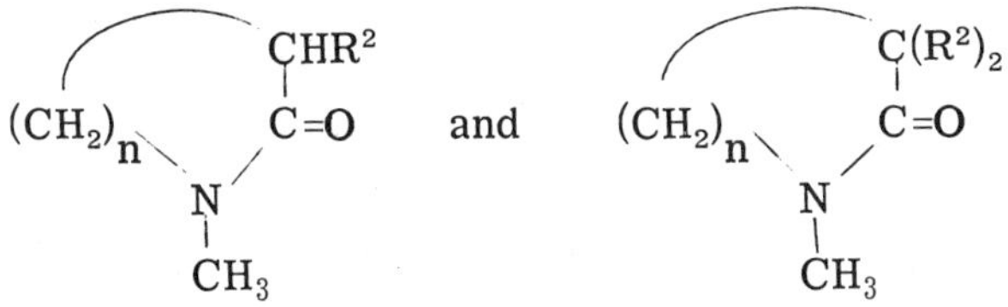

$$R = H; \quad R^1 = CH=CH_2; \quad R^2 = CH_2CH=CHCH_3$$

$$R = H; \quad R^1 = C(CH_3)=CH_2; \quad R^2 = CH_2CH_2CH(CH_3)_2$$

$$CH_2CH(CH_3)C_2H_5$$

(after hydrogenation)

$$R = H; \quad R^1 = C_6H_5; \quad R^2 = CH_2CH_2C_6H_5$$

$$R = H; \quad R^1 = Si(CH_3)_3; \quad R^2 = CH_2CH_2Si(CH_3)_3$$

$$R = CH_3; \quad R^1 = C_6H_5; \quad R^2 = CH_2CH(CH_3)C_6H_5$$

Although DMSO is used extensively as an aprotic solvent, the hydrogens in DMSO are quite labile. DMSO under the influence of base can react with a variety of conjugated dienic hydrocarbons, arylalkenes, and aromatic and heteroaromatic ring compounds (34). The net result of these reactions is the formation of methylated analogs of the starting hydrocarbons.

Reaction of Olefins with Miscellaneous Compounds

Ethylene reacts with salts of saturated carboxylic acids containing at least one hydrogen atom attached to the α-carbon atom (35). The ethylation which is catalyzed by alkali metals and their derivatives and which occur at 150-250° under ethylene pressure, can be presented as follows:

$$RCH_2C\begin{smallmatrix}O\\\\OM\end{smallmatrix} + CH_2=CH_2 \xrightarrow{M} RCH\begin{smallmatrix}C_2H_5\\\\CO_2M\end{smallmatrix} + RC\begin{smallmatrix}(C_2H_5)_2\\\\CO_2M\end{smallmatrix}$$

(R = H, alkyl, cycloalkyl. M = alkali metal)

Aldimines, which can be synthesized in good yields by the condensation of primary amines with aldehydes, undergo a facile reaction with isoprene and vinylaromatic hydrocarbons in the presence of alkali metals (36). The alkenylation of aldimines with isoprene is made at atmospheric pressure in benzene solvent. Almost quantitative yields of mono- and diadducts are obtained. The alkenylation with isoprene can be presented as follows.

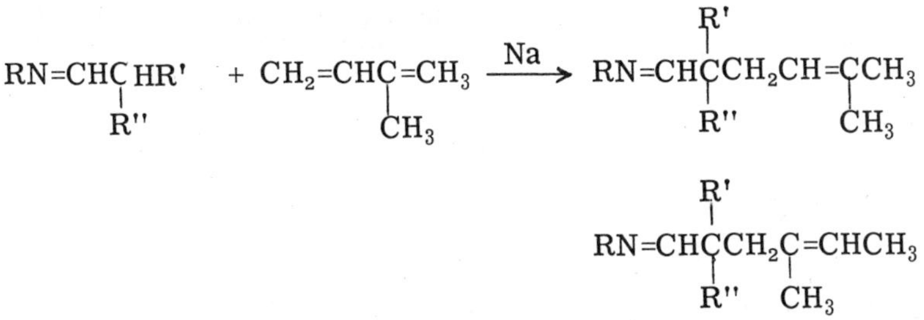

$$RN=CH\underset{\underset{R''}{|}}{C}HR' \ + \ CH_2=CH\underset{\underset{CH_3}{|}}{C}=CH_3 \ \xrightarrow{Na} \ RN=CH\underset{\underset{R''}{|}}{C}CH_2CH=\underset{\underset{CH_3}{|}}{C}CH_3$$

$$RN=CH\underset{\underset{R''}{|}}{C}CH_2\underset{\underset{CH_3}{|}}{C}=CHCH_3$$

When R' is hydrogen the monoadducts can undergo further reaction with isoprene to form dialkenylated aldimines.

Intramolecular Cyclization Reactions

Potassium hydride catalyzes cyclization of cycloocta-dienes at 190° to form 84-94% of 31 (37), while 1, 4, 7-cyclo-nonatriene undergoes in the presence of potassium tert-butoxide in dimethyl sulfoxide double cycloisomerization to 32.

31   32

ω-Phenylalkenes undergo cyclization and cyclialkylation at 185° in the presence of either potassium or cesium, i.e.,

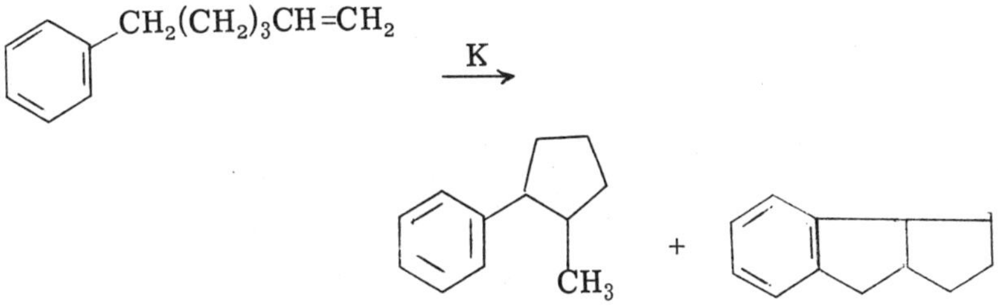

$$\underset{}{}CH_2(CH_2)_3CH=CH_2$$

$$\xrightarrow{K}$$

+

cis and trans   12%

66%

6-(3-Pyridyl)-1-hexene in the presence of either sodium or potassium forms 33 and 34 in good yields. 33 Is the preferred tricyclic compound (40).

33   34

Conclusion

Base-catalyzed carbon-carbon addition reactions are of synthetic importance as they permit the production of difficult to prepare hydrocarbons and related compounds in good yields by a one-step procedure. The reactions proceed under relatively mild conditions and in the presence of either dispersed alkali metals or alkali metals deposited on supports. In the case of olefins containing an activated double bond the use of t-BuOK in DMSO can be used as an effective catalyst.

Literature Cited

(1) (a) Pines, H., Vesely, J. A., and Ipatieff, V. N., J. Amer. Chem. Soc. (1955) 77, 347;
(b) Pines, H. and Eschinazi, H. E., ibid. (1955) 77, 6314;
(c) Pines, H. and Haag, W. O., J. Org. Chem. (1958) 23, 328.
(2) (a) Pines, H., Acct. Chem. Research (1974) 7, 155;
(b) Pines, H., Synthesis (1974), 309.
(3) Reggel, L., Friedman, S., and Wender, I., J. Org. Chem. (1958) 23, 1136.
(4) Schriesheim, A., Hofmann, J. H., and Rowe, C. A., J. Amer. Chem. Soc. (1961) 83, 3731.
(5) Pines, H., Vesely, J. A., and Ipatieff, V. N., J. Amer. Chem. Soc. (1955) 77, 554.

(6) For general discussion of sodium anthracene complexes see K. Tamaru, Advan. Catal. Relat. Subj. (1969) 20, 327.

(7) Mark, V. and Pines, H., J. Amer. Chem. Soc. (1956), 78, 5946.

(8) Shaw, A. W., Bittner, C. W., Bush, W. V., and Holzman, G., J. Org. Chem. (1965) 30, 3286.

(9) Hambling, J. K., Chem. Brit. (1969) 5, 354.

(10) Takabe, K., Katagiri, T., and Tanaka, J. Bull. Chem. Soc. Japan (1972) 45, 2662.

(11) (a) Suga, K., Watanabe, S., Watanabe, T., and Kuniyoshi, K., J. Appl. Chem. (London) (1969) 19, 318;
    (b) Watanabe, S., Suga, K., and Fujita, T., Synthesis (1971) 375.

(12) Zwierzak, A. and Pines, H., J. Org. Chem. (1963) 28, 3392.

(13) Brown, W. G., J. Amer. Chem. Soc. (1968) 90, 1916.

(14) Zwierzak, A. and Pines, H., J. Org. Chem. (1962) 27, 4084.

(15) Kolobielski, M. and Pines, H., J. Amer. Chem. Soc. (1957) 79, 1698.

(16) Shabtai, J. and Pines, H., J. Org. Chem. (1964) 29, 2408.

(17) Shabtai, J. and Pines, H., J. Org. Chem. (1965) 30, 3854.

(18) Shabtai, J., Ney-Igner, E., and Pines, H., J. Org. Chem. (1975) 40, 1158.

(19) Pines, H. and Schaap, L., J. Amer. Chem. Soc. (1958) 80, 3076.

(20) Eberhardt, G. G., J. Org. Chem. (1964) 29, 643.

(21) Stipanovic, B. and Pines, H., J. Org. Chem. (1969) 34, 2106.

(22) Pines, H. and Mark, V., J. Amer. Chem. Soc. (1956) 78, 4318.

(23) Eberhardt, G. G. and Petersen, J. J., J. Org. Chem. (1965) 30, 82.

(24) Pines, H. and Sih, N. C., J. Org. Chem. (1965) 30, 280.

(25) Watanabe, S., Suga, K., and Fujita, T., Synthesis (1971) 3, 375.

(26) (a) Pines, H. and Wunderlich, D., J. Amer. Chem. Soc. (1958) 80, 6001;
    (b) Shabtai, J. and Pines, H., J. Org. Chem. (1961) 26, 4225;
    (c) Shabtai, J., Lewicki, E., and Pines, H., J. Org. Chem. (1962) 27, 2618.

(27) (a) Profft, E. and Schneider, F., Arch. Pharm. (1955)
        289 99.
     (b) Pines, H. and Wunderlich, D., J. Amer. Chem. Soc.
        (1959) 81, 2568.
(28) Pines, H. and Notari, B., J. Amer. Chem. Soc. (1960)
        82, 2209.
(29) (a) Pines, H. and Kannan, S. V., Chem. Commun. (1969)
        1360.
     (b) Kannan, S. V. and Pines, H., J. Org. Chem. (1971)
        36, 2305.
(30) (a) Pines, H. and Notari, B., J. Amer. Chem. Soc.
        (1960) 82, 2209.
     (b) Chumakov, Y. I. and Ledovskikh, Ukr. Chim. Zh.
        (1965) 31, 506.
     (c) Pines, H. and Oszczapowicz, J. Org. Chem. (1967)
        32, 3183.
     (d) Stalick, W. M. and Pines, H., J. Org. Chem. (1970)
        35, 415.
     (e) Pines, H. and Sartoris, N. E., J. Org. Chem. (1969)
        34, 2113.
     (f) Sartoris, N. E. and Pines, H., J. Org. Chem. (1969)
        34, 2119.
     (g) Oszczapowicz, J. and Pines, H., J. Org. Chem.
        (1972) 37, 2799.
(31) Stalick, W. M. and Pines, H., J. Org. Chem. (1970)
        35, 422.
(32) (a) Pines, H. and Stalick, W. M., Tetrahedron Lett.
        (1968) 3723.
     (b) Pines, H., Stalick, W. M., Holford, T. G., Golab,
        J., Lazar, H., and Simonik, J., J. Org. Chem.
        (1971) 36, 2299.
(33) Pines, H., Kannan, S. V., and Simonik, J., J. Org.
        Chem. (1971) 36, 2311.
(34) (a) Argabright, P. A., Hofmann, J. E., and Schriesheim,
        A., J. Org. Chem. (1965) 30, 3233.
     (b) Feldman, M., Danischefsky, S., and Levine, R.,
        J. Org. Chem., (1966) 31, 4322.
     (c) Russell, G. A. and Weiner, S. A., J. Org. Chem.
        (1966) 31, 248.
     (d) Nozaki, H., Yamamoto, Y., and Noyori, R.,
        Tetrahedron Lett. (1966) 1123.
(35) Schmerling, L. and Tockelt, W. G., J. Amer. Chem.
        Soc., (1962) 84, 3694.

(36) (a) Martirosyan, G. T. , Kazaryan, A. Ts. , and
        Mirsaryan, So. , Arm. Khim. Zh. (1973) 26, 569.
     (b) Kazaryan, A. Ts. and Martirosyan, G. T. , Arm.
        Khim. Zh. (1972) 25, 861.
(37) Slaugh, L. , J. Org. Chem. (1967) 32, 108.
(38) Wathley, J. W. H. and Winstein, S. , J. Amer. Chem.
     Soc. (1963) 85, 3716.
(39) Pines, H. , Sih, N. C. , and Lewicki, E. , J. Org. Chem.
     (1965) 30, 1457.
(40) Pines, H. , Kannan, S. V. , and Stalick, W. M. , J. Org.
     Chem. (1971) 36, 2309.

13

Mixing in Emulsion-Reaction Types of Liquid-Liquid Processes

J. Y. OLDSHUE

Mixing Equipment Co., Inc., 135 Mt. Read Blvd., Rochester, NY 14611

This discussion is aimed primarily at alkylation processes, which are characterized by two-phase emulsions. The traditional process includes a hydrocarbon phase in contact with a sulfuric acid catalyst to form the environment for proper alkylation. In the petroleum industry, the desired product is a high octane alkalate, but other alklation processes have other desired products.

There are several different ways of running full scale alkylation processes. One of the earlier systems was the use of a "cascade reactor", there was a sulfuric acid catalyst in each stage, which did not flow between stages, and the hydrocarbon was allowed to flow in each stage in series. A mixing device was used to form a two-phase emulsion, which then must settle in each stage so that clear hydrocarbon can flow into the next stage. Mixers used included jets, submerged pumps, and mixing impellers.

Another type of cascade continuous flow system did not try to settle in each stage, but allowed emulsion to flow co-currently through the system.

A third type used a high energy side entering propeller unit, which carried out the process in a single stage.

This paper analyzes the role that mixing can play in this reaction, and tries to divide "mixing" down into its several component fluid mechanic parts, which is helpful in analyzing process results.

In order to orient the reader, Table I illustrates the steps that are desirable in designing a fluid mixer. Under I, we look at the mechanics of a flow from impellers, what kind of fluid regimes are required by the alkylation process, and how do we scale up and scale down the operation.

Under II, impeller power characteristic curves hold true whether we are doing the desired process job or not. It is essential to have data on various impellers in these areas, but this data is largely available and it is primarily a matter of specifying the specific gravity and viscosity of the various

224

TABLE I

ELEMENTS OF MIXER DESIGN

I PROCESS DESIGN   1 - FLUID MECHANICS
                              OF IMPELLERS

                          2 - FLUID REGIME
                              REQUIRED BY PROCESS

                          3 - SCALE UP; HYDRAULIC
                              SIMILARITY

II IMPELLER POWER   4 - RELATE IMPELLER
   CHARACTERISTICS   HP, SPEED & DIAMETER

III MECHANICAL DESIGN   5 - IMPELLERS

                          6 - SHAFTS

                          7 - DRIVE ASSEMBLY

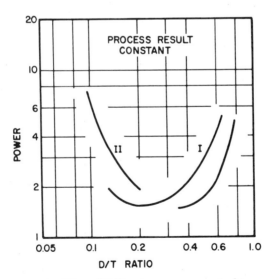

Figure 1.   *Effect of* D/T *ratio on power required for a
given process result*

phases and emulsion that the impeller will see during various
parts of the process operation.

In III, mechanical design is essential for a satisfactory
operation, but this part of the design is not treated in this process oriented paper.

Fluid Mechanics of Flow From Impeller

By way of review, all the power that we apply to the mixing
impeller produces a pumping capacity, Q, and a velocity head, H,
within the discharge from the impeller, such that power P, is proportional as shown below.

$$P \propto QH$$

Other relationships, such as:

$$P \propto N^3 D^5$$

where N is impeller speed and D is impeller diameter, as well as:

$$Q \propto ND^3$$

allow us to calculate the way that we can vary the flow to impeller head ratio at a given power level with various impeller sizes
and speed. It turns out that if we combine the previous three
equations, that we arrive at a fourth,

$$Q/H \propto D^{4/3}$$

This equation essentially tells us that the large diameter impeller running at slow speed at a given power level gives us a high
volume of fluid pumped, and a relatively low impeller head, while
a small impeller running at high speed gives us less flow and a
higher level of impeller head.

The impeller head is related to various fluid shearing phenomena going on in the tank, and as such, is an important element
of emulsion and reaction processing.

In general, every process has an optimum combination of flow
and impeller head. For many processes, the particular impeller
type used to generate that flow and impeller head is relatively
unimportant, and two impellers will achieve the same process result when operated at their optimum geometry to give this flow to
head ratio for the particular process requirement. Fig. 1 illustrates this phenomena with two different impellers producing the
best process result at different actual impeller diameters.

The whole question of impeller shear effects has been developing rapidly in the last few years, and it will be helpful to
summarize the role that these play in the alkylation process.
Using for illustration the radial flow impeller, Fig. 2 shows the
velocity profile based on average velocities at a point coming off
the blades of an impeller. As shown in the diagram, by taking the
slope at any point, we obtain the velocity gradient, which is the
shear rate that exists at that point. This shear rate can be calculated at any point in the mixing tank if we will make various
velocity measurements and establish the velocity profile at any
point.

Actually, the fluid shear rate must be multiplied by the vis-

TABLE II

SHEAR RATE TO RUPTURE A GIVEN DROPLET SIZE

μ CP	$\mu' = 80$ Cp	
	SHEAR RATE SEC. $^{-1}$	SHEAR STRESS IN CONTIN. PHASE g/cm^2
80	93	0.08
240	38	0.09
370	22	0.08
745	16	0.12
1175	14	0.17

μ = continuous phase viscosity

μ' = discontinuous phase viscosity

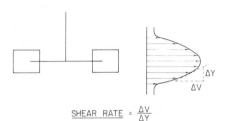

$$\text{SHEAR RATE} = \frac{\Delta V}{\Delta Y}$$

Figure 2.   Schematic of velocity flow pattern from radial flow turbine, showing shear rate calculation

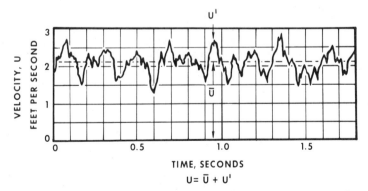

TIME, SECONDS

$$U = \bar{U} + U'$$

Figure 3.   Typical velocity vs. time from a high frequency response probe measuring velocities at a point from an impeller mixer

cosity, to give the fluid shear stress which carries out our process result, in accordance with the equation below.

Fluid shear stress = μ (fluid shear rate)

An article by Karam (1) gives typical data to illustrate the difference between shear rate and shear stress. Table II is extracted from cross plots of their data, showing the shear rate required with different continuous phase viscosities and one dispersed phase viscosity to break up a second fluid of the same size droplet. This shows that the shear stress in grams per centimeter squared is the basic parameter and the viscosity and shear rate are inversely proportional to give the required shear stress.

Looking at the profile from the average velocity at a point, there are 4 regions in the tank which are of particular interest. These give us the following points of pertinent definitions:

1. The average impeller zone shear rate.
2. The maximum impeller zone shear rate.
3. The average tank zone shear rate.
4. The minimum tank zone shear rate.

The average velocity gradients throughout the tank have been defined by several different methods. Estimates from our laboratory indicate that these velocity gradients are an order of magnitude less than the average velocity gradient around the impeller zone.

The minimum velocity gradient of the tank has been estimated as 1/4 to 1/3 of the average velocity gradient throughout the entire tank.

If a flow in the tank is turbulent, either because of high power levels or low viscosity, then a typical velocity pattern at a point would be illustrated by Fig. 3. The velocity fluctuation μ' can be changed into a root mean square value (RMS), which has great utility in estimating the intensity of turbulence at a point. So in addition to the definitions above, based on average velocity point, we also have the same quantities based on the root mean square fluctuations at a point. We're interested in this value at various rates of power dissipation, since energy dissipation is one of the major contributors to a particular value of RMS μ'.

Looking at the average velocity at a point, Fig. 4 gives the result that the maximum shear rate around an impeller increases with impeller diameter at a given RPM, while the average shear rate around the impeller remains essentially a constant with impeller diameter.

This yields a further relationship, as shown in Fig. 5, which shows that typically on scaleup, maximum shear rates increase, while average shear rates decrease. These now are shear rates that affect large scale particle, variously estimated at 100 to 200 microns and larger.

It is impossible to keep the ratio of all the pertinent mixing parameters constant in scaleup, so that we must pick and choose the mixing parameters and shear rates that are important for a given type of process.

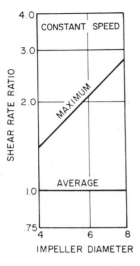

Figure 4. Data showing that maximum shear rate around an impeller increases, while average shear rate remains constant with different diameters having the same RPM

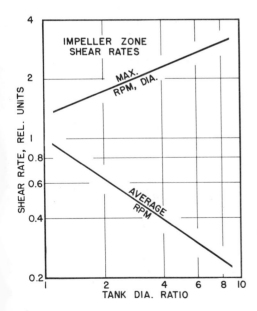

Figure 5. Effect of scale-up on maximum and average impellers zone shear rates

On the other hand, the turbulent velocity fluctuations RMS
μ' affect microscale processes, again variously estimated at 100
to 200 microns or less, on down to possibly the mixing length or
possibly on down to the scale of molecular reaction.

Turbulent Velocity Fluctuations

Experimental data reported by J. Y. Oldshue, (6) were obtained with a hot wire velocity meter. Details of that equipment are
included in the appendix to this report. Additional data from
this equipment are shown on Figs. 6, 7 and 8. Fig. 6 shows some
of the data which give the ratio of the RMS velocity fluctuation
to the average velocity at that point as a function of position
above and below the impeller centerline. These data are for a 4"
diameter, 6 flat blade turbine at 259 RPM. This indicates the
ratio of fluctuating velocity to average velocity at a point can
be as high as 0.75 and is generally in the range of 0.5 to 0.7.
These measurements were made approximately 3/4" from impeller
periphery. Additional measurements made in other parts of the
tank are shown in Fig. 7. These indicate that the velocity fluctuation in other parts of the tank in the range of 0.05 to 0.15
of the average velocity which is more typical of the turbulence
behind grids and in pipelines.

Other background on turbulent velocity parameters is given
by Cutter (3) who calculated several parameters from photographs
of flow patterns in a mixing vessel. He found that the RMS velocity fluctuation was primarily related to energy dissipation. His
measurement scale in making these photographic studies could be
quite different from the scale velocities from the hot wire velocity meter reported here, so that a direct comparison of these results with those of Cutter has not been made in this paper.

It since appears that the frequency and size of the turbulent
velocity fluctuations are more compatible with phenomena on a molecular micron size scale in a mixing vessel. For example, in a
publication Paul & Treybal (4) shows the yield of a given reaction which had several alternate paths was determined by the RMS
velocity fluctuations at the feed point. Paul used data from
Schwartzberg & Treybal (5) and Cutter to calculate RMS velocity
at the feed point.

Another parameter which can be measured by the hot wire velocity meter technique is a dissipation length.

$$\lambda = \text{DISSIPATION LENGTH}$$

$$\frac{1}{\lambda^2} = \frac{1}{u^2}\left(\frac{du}{dx}\right)^2$$

$$u = \text{VELOCITY}$$

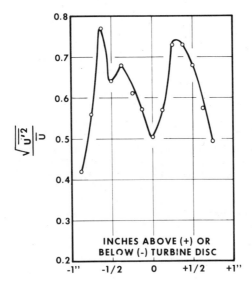

Figure 6.   The ratio of RMS velocity fluctuations to average velocity at different distances above and below the impeller centerline

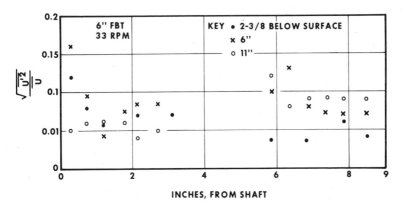

Figure 7.   Effect of distance from impeller shaft on the ratio of RMS velocity fluctuation to the average velocity at a point

This is helpful in indicating the size of the turbulent eddies. λ is a measure of, although not equal to, the mean size of the smaller eddies which are dissipated into energy. Fig. 8 gives some experimental values. At the moment, λ values are not being used to quantitatively analyze processes in the mixing vessel.

Some Practical Concerns On Scaleup Of Alkylation

Alkylation needs pumping capacity to maintain a uniform mixture of components within each phase as well as a uniform distribution of dispersed phase in the continuous, and the dispersion step itself requires a certain amount of shear stress to create interfacial area for mass transfer. There is no way of knowing what levels of each are needed without running some experiments.

If this is coupled with the various kinds of continuous flow phenomena, and the requirement for settling of the various phases in one or more stages of the system, there comes out to be some very definite optimum mixer combinations for a given process.

For example, if the emulsion must settle in each stage, so that a clear hydrocarbon phase can be passed on to the subsequent stage, then the optimum requirement for mixer variables becomes quite unique. As the speed of the given impeller is increased, the interfacial area is increased, better mass transfer is produced, possibility of better uniformity is available, but difficulty in settling a clear hydrocarbon phase is increased. There is a definite optimum power level, D/T ratio and other parameters, that have been found to exist in these kinds of processes.

In general, the relatively large diameter mixing impellers with relatively high pumping capacity and lower levels of fluid shear have been found to be more effective than higher shear devices, such as high speed pumps or high velocity jet nozzles.

If the emulsion is to flow unsettled from stage to stage, then there is no concern about settling until the final product settler is reached. This means that at various power levels, different combinations of variables are optimum. For example, at a low total hp level, large D/T ratios are required to achieve the pumping and blending required to maintain a uniform emulsion. As power levels are increased, the total productivity of the process usually increases, and smaller D/T's are more important because a level of pumping action at a higher power level can be achieved with smaller D/T ratios and still give adequate blending. As power levels are increased further, more of the energy can be put into shear rates, since a smaller proportion of the power is needed to achieve uniformity of blending and emulsification. At some point, the improvement in process yield with power begins to reach a peak and fall off, and at that point, economic conditions enter into the picture and dictate practical levels of power.

If consideration is being given to a single stage reactor, then high power levels are used to go along with coil and jacket type cooling, and power levels are often dictated by the necessi-

ty for heat transfer and mass transfer, rather than subsequent
settling and stage operation.

Some Considerations In Scaling Up Or Scaling Down Alkylation Studies

Table III gives the change in various mixing parameter ratios for 4 different scaleup calculations. What this shows is
that we cannot control the ratio of every individual fluid mechanics parameter. It also indicates that it is better to use
correlating parameters to find out how they change on scaleup,
rather than searching for a constant scaleup parameter for every
conceivable mixing process. There is no assurance that there is
a constant scaleup parameter for any given process job we are
studying. Therefore, pilot plant studies should be directed toward establishing a controlling factor in the process, so that
its control on scaleup can be more closely monitored.

Referring to this table, it is seen that if constant power
per unit volume is used as a parameter, then maximum impeller
zone shear rates go up while the pumping capacity system goes
down. Another column shows that a constant impeller tip speed,
the power per unit volume goes down drastically and the pumping
capacity goes down still further.

In general, small scale models have too much pumping capacity and too little impeller zone shear rate to be typical of the
performance to be expected in a full scale prototype. Thus, it
is normally necessary to purposely scale down the pumping capacity, and increase the shear rate on small scale equipment to
more closely duplicate the performance which will be expected on
full size units.

In alkylation, this means that the large size tank will normally have higher shear stresses, and finer bubbles produced
around the impeller, but also a much greater tendency for coales-
cense to occur in the lower shear zones in the tank. The pumping
capacity per unit volume of the full size unit will tend to be
much less than it is in the small size unit and therefore, the
uniformity of reagents, the blending of these materials, and the
stability of the emulsion will be less than it would be on the
pilot scale. Non-geometric designs must often be used.

Some Considerations About Pilot Plants

There is a minimum size pilot plant for two phase processes.
For example, Fig. 9 shows an impeller that, let's say, is 2 centimeters in blade width. The jet from the impeller gives a macroscale shear rate of 10 at a given speed, while the shear rate
across a quarter centimeter distance is 9-1/2, across a 0.5 centimeter distance is 7.0, across half the blade width is 5, and
across the entire impeller is zero. Thus, a bubble that is 2-1/2
centimeters in size will see no shear rate, while a micron sized

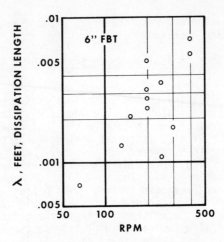

Figure 8.   The effect of RPM on the dissipation length in a mixing vessel

TABLE III

PROPERTIES OF A FLUID MIXER ON SCALE UP

PROPERTY	PILOT SCALE 5 GALLONS	PLANT SCALE 625 GALLONS			
POWER (P)	1.0	125	3125	25	0.2
P/VOL.	1.0	1.0	25	0.2	0.0016
SPEED (N)	1.0	0.34	1.0	0.2	0.04
TURBINE DIA(D)	1.0	5.0	5.0	5.0	5.0
PUMPING CAPACITY (Q)	1.0	42.5	125	25	5.0
Q/VOL.	1.0	0.34	1.0	0.2	0.04
ND	1.0	1.7	5.0	1.0	0.2
$\dfrac{ND^2\rho}{\mu}$	1.0	8.5	25.0	5.0	1.0

bubble will see a shear rate of 10.

We cannot let the physical size of the impeller blade in the pilot scale get out of proportion to the particle we are trying to disperse and the bubble we are trying to create.

A way to evaluate the effect of macroscale maximum and average impeller zone shear rates, and the microscale fluctuating turbulent intensity forces, is to build a series of different impellers as shown in Table IV. Merely for illustration, 5 impellers are shown. The first impeller is a standard blade width, run at a diameter of 1 and a speed of 1, to yield a maximum shear rate of 1, an average shear rate of 1, and a fluctuating shear rate of 1 on a relative basis. There is no relationship between the values of 1 for the three shear rate quantities. This impeller would be run and the process result evaluated. Then impeller number 2 would be used, which is a smaller diameter, higher speed impeller of the same blade proportions. It gives the shear rate properties as shown on the table. A 3rd, 4th and 5th impeller are used, which vary the various shear quantities. By observing the effect on process result, this allows a general interpretation of the role of various macro and micro scale shear rates in the process to be evaluated, and give some idea of the kind of scaleup performance to be expected with various kinds of scaleup parameters.

Some Comments About Liquid-Liquid Dispersion

One of the key points is to know which phase is continuous. If there is an ambivalent range, which means that either phase can be continuous, then the phase that the impeller starts in will normally be the continuous phase and it will suck in the dispersed phase into the zone. Fig. 10 illustrates the ways the different phases can be dispersed, and also shows the method of dispersing a heavy phase and light phase with the impeller still being at the bottom portion of the tank.

Emulsion break time can be a measure of the degree of tightness of the emulsion, and Figs. 11 and 12 show data obtained under various kinds of experiments illustrating the effect of different mixing variables on emulsion break time for a particular two phase system. This kind of data does give an indication of the settling times to be expected in the settling stage following a continuous system. A general truism says that the first phase will settle with more clarity than will the continuous phase. The continuous phase normally has a haze that is very difficult to settle, while the dispersed phase, once it settles, is usually much more clear.

Some Considerations About Instantaneous Concentrations At The Feed Point

Figs. 13 and 14 show the average velocity at a point as de-

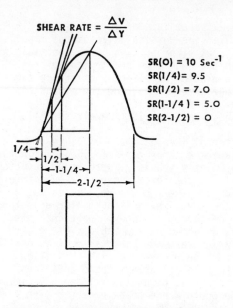

Figure 9. Schematic of different shear rates
as a function of different size particles in the
flow stream

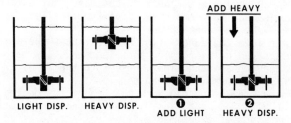

Figure 10. Method of controlling which phase is dispersed
in batch extractions

IMPELLER			SHEAR RATE		
D	N	Dω/D	MAX	AVG	FLUC
1	1	S	1.0	1.0	1.0
0.7	1.8	S	1.3	1.8	1.0
1	1.3	N	1.3	1.3	1.0
1	1	N	1.0	1.0	0.5
0.8	1.3	S	1.0	1.3	0.7

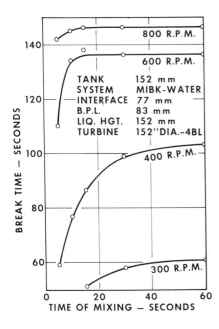

Figure 11.   *Effect of mixing time and RPM on break time for a liquid–liquid emulsion*

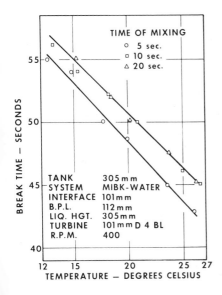

Figure 12.   *Effect of temperature on the break time of a liquid–liquid emulsion*

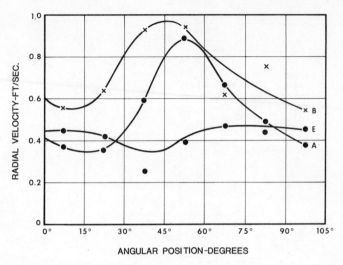

Figure 13. *Effect of blade angle on the velocity from a flat blade turbine*

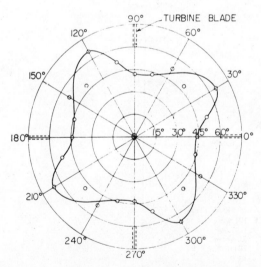

Figure 14. *Effect on the angle between the blades on the average velocity at a particular position in the discharge stream from a radial flow impeller*

termined by the angle between the blades. What this shows is
that the velocity at a point fluctuates about 50% from the average, as is also shown by the RMS values discussed previously.
     If a chemical is injected at a constant rate, then the molar
concentration ratio at that point is actually fluctuating over a
factor of 2. If reaction rates are fast enough so that this
would affect the path that a reaction would take, this could
affect the distribution of reaction products in a way that would
be either good or bad, depending upon circumstances. For example,
in neutralizing a sensitive chemical or biological product, adding acid or base at a constant rate into the eye of the impeller,
can cause the pH to fluctuate markedly at that point, and is
thought to be one of the reasons for the decrease in yield from
a theoretical value when trying different types of introduction.
     If a uniform concentration of a miscible liquid is more de-
sirable than the high intensity turbulent dispersion of these materials in the stream, then it may be well to have injection
points out in a more uniform, less fluctuating environment. Thus
evaluation of injection point conditions can be very critical in
reactions that may take different paths, depending upon chemical
concentrations and fluid mechanics variables.
     In the paper by Paul & Treybal,they showed that the reaction
products of a competitive, consecutive second order reaction do
depend on the RMS fluctuation value at the injection point.

Summary

     Mixers can be designed to increase pumping capacity at a con-
stant level of impeller shear rate, which would increase the total horsepower of the system, they can be designed to increase
pumping capacity at the expense of fluid shear rate variables, or
they can be designed to increase fluid shear rate variables at
the expense of pumping capacity.
     To determine which of these would be more advantageous,
there has to be some information or speculation on where a given
process result stands relative to these variables. On a full
scale unit, it is usually very difficult to make significant
changes in mixing variables, so that it is seldom possible to
assess possible improvements in increasing horsepower, level, the
torque in the mixer drive to change other variables, or whether
it is worth making major changes in geometry, feed introduction,
etc.
     If small scale experiments are to be used for evaluation
purposes, then they must be designed to have blend time, pumping
capacity and shear rate levels relatively similar to full scale
installation. This normally means a non-geometric model of scale-
down to control various mixing parameters.
     The small scale model must be of the same general type as
full scale with regards staging, continuous flow and other variables. The impeller must not be allowed to get out of proportion

in terms of physical size to the bubbles and droplets involved in the full scale process.

To shed most light on the effect of mixing, the usual order of variables includes the following:

1. Vary mixer speed, which varies both pumping capacity and all of the shear rates. If this does not change the process result, then the probability of other variables affecting it are relatively small.
2. Making a change in impeller size to tank size ratio of a given type impeller will normally tell whether pumping capacity or the entire variety of shear rates is significant in the process result.
3. If it is desired to separate out macroscale and microscale shear rate effects, then a change in the internal geometry of an impeller is necessary to change the relative proportions of these two quantities.

Symbols Used

D   Impeller diameter

D/T   Impeller diameter-to-tank diameter ratio.

D_ω Impeller blade width.

$\frac{ND^2\rho}{m}$ Reynolds number, ratio of inertia force to viscosity force, dimensionless.

P   Total power.

$\frac{Pg}{\rho N^3 D^5}$ Power number, ratio of applied force to graviation force, dimensionless.

Q   Volumetric fluid displacement of impeller.

RMS   Root mean square, $\sqrt{\overline{\mu^2}}$

T   Tank diameter

u   Velocity in x direction.

$\bar{u}$ Mean velocity, ft./sec.

u'   Fluctuating velocity.

Z   Liquid depth.

λ   Dissipation length. $\left[\dfrac{1}{\lambda^2}=\dfrac{1}{u^2}\overline{\left(\dfrac{du}{dx}\right)^2}\right]$

μ   Continuous phase viscosity.

μ´   Discontinuous phase viscosity

ρ   Density of fluid or solid.

Literature Cited

1 Karam, H. J., and Bellinger, J. C.: "Deformation and Break-up of Liquid Droplets in a Simple Shear Field", Ind. Engng. Chem., Fundamentals 7 (4), 576 (1968).

2 Oldshue, J. Y.: "Fermentation Mixing Scale-up Techniques", Biotech, and Bioeng. VIII (1),3 (1966).

3 Cutter, L. A.: "Flow and Turbulence in a Stirred Tank", A.I.Ch.E.J 12 (1), 35 (1966).

4 Paul, E. L., and Treybal, R.E.: "Mixing and Product Distribution for a Liquid-Phase, Second-Order, Competitive-Consecutive Reaction", presented at 62nd Annual Meeting A.I.Ch.E., Washington, D. C. (November 1969).

5 Schwartzberg, H. E., and Treybal, R. E. : "Fluid and Particle Motion in Turbulent Stirred Tanks", Ind. Engng. Chem. Fundamentals 7 (1), (1968).

6 Oldshue, J. Y.: "The Spectrum of Fluid Shear in a Mixing Vessel", Chemeca '70, Butterworth (1970).

14

Improved Mixing in Alkylation

D. E. ALLAN and R. H. CAULK

Exxon Research and Development Laboratories, P. O. Box 2226, Baton Rouge, LA 70821

Turbulent transfer processes are important in many
petroleum and petrochemical operations. For example, turbulent
mixing in stirred tanks affects both yield and product quality
in such chemical conversions as alkylation and vinyl chloride
polymerization; it influences the rate in physical conversions
such as crystallization by dilution chilling. Mixing efficiency,
as measured by process results, is tied to the type and degree
of turbulence present in the reactor. This paper will show
that the efficiency of mixing may be improved through a better
understanding of turbulent flow field coupled with an understanding of how flow affects the process. This study was an
attempt at combining turbulent flow studies, studies on model
systems and pilot plant data in order to understand the mixing
processes involved.

The paper is in two parts. First is a study of turbulent
flows generated by radially discharging impellers. It has been
asserted that mixing is a flow phenomena, hence, the first step
in gaining an understanding of mixing is to catalog flow
characteristics generated by common impellers. The second part
is a detailed comparison of the performance of two impellers
that generated significantly different flows in a pilot plant
test which studied the alkylation of butylenes.

Summary

This paper discusses a way to improve mixing, hence
product quality, in acid catalyzed alkylation without increasing the power consumption. It is essentially divided into two
parts: (1) a basic study of the fluid mechanics of impeller
driven flows which has yielded a new impeller design, and
(2) an alkylation mixer comparison with sulfuric acid catalyst
done in an alkylation pilot plant.

Looking at mixing as a turbulent flow generated by an
impeller, a study of the flows generated by conventional and
novel radially discharging impeller styles was done at Exxon's

Corporate Research Laboratories. Modern laser velocimetry
techniques were employed. Local average velocities and turbulent intensities were obtained in the discharge from the
impellers. Average velocities varied with power number for
the various styles as expected and this means the most commonly
used flat blade turbine generated the highest average velocities. The significant result was that turbulent intensities
were similar for all conventional impellers. One novel impeller generated intensities over twice as high in the fluid near
the impeller. Mixing literature suggests an improvement in
processes such as alkylation based on this flow difference.
    A mixer comparison with sulfuric acid catalyst was carried
out in the Exxon Research and Development Laboratories alkylation pilot plant. The novel impeller was compared to the
conventional flat bladed turbine at constant power levels
using refinery olefin and isobutane streams. Analysis of the
product quality-acid composition data for the pilot plant
comparison showed that the octane incentive for the new mixer
was related to acid strength and varied between 0.4 and 0.25
MON in the commercial operating region (96-90% H_2SO_4). In
addition to the quality comparison, catalyst life effects were
calculated and compared for both mixers. These data showed
that the novel impeller required somewhat lower acid replacement rate than the flat bladed turbine at high acid strengths
(98-94% H_2SO_4). Calculated acid replacement rates for both
mixers were equivalent at lower acid strengths (94-90% H_2SO_4).
Interpretation of these pilot plant results in light of the
mixing studies indicates that the key role of the novel impeller in alkylation is to generate additional emulsion surface
area.

The Impellers and The Flows They Generate

    Flow characteristics in a mixing vessel can influence
process performance. The impeller is a device which imparts
motion to the medium in which it operates. The characteristics
of the flow which are of greatest interest are the mean fluid
velocity at all points within the fluid and the turbulent
fluctuations superimposed on the mean velocity. Paul and
Treybal (1) have discussed how the detailed flow characteristics can influence process performance. This paper will show
how impeller style can influence the flow characteristics.
    The impellers used in this study are shown in Figure 1.
They are both of the radially discharging type which is
characterized by a strong radial jet of fluid that moves out
to the vessel walls while entraining fluid from above and below.
Near the wall it splits into two circulation zones. The flat
blade turbine (FBT) is commonly used in industry and consists
of a disc with several paddles fixed normal to the disc which
serve to generate the radial flow. The novel impeller design

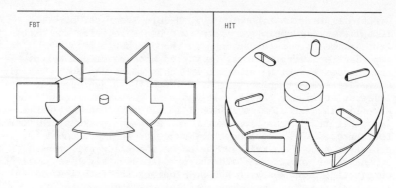

Figure 1. Impeller styles compared in alkylation pilot plant

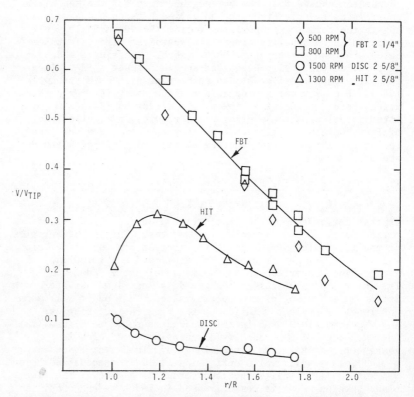

Figure 2. Average radial velocity along centerline of jet produced by impeller vs. radial position

(HIT)* has two discs that limit the flow into the impeller. It
also has several pairs of paddles fixed normal to the discs
which generate the radial flow.

Earlier work has used several methods to measure turbulent
flow parameters in the discharge stream of an impeller and
these are summarized by Gunkel and Weber (2). Only the FBT
impeller style was used in all earlier studies.

In this work, measurements of the mean velocity and the
turbulent fluctuations in the flows generated by the two impellers were confined to the center line of the radial jet since
this is the zone where the flow is most strongly influenced
by the impeller style. Turbulent flow parameters were determined using a DISA Electronics Laser Anemometer System. As the
name implies, this is an optical device that measures the
instantaneous velocity at a point. The technique is linear
which allows accurate flow measurements at very high turbulence
levels such as found in mixing vessels. Since there is no probe,
there are no flow disturbances and the measurement is independent of fluid properties. The anemometer used in this work was
sensitive to the direction of the instantaneous velocity.

A glass mixing vessel of 0.20 m diameter was used for the
study. It was enclosed in a square plexiglass tank that was
also filled with the same liquid as in the mixing vessel
in order to minimize refraction of the light beams. The liquid
level was 0.20 m and the impellers were 0.10 m from the bottom.
Impellers of approximately 0.064 m diameter were used to
minimize wall feedback effects. Water was the working fluid
for the results reported here.

The instantaneous velocity, V_i, in a turbulent flow
can be expressed as

$$V_i = \bar{V}_i + U_i \ (i = 1,2,3)$$

where $\bar{V}_i$ is the time mean velocity and U_i is the instantaneous
fluctuating velocity in the i direction. By definition, the
average fluctuating velocity, $\bar{U}_i$, is zero. The relative
intensity of turbulence, I, is defined as

$$I_i = \frac{\sqrt{\overline{U_i^2}}}{\bar{V}_i}$$

where $\overline{U_i^2}$ is the mean of the square of the fluctuating velocity.
Figures 2 and 3 are measurements of the radial component of
velocity along the center line of the strong radial jet issuing
from the different impeller styles versus the radial position
(r) normalized by the impeller radius (R).

*High intensity turbine

The variation of mean radial velocity with distance from
the impeller tip is shown in Figure 2. When the velocity is
normalized using the impeller tip speed (V_{TIP}) a relation is
obtained for each impeller style for various rotational speeds.
The mean radial velocity is a measure of impeller pumping
capacity. As expected, the various styles show grossly different
characteristics in terms of pumping capacity.

In all cases the mean radial velocity decreases with
increasing radial distance (except for the HIT impeller near
the tip). This is due to two things: (a) in a cylindrical
geometry, the cross sectional area for flow increases with
radial distance, and (b) the radial jet entrains fluid from
above and below, hence velocity must decrease to conserve
momentum.

The distribution of the turbulent intensities for the
radial component of velocity is plotted in Figure 3 as a function of radial position. Turbulent intensity is a measure of
the energy partitioning between the fluctuating velocity to
which local mixing is attributed and mean velocity which relates
to pumping or blending. The FBT ranges from 0.35 near the
impeller to 0.60 toward the wall. The increase in intensity
with distance is due to the mean velocity decreasing more
rapidly than the fluctuating velocity. Intensity is seen to be
independent of impeller speed. That is, once fully developed
turbulent flow is achieved, additional energy input is divided
in a constant manner between flow and turbulence.

The new impeller style (HIT) generates a different profile
and is characterized by a zone of very high turbulent intensity,
up to 0.95, near the impeller tip. These higher turbulence
levels mean more energy is being put into fluctuations (local
mixing) at the expense of pumping. The very high levels also
indicate that instantaneous flow reversals are occurring (this
condition could only be observed using a Laser Anemometer).
Negative instantaneous velocities, when made part of an average
velocity, reduce the time mean near the impeller tip. This
explains the decreasing portion of the mean velocity profile
for the HIT impeller as we move toward the impeller. That is,
the new impeller generates a "washing-machine" action where
some of the fluid ejected is periodically sucked back into the
impeller.

Results on the FBT agree with earlier studies of Mujumdar
et al (3) who used a hot wire anemometer for their measurements.
Recent mixing literature (1, 4) suggests that the turbulent
flow characteristics of the novel HIT impeller should be of
value in alkylation.

Experimental Alkylation Equipment

The experimental equipment used to obtain the mixer alkylation data described in this paper is a pilot plant at the Exxon

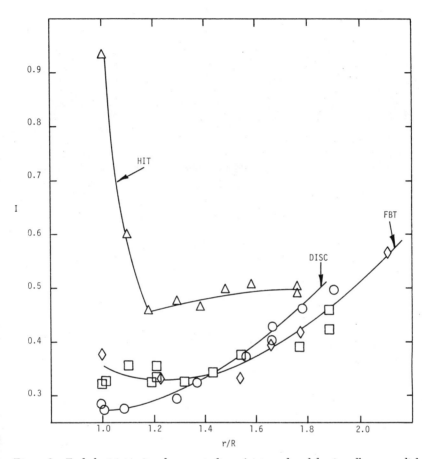

Figure 3. *Turbulent intensity along centerline of jet produced by impeller vs. radial position*

Research and Development Laboratories (ERDL) in Baton Rouge, La.
This unit was designed to be similar in performance to the
stirred alkylation units currently in commercial use. Although
it was not possible to maintain exact mixing similarity, this
unit maintained similitude in the more important scale-down
parameters.

A condensed version of the ERDL alkylation pilot plant flow
plan is shown in Figure 4. Olefin and isobutane feed streams
are separately pumped to the unit from large feed storage
vessels with Lapp Pulsafeeder diaphragm pumps and metered with
turbine flow meters. The streams are then combined, caustic
scrubbed, water washed and dried with molecular sieves before
being sent to the reactor. The combined feed stream is then
injected into the acid-hydrocarbon emulsion in the stirred
reactor vessel. In order to maintain a constant temperature
environment both reactor and settler are coolant jacketed.
After reaction, the acid-hydrocarbon emulsion passes from the
reactor to a baffled settler where a phase separation takes
place and settled acid is pumped back to the reactor. The
settled hydrocarbon phase is caustic scrubbed to remove any
entrained acid and sent either to a material balance accumulator
or to a product accumulation system.

Information that has been obtained from this pilot plant
includes: (1) alkylate quality at simulated commercial
operating conditions, (2) alkylate yield, (3) isobutane
consumption, (4) catalyst life, (5) maintenance of product
quality with catalyst life, and (6) selectivity data (which
contributes to the determination of reaction rate constants).

Mixer Alkylation Comparison - Pilot Plant Parameters

In order to get the most meaningful comparison between
the two mixer styles in an alkylation environment, two back-
to-back dying acid runs* (each of about two weeks duration)
were made in the ERDL alkylation pilot plant at the nominal
test conditions shown below.

Reactor Parameters

Reactor Temperature   - 40°F
Olefin Space Velocity - 0.15 vol. olefin/hr/vol. H_2SO_4
H_2SO_4 in Emulsion   - 60 vol. %

*In a dying acid run one begins with fresh acid in the reactor
and without spent acid withdrawal or fresh acid makeup allows
the catalyst diluents (acid soluble oil and water) to build up
with time. This type of run can be contrasted to a steady
state run where spent acid is purged and fresh acid added so
as to keep the diluent level constant.

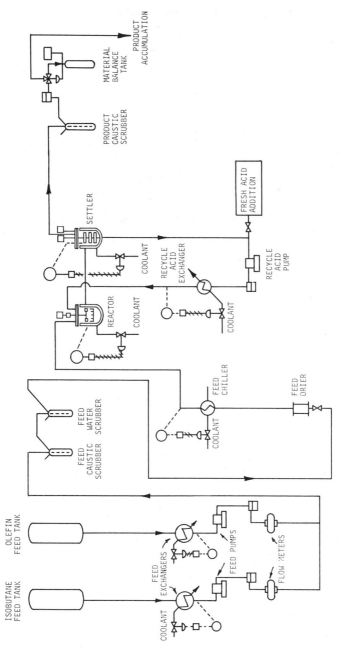

Figure 4. ERDL alkylation pilot plant

Mixing Parameters

Mixing Devices	- Flat bladed turbine, novel high intensity turbine
Mixer Speed	- 170 RPM (FBT), 245 RPM (HIT) (i.e., constant power)
Mixer Size	- Both mixers are 7" in diameter
Jet Reynolds No.	- 3245

The reactor parameters shown above are typical for commercial stirred alkylation units in H_2SO_4-butylene service. It is important to emphasize that the mixer comparison was made at constant power.

The olefin and isobutane feed streams used in the mixer alkylation comparison were obtained from Exxon Company USA's Baton Rouge Refinery. The olefin stream was catalytically cracked butenes while the isobutane stream was obtained as the overhead from a deisobutanizer tower. The composition of these streams is shown below:

Feed Stream	Olefin	Isobutane
Composition, Vol. %		
Isobutane	37.1	85.6
Olefin		
$C_3^=$	0.1	
$C_4^=$	28.0	} 0.1
$C_5^=$	3.6	
Total	31.7	0.1
Other Paraffins	31.2	14.3

Alkylation Pilot Plant Results

Previous investigators such as Albright et al (5) have shown that alkylation catalyst composition has a profound effect on product quality and that quality comparisons must be made at constant catalyst composition. In addition to measurement of product quality differences, one of the objectives of the mixer comparison was to look at possible differences in catalyst composition between the two mixer styles and the subsequent effect on catalyst life.

The effect of catalyst age (measured as hours on olefin) on the concentration of H_2SO_4* in the catalyst phase is shown in the top part of Figure 5. The H_2SO_4 concentration for both mixers decreased smoothly and nearly linearly with increasing catalyst age and there seemed to be no significant differences

---

*H_2SO_4 concentration in the catalyst was determined by NaOH titration.

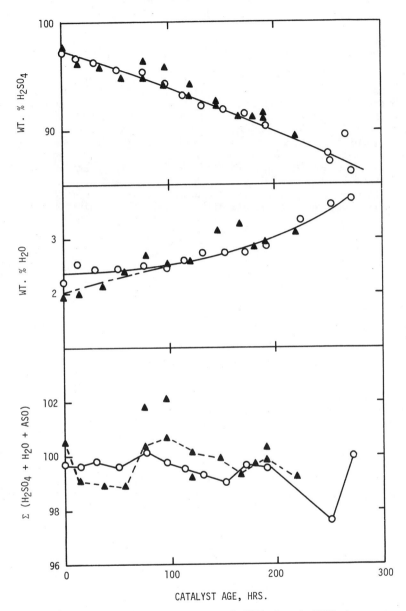

Figure 5.   Catalyst composition vs. age.   ○, HIT mixer;   ▲, FBT mixer.

between them. The starting acid for the HIT mixer study was
97.9% H_2SO_4 while the discharge acid was 86.0% H_2SO_4. The fresh
or starting acid for the FBT study was 97.3% H_2SO_4 while the
discharge acid strength was 89.6% H_2SO_4. The reason for the
difference in discharge strengths was that the FBT study was
terminated at a lower catalyst age. The second part of Figure
5 shows the effect of catalyst age on the catalyst water
concentration[φ] for both mixers. These data show that the water
content varied smoothly from about 2 wt. % for the fresh acid
to about 3-3.5 wt. % for the discharge catalyst. Although
there was a slight difference in water content between the
mixers initially, the water data for the most part are quite
similar. Since the total feed to the ERDL pilot plant was dried
to a level of 1-2 ppm H_2O, this water buildup was due to hydrocarbon oxidation by the sulfuric acid. Finally, the effect of
catalyst age on acid soluble oil[Δ] level in the catalyst for the
mixer comparison is shown in Figure 6. These data show that,
as anticipated, acid soluble oil built up steadily with
increasing catalyst age reaching a level of 10 wt. % for the
HIT mixer discharge acid and ∿7 wt. % for the FBT mixer discharge
acid. For all practical purposes both mixers showed about the
same variation of acid soluble oil content with catalyst age.
There are two interesting features with regard to the data in
Figure 6 that should be noted. The first point is that both
runs show a finite acid soluble oil level in the acid before the
olefin stream was started (as measured by catalyst age). This
low level was due to feeding isobutane over the fresh H_2SO_4
prior to starting olefin feed and shows that significant acid
soluble oil components can be formed from compounds other than
dienes, olefins and/or sulfur at low temperatures. The other
interesting point with regard to the Figure 6 data is that the
HIT data become non-linear after about 225 hours on olefins.
This seems to indicate an "acid runaway" condition* and more will

---

[φ]Water concentration in the catalyst was determined by titration
with fuming H_2SO_4 (oleum) after Albright et al (5).

[Δ]Acid diluent commonly referred to as red oil or sludge which
is formed primarily in the reactor (emulsion residence time
in the settler is an order of magnitude less than the reactor).
Although feed contaminants such as dienes or sulfur are
important red oil precursors, it can also be formed from the
olefin, isobutane and alkylate.

*"Acid runaway" is a term commonly used in commercial alkylation
operations and is characterized by a condition of low acid
strength and rapid build up of diluents. In certain situations
an acid runaway may force olefin feed to be cut out since
fresh acid usually cannot be added fast enough to counteract
the rapid diluent formation.

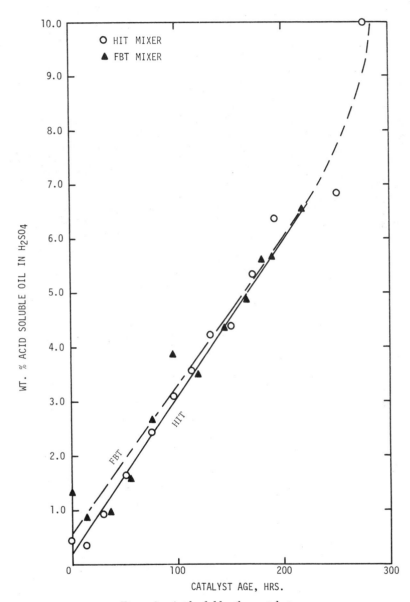

Figure 6.   Acid-soluble oil vs. catalyst age

be said about it in the discussion on catalyst life.

The consistency of the raw catalyst composition data can be seen by looking at the bottom part of Figure 5. This is a plot on the sum of wt. % H_2SO_4, wt. % H_2O and wt. % acid soluble oil for each of the acid analyses in the mixer comparison study. The absolute difference between this sum and 100 gives a measure of the consistency of the acid analyses. Except for the one point at about 250 hrs. (and the analyses that were rerun at $\sim$75-100 hrs.), the sums in Figure 5 for the mixer comparison runs vary between 99% and 101% with most sums being closer than 1%. If it is considered that these sums are formed from the results of three independent analyses, it can be said that the quality of the acid composition data is very good.

One of the most important facets of the mixer comparison was the effect of the mixer type on alkylate quality. As mentioned previously, since acid composition does have a profound effect on quality it is necessary to compare mixing effects at constant acid composition. For this study, the acid component which seemed to play the key role in this composition/quality effect is the acid soluble oil. This component, which is characteristic of strong acid processes, is a complex mixture in which cyclic conjunct polymers predominate (6). Since alkylate quality does vary markedly with acid soluble oil content, many theories on its role in quality enhancement have been proposed such as a hydride transfer agent, surfactant, or agent to increase isobutane solubility.

Therefore, since the catalyst water content was relatively constant, acid soluble oil level has been used as a primary indicator of acid composition to compare the product qualities of the two mixers. A comparison of octane product quality for the two mixers is illustrated in Figure 7 where the data are plotted as GC calculated C_6^+ MON (cl)* versus wt. % acid soluble oil in the catalyst. Although initial product quality was poor, Figure 7 shows that the MON went through a broad maximum for both mixers from 2 to 4 wt. % acid soluble oil. These data show that the HIT mixer gave a product quality incentive over the flat bladed turbine and that the magnitude of this incentive varied between 0.4 and 0.25 MON in the commercial operating region (2-7 wt. % acid soluble oil or 96-90 wt. % H_2SO_4).

As far as the general shape of the curves is concerned, the presence of maxima indicates that at low concentrations acid soluble oils interact directly with the alkylation reactions and have a beneficial effect toward product quality while at high concentrations the diluent effect on acid strength outweighs any possible benefits.

---

*Clear motor octane number

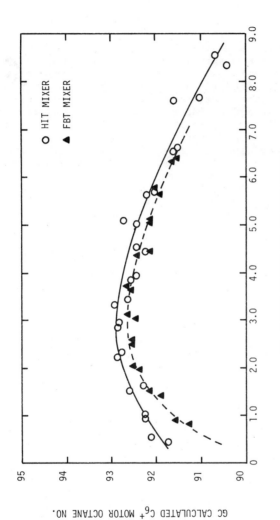

Figure 7. *Product quality as a function of acid-soluble oil level*

In addition to product quality the economics of the
alkylation process dictate that catalyst deactivation must be
fully considered when evaluating potential process improvements.
Catalyst deactivation in alkylation is characterized by the
formation of diluents that lower acid strength and have a deleterious effect on product quality beyond a certain level. These
diluents can either be absorbed directly in the acid such as
water in the hydrocarbon feed or be formed from feed components.
The primary diluent formation route is polymerization of these
feed components (especially dienes and sulfur compounds) to acid
soluble oils. In addition, oxidation of the feed and/or
alkylate hydrocarbons by the acid may occur which forms water
and tends to dilute the acid. In a commercial alkylation operation the diluents formed are continuously purged by withdrawing
spent acid and sending it for regeneration. Fresh or replacement acid is added back at a rate that replaces the acid lost
by purging and compensates for any internal losses such as
oxidation. Due to the importance and costs associated with
catalyst life, deactivation rates for both mixers were measured
and are discussed in the following paragraphs.

The quantitative measure of catalyst life in alkylation
is the pounds of diluent (i.e., acid soluble oil and H_2O)
generated per gallon of olefin (pure) fed. This factor when
appropriately combined with acid losses (i.e., oxidation) and
fresh and spent acid concentrations, gives the fresh acid
replacement rate which is a direct measure of the regeneration
cost for any particular alkylation process. In order to calculate values for the acid replacement rates, it was necessary
to formulate a "dying acid" model for the pilot plant reactor-
settler system.* This model assumed the reactor-settler system
to be a perfectly mixed tank and took into account acid soluble
oil generation and H_2SO_4 oxidation. Using data on acid composition, additions/deletions (sampling and losses) and catalyst
inventories, the model was applied incrementally to the data
to calculate diluent and oxidation parameters. These parameters
were then used to calculate equivalent acid replacement rates
for each mixer.

The calculated equivalent acid replacement rates for each
mixer are shown in Figure 8 as functions of acid strength

---

*In a "dying acid" type of experiment with no additions/with-
drawals, the fresh acid replacement rate has no real meaning.
Values calculated for the mixer comparison data are called
equivalent replacement rates. These equivalent replacement
rates are the rates necessary (i.e., for a given diluent formation rate) to maintain equilibrium in a hypothetical steady
state alkylation unit which uses fresh acid at 98% titratable
acidity and spends its acid at the acid strength in question.

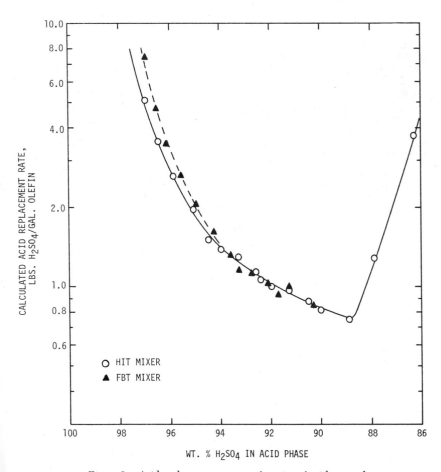

Figure 8.   Acid replacement rate as a function of acid strength

and as expected decrease with decreasing H_2SO_4 concentration
down to about 90 wt. % H_2SO_4. These data indicate that the HIT
mixer showed a lower acid replacement rate in the high acid
strength portion of the commercial operating region (i.e.,
98-94 H_2SO_4).* The difference in acid replacement rates seemed
to be the greatest with the fresh acid and progressively
narrowed with decreasing acid strength. In the lower acid
strength portion of the commercial operating region (i.e., 94-
90% H_2SO_4), the acid replacement rates for both mixers were
essentially equivalent. Below 90% H_2SO_4, data from the HIT
mixer run graphically showed how quickly replacement rates can
rise when the acid strength gets below a critical value (i.e.,
"acid runaway").

Interpretation of Mixing and Alkylation Results

Since data are available on the flows generated by the two
impellers, as well as on process performance some speculations
can be made on the role of the novel HIT impeller in alkylation.
A summary of the relative flow characteristics and alkylate
quality are given in the table below.

Impeller	Power	Flow at Olefin Addition Point	Turbulence	Emulsion Surface Area	Δ MON @ 3% Acid Soluble Oil
FBT	1.0	1.0	1.0	1.0	--
	2.0	1.26	1.26	1.3	0.3
	3.0	1.45	1.45	1.6	0.7
HIT	1.0	0.59	1.45	1.2	0.25

Results on operating the FBT at higher power levels are from
other pilot plant runs. The Δ MON is at the optimum acid
soluble oil level of about 3.0 wt. %.

High flow of emulsion generated by the impeller should
disperse the olefin feed more quickly throughout the reactor
and thereby improve quality. Comparing the FBT and HIT impellers at constant power, the flow is 40% lower for the HIT
impeller. Conventional wisdom would predict the product quality
to be lower. Hence impeller pumping is not the key variable.

Micromixing and the subsequent rapid dispersion of the
olefin should be increased by increasing the turbulence at the
point of injection of olefin. At constant power the HIT
impeller has turbulence levels that are 1.45 times as high. For
the FBT to reach these levels, its power must be increased by a

*This difference in replacement rates for the two mixers is due
to: (1) a small difference in the effect of catalyst age on
acid soluble oil level (see Fig. 6) and (2) about a 10% difference in acid inventories for the two runs.

factor of three. At these high power levels the product quality
is improved by 0.7 MON. Since this is twice the improvement
obtained by the HIT impeller, this suggests that turbulence
alone (and its influence in dispersing olefin) is not the key
variable.

The combination of reducing the flow and increasing the
turbulence level has been shown by Brown and Pitt (4) to decrease
emulsion drop size and, in systems as we have here with a fixed
amount of dispersed phase, increase the emulsion surface area.
The HIT impeller has been shown to generate emulsions with 20%
more surface area than did the FBT when compared at the same
input power level. This is in line with Brown and Pitt's correlations if the observed flow characteristics for the HIT
impeller are used. To increase the emulsion surface area by
this amount using the FBT, the power level must be approximately
doubled. When the FBT power level is doubled, the improvement
in product quality is in line with that obtained with the HIT
impeller. This suggests the key role of the impeller in
alkylation is to generate emulsion surface area.

The bigger increases in alkylate quality at higher acid
strengths due to improved mixing is likely due to an increased
importance of mixing in stronger acid environments. At present
insufficient data exist to be sure if the role of the impeller
is the same in this region.

Literature Cited

1. Paul, E. L. and Treybal, R. E., AIChEJ (1971), 17, 718.
2. Gunkel, A. A. and Weber, M. E., AIChEJ (1975), 21, 931.
3. Mujumdar, A. S., Huany, B., Wolf, D., Weber, M. E., and
   Douglas, W. J. M., Can. J. Chem. Eng. (1970), 48, 475.
4. Brown, D. E. and Pitt, K., Chem. Eng. Sci. (1974), 29,
   345.
5. Albright, L. F., et al, Ind and Eng Chem-Proc Des and Dev.
   (1972), 11 (No. 3), 446.
6. Miron, S. and Lee, R. J., J. Chem. Eng. Data (1963), 8
   (No. 1), 150.

15

Computer Modeling of Alkylation Units

J. C. KNEPPER* and R. D. KAPLAN
Amoco Research Center, Naperville, IL 60540

G. E. TAMPA**
Amoco Oil Refinery, Whiting, IN 46394

Computer simulation (1,2) is a powerful tool for improving the design and operation of process units. Such techniques have been used to design units, to evaluate proposed process modifications, to compare operations between units, to identify declining unit performance, to optimize current performance, and ultimately to control unit operation. Within the petroleum industry, a catalytic cracking simulation model (3) is being used to predict and guide unit operation.

Amoco has now applied this approach to sulfuric acid alkylation units. Gilmour (4) had previously developed an alkylation unit model, using PACER, a general purpose simulator. He presented no quantitative results of model applications, however. Sauer et al. (5) also developed an alkylation unit simulation, but they did not use a rigorous approach and indicated that "the similarity to alkylation is not complete because several simplifying assumptions are made in describing the process and in using the correlations (6)."

In contrast, the Amoco model is more fundamental and is based on extensive laboratory and commercial test data that accurately reflect unit operations. Because Stratco and cascade units account for over 80% of the alkylation capacity of Amoco refineries, we developed complete computer models for each type.

Unit Description

Flow diagrams of Stratco and cascade alkylation units, the two which were modeled, are shown in Figures 1 and 2. Both types

---

*Present address is Rio Blanco Oil Shale Project, 9725 E. Hampden Ave., Denver, Colorado 80231.
**Present address is Standard Oil (Indiana), 200 E. Randolph Drive, Chicago, Illinois 60680.

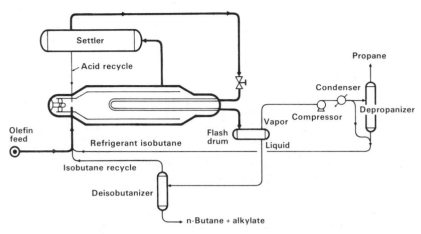

Figure 1.   Stratco effluent-refrigeration alkylation unit

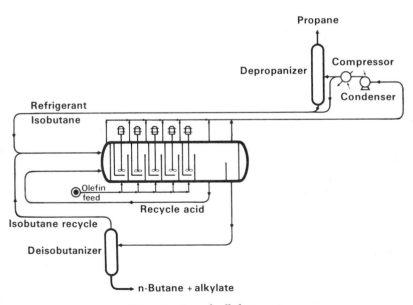

Figure 2.   Cascade alkylation unit

typically contain multiple reactors.

In the Stratco effluent refrigeration unit, mixed olefins and isobutane flow in parallel to the shell side of each reactor, where the exothermic alkylation reactions take place in a hydrocarbon-in-acid emulsion. The reactor effluent flows to a settler where acid and hydrocarbon phases are separated. The acid is then recycled to the shell side of the reactors. A back pressure valve in the hydrocarbon effluent line from the settler is used to maintain total liquid operation in the reactor and settler. Downstream of the valve, the pressure is reduced. Some hydrocarbon vaporizes, cooling the remaining material. This stream is then flashed through heat exchange tubes to control reaction temperature. The liquid-vapor hydrocarbon effluent then flows to a pressure-controlled flash drum. Liquid from the drum is sent to a deisobutanizer, which produces an isobutane recycle stream overhead and a crude alkylate bottoms product. Flash drum vapors are compressed and condensed. Then a part of this stream is depropanized. The depropanizer bypass and bottoms product form the liquid refrigerant recycle stream, which returns to the reaction zone. The unit also includes interstream heat exchangers that improve the energy efficiency of the process.

In the cascade unit, the olefin feed flows in parallel to several mixed zones within each reactor shell, while refrigerant isobutane, isobutane recycle and acid flow in series through these zones. Compared with single-stage reactors handling equivalent feed streams, this series of stirred reactors achieves a higher average isobutane concentration. Temperature is determined by reactor pressure control, which sets the extent of vaporization of the hydrocarbons in the reaction zones. The isobutane recycle and refrigerant isobutane streams are produced similarly in both Stratco and cascade units.

Need For Models

Simulation models are needed to evaluate commercial unit performance because of the strong emphasis on recycle within the process, as is shown in Figures 1 and 2. Analysis of such flow patterns require tedious, time-consuming, trial-and-error calculations that are best done by computer.

Models can be used to balance the key elements of profitability, energy versus product quality. For example, good reactor performance is favored by low temperatures and high isobutane concentrations in the reaction zones. Refrigeration systems serve the double role of controlling reactor temperature and keeping isobutane concentration high through efficient removal of propane from the system. These results are obtained at the expense of compressor energy. The deisobutanizer also maximizes the concentration of isobutane in the reactor system, but only at the expense of energy.

Simulation models are also particularly useful for assessing the economics of unit maintenance and proposed modifications. Deisobutanizer condenser performance is a good example. As these units age and their heat transfer capability decreases, there is a decline in the rate and/or purity of the isobutane recycle stream. With a comprehensive simulation model, the evaluation of the economics of a turnaround can be made within a matter of days by using actual measured and design heat transfer coefficients.

Description of Models

The options of using available general purpose process flow-sheet simulators, or formulating special purpose models were considered. The latter was chosen to satisfy the need for detailed, accurate, and flexible models, which are suitable for routine use.
The components of our models are:

- Data input

- Vapor-liquid equilibrium constant and enthalpy value calculation routines

- Dew point, bubble point, and flash calculations

- Stream enthalpy and temperature from enthalpy routines

- Heat exchanger and distillation column simulations

- Convergence techniques for trial-and-error calculations

- Reaction correlations

- Process economics

- Optimization routine

- Data output

Thermodynamic calculations are used to evaluate vapor-liquid equilibrium constants, enthalpy values, dew points, bubble points, and flashes. Established techniques simulate the heat exchangers and distillation columns, and handle convergence and optimization.
Reaction correlations relate alkylate octane and yield, acid usage, and isobutane consumption to feed composition and process variables. These correlations, determined from pilot plant and

commercial tests, are of the form:

Response = f (olefin feed composition, temperature, space velocity, isobutane concentration,
acid strength, mixing conditions, etc.)

Figures 3 and 4 show typical correlations. The effect of
propylene concentration in the olefin feed on acid consumption is
shown in Figure 3. There is a critical propylene concentration
below which acid consumption increases at a low, linear rate and
above which the increase is more drastic. The various symbols
reflect a number of different pilot plant and commercial test results. Figure 4 presents pilot plant data showing the effect of
isobutane in the reactor effluent on research octane number. A
number of other alkylation correlations have appeared in the lit-
erature ($\underline{7}$, $\underline{8}$, $\underline{9}$, $\underline{10}$, $\underline{11}$, $\underline{12}$).

Model structure is shown in Figure 5. Process variables,
unit constants (such as heat transfer coefficients), and feed
streams are described on input or as selected by the optimization
routine. Then, heat and material balances are performed using an
assumed alkylate yield and isobutane consumption. These results
form a set of reaction conditions which are used in correlations
to calculate reactor performance. The heat and material balance
calculations are repeated if reactor performance differs significantly from that used in the previous calculation. Operating
incentives are then computed and may be used in the optimization
routine to select new values of the optimization variables.

The accuracy of the models has been validated by comparison
with commercial operating data. Unit performance data are used
to adjust reaction correlations to base point levels.

Examples of Model Applications

The models have been applied to analyze various operations.
Four examples of model usage are:

1) To identify profitable unit modifications.
2) To compare performance between units.
3) To evaluate optimal unit capacity.
4) To determine optimal deisobutanizer operation.

Unit Modifications.

Model simulations are useful for evaluating the benefits of
proposed configurational changes in units. For example, simulations suggested that modifications to the internals of cascade
units would improve performance. After these changes had been
made in an Amoco commercial unit, weekly average acid con-
sumptions as low as 0.2 lb/gal of alkylate were recorded.

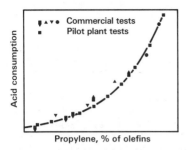

Figure 3. Effect of propylene concentration on acid consumption

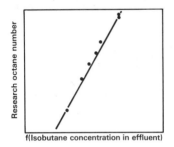

Figure 4. Effect of isobutane concentration on research octane number

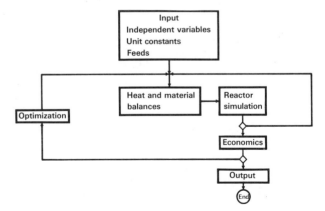

Figure 5. Model structure

Comparison of Performance.

The models may be used to compare performance between several units or unit types. For example, a week of production data from the above-mentioned modified cascade unit was compared with simulated Stratco performance. The agreement between Amoco's Stratco simulation model predictions and material balance and performance predictions provided by the Stratford Engineering Corporation had been confirmed earlier. Model adjustments allowed a comparison at constant operating conditions to be made. Table I compares the performance of the modified cascade unit with the Stratco simulation model predictions. Acid consumption is 36% lower in the modified cascade unit. The Stratco unit does show a slight research octane advantage over the cascade. However, overall economics favor the modified cascade in this case.

Optimal Unit Capacity.

The simulation models can easily be used to identify optimal unit capacity. In this example, feed rates to the cascade unit were varied in such a way that composition remained constant. Maximum refrigeration and deisobutanizer reboiler duty were also specified. Variation in olefin feed composition with changes in feed rate could also have been simulated. Figure 6 shows the effect of increasing feed rate on average reactor isobutane concentration and temperature, and on unit operating incentives (profit). The average isobutane concentration decreases significantly as feed rate increases; but the average reactor temperature remains relatively constant at 46°F until a butane-butylenes feed rate of about 1000 B/H is reached. Thereafter, temperature rises sharply. This temperature pattern results from the imposition of dual limitations: first, minimum allowable reactor zone temperature, then minimum permissible compressor suction pressure. As feed rate increases, the combination of higher space velocity, lower isobutane concentration and, eventually, higher temperature results in lower octanes and alkylate yield per barrel of olefin feed, and in higher acid consumption. However, the lower part of the figure shows that the value of the additional product overcomes these disadvantages until the optimal feed rate of 1050 B/H is reached. Alkylating incremental feed above this point is not economical.

Optimal Deisobutanizer Operation

Deisobutanizer operation is a key factor in the performance of an alkylation unit. The deisobutanizer fulfills two main functions; separating alkylate from lighter components, and maintaining a high concentration of isobutane in the reactor. The Stratco model was used to search for the combination of energy costs and alkylate quality giving maximum profit. Simulation model predic-

Table I

Comparison of modified cascade operation with Stratco performance

Conditions	Modified Cascade	Stratco Effluent Refrigeration
Propylene in olefin feed, vol%	11.8	
Average isobutane concentration in reactor, vol %	69.6	
Space velocity, V/V/hr	0.27	
Temperature, °F	50.0	
Spent acidity, Wt%H_2SO_4	91.5	
Research octane	95.0	95.2
Acid consumption, lb 98.5-91.5 wt% H_2SO_4/ gal. C_5+ alkylate	0.369	0.580

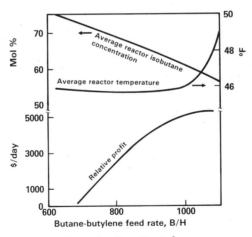

Figure 6. Capacity study

tions were made for given olefin and make-up isobutane stream
rates and compositions to show the effect that deisobutanizer
reboiler duty and reflux ratio have on relative profit. The results are shown in Figure 7, where the elliptical curves are constant profit contours. The larger the profit, the smaller the
ellipse.

Increasing recycle rate at constant reboiler duty initially
leads to improved performance because of higher reactor isobutane
concentrations. Further increases in recycle rate will eventually have a negative effect on performance because of decreased
isobutane concentration in the recycle as well as other effects.

Increasing reboiler duty at constant recycle leads to increased recycle purity and a higher reactor isobutane concentration. As a result, performance improves. Further increases in
reboiler duty will continue to improve octane and decrease acid
consumption. Eventually, the value of these improvements will not
be sufficient to pay for the increased reboiler energy usage.

A section drawn through Figure 7 at a constant isobutane recycle rate is shown in Figure 8, for which the abscissa is now
reflux ratio (reflux/isobutane recycle). The optimum corresponds
to the value of reboiler duty giving the maximum profit for the
specified recycle rate. As the reboiler duty (or reflux rate)
decreases, the profit drops sharply. No reflux corresponds to
the operation of the deisobutanizer as an "isostripper". Figure 8
shows that "isostripper" operation significantly reduces unit
profitability.

The effect of reduced reflux rate on Motor octane is shown in
Figure 9. As "isostripper" operation is approached, octane drops
markedly because of sharply reduced reactor isobutane concentration.

Conclusions

Complete simulation models have been formulated for cascade
and Stratco sulfuric acid alkylation units; and studies have
confirmed the accuracy of the models. Application studies include
cases in which model usage identified profitable unit modifications, determined optimal unit capacity and optimal distillation
tower operation, and compared the performance of cascade and
Stratco units. "Isostripper" deisobutanizer operation was determined to be relatively unprofitable for sulfuric acid alkylation
units; and acid consumption on a modified cascade unit was found
to be 36% below that expected for a Stratco unit. The examples
presented suggest the broad applicability of the simulation
models for improving alkylation unit operation. Use of the models
not only pinpoints areas where significant improvements are possible, but also quantifies incentives needed to get them imple-
mented quickly.

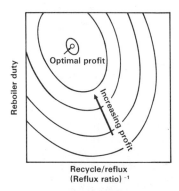

Figure 7. *Effect of deisobutanizer operating variables on unit profitability*

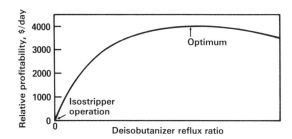

Figure 8. *Effect of deisobutanizer reflux ratio on profitability at a given isobutane recycle rate*

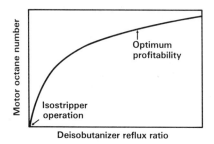

Figure 9. *Effect of deisobutanizer reflux ratio on motor octane number*

270INDUSTRIAL AND LABORATORY ALKYLATIONS

Acknowledgement

We gratefully acknowledge constributions made toward the
development of these models by B. E. Brown, G. W. Elmer,
B. S. Kennedy, S. J. Choi, and the late W. F. Pansing.

Abstract

Computer simulation models have been formulated for cascade
and Stratco sulfuric acid alkylation units. These complete
models incorporate mathematical descriptions of all the interact-
ing parts of the units, including reactors, distillation columns,
compressors, condensers, and heat exchangers. Examples illus-
strate diverse model applications. These include identifying
profitable unit modifications, comparing cascade to Stratco performance, evaluating optimal unit capacity and determining optimal deisobutanizer operation.

Literature Cited

1) Crowe, C.M., Hamielec, A.E. Hoffman, T.W. Johnson, A.I. and
Woods, D.R., "Chemical Plant Simulation," Prentice-Hall,
Inc., Englewood Cliffs, N.J. (1971).
2) Franks, R.G.E., "Modeling and Simulation in Chemical Engineering," Wiley-Interscience, New York (1975).
3) Wollaston, E.G., Haflin, W.J. Ford, W.D., and D'Souza, G.J.,
Oil and Gas Journal (1975) 73, (38), p. 87.
4) Gilmour, R.H., British Chemical Engineering, (March 1969),
14, p. 315.
5) Sauer, R.N., Colvile, A.R., Jr., and, Burwick, C.W., Hydrocarbon Processing and Petroleum Refiner (February 1964),
43, p. 84.
6) Ibid, p. 86.
7) The Phillips Petroleum Company, "Hydrofluoric Acid Alkylation", p. 117-128 (1946).
8) Mrstik, A. V., Smith, K.A., and Pinkerton, R.D., "Advances in
Chemistry Series, V," p. 97, American Chemical Society.
9) Payne, R.E., Petroleum Refiner (September 1958), 37, p. 316.
10) Putney, D.H., in "Advances in Petroleum Chemistry and Refining II," p. 315, Kobe, K.A., and McKetta, J.J.,Jr.,
Editors, Interscience Publishers, Inc., New York (1959).
11) Jernigan, E.C., Gwyn, J.E., and Claridge, E.L., Chem. Eng.
Prgr. (1965), 61, p. 94.
12) Van Zoonen, D., World Petroleum Congress, Mexico City,
Paper No. P.D. 17 (1967)

Sulfuric Acid Requirements for Industrial Alkylation Plants

ORLANDO WEBB

Rt. 4, F-12, Lee's Summit, MO 64063

The most common utilization of the light olefins produced
from catalytic cracking and coking operations is as feedstock
for alkylation units for reaction with isobutane. The alkylate
product has high research and motor octane numbers and is used
in motor gasoline blending or, to a limited extent, rerun and
used as the base stock for manufacturing aviation gasoline.

Of the 125 commercial alkylation units operating in the
United States today, 59 employ concentrated sulfuric acid as the
catalyst. Total alkylate production is over 850,000 barrels per
stream day, and of this domestic total the daily capacity of the
sulfuric acid alky units is in excess of 500,000 barrels (1). To
produce a satisfactory product and avoid corrosion and other
problems the titratable acidity of the acid catalyst must be
maintained at a relatively high level, normally above 88%
(equivalent wt. % H_2SO_4). In most plants this is accomplished by
continuous addition of fresh sulfuric acid of 98.0 to 99.5%
H_2SO_4 strength and continuous discard of spent acid of the minimum acceptable strength. Figure 1 is a block diagram showing the
general process arrangement. The discard acid is either returned
to the acid supplier, or to a much lesser extent, utilized elsewhere in the refinery.

Purchased acid costs have increased in recent years from
nominally $20/ton to $30-40/ton in many locations. Typically,
the current cost of make-up acid ranges from a low of about
$0.10 to highs of over $1.00 per barrel of alkylate. It is one
of the major operating expenses for a sulfuric acid alkylation
unit. The purpose of this paper is to discuss the factors that
affect fresh acid requirements and to point out areas that warrant
close operator attention.

Plant Evaluation Considerations

Before considering specific acid reactions and theories, one
should pause briefly and take note of the operating procedure
used on a typical commercial unit and the effect this can have on

the apparent plant performance, irrespective of feed impurities,
side reactions, etc. First, recognize that the volume of the
acid catalyst in the reaction section of the plant is very large
compared to both the fresh acid make-up and the discard acid
flow rates. For example, there may be as much as 1,800 barrels
of acid inventory in a 17,000 barrel per day unit, and the fresh
and discard acid flow rates may be only roughly 30 barrels per
hour. The configuration of the large settlers used for separating and recycling the acid usually is such that even a small
change in the acid level, say one inch, may represent 45 minutes
of total fresh acid flow. Thus the maintaining of a constant
inventory in the plant is a difficult operating goal to satisfy.
Plant performance evaluations should take into account the possible effect of inventory changes.

Second, even assuming that an essentially constant volume of
acid catalyst is maintained, the operator must decide what the
optimum discard acidity should be and attempt to maintain it by
making adjustments in fresh acid and discard acid flow rates. If
the discard acidity is allowed to increase, excessive fresh acid
is used. If it is allowed to decrease, alkylate quality
declines, undesirable side reactions increase, and acid usage
increases·. A decrease in average system acidity of 1% typically
reduces alkylate quality by about 0.4 octane numbers. In certain
situations, acidity may drop below the point where the acid is no
longer an effective alkylation catalyst but continues to absorb
olefins. Chain polymerization reactions can also be initiated.
This is accompanied by rapid temperature rise and a sharp further
decline in titratable acidity. In such situations, operators
commonly resort to very large acid additions in an effort to
bring the entire volume of acid up in strength. Recognizing the
enormous acid inventory in the system, it is obvious that such
upsets, in addition to changes in acidity at the beginning and
end of an operating period, must also be taken into account in
making a factual plant performance evaluation of average acid
usage.

Third, and now assuming essentially constant volume and
acidity are maintained, there is the problem of determining an
accurate discard acid analysis. Discard acid samples are viscous
and can contain appreciable entrained hydrocarbon. The hydrocarbon phase must be completely removed by centrifuging the
sample if the analysis is to have any meaning or reproducibility.
Most analytical laboratories report the titratable acidity of the
acid portion of the sample. This is calculated from the amount
of sodium hydroxide used in neutralizing the sample in a standard
titration procedure. While the result is expressed as wt.% H_2SO_4
it is important to recognize that the discard liquid is not a
simple sulfuric acid solution, but is actually a very complex
mixture of sulfuric acid, weak organic acids, sulfonic acids,
sulfides, acid soluble polymeric olefin compounds, carbon, water,
and probably other compounds depending on feed composition and

impurities. The nature and relative amounts of the acid diluting compounds can vary and have a pronounced effect on the make-up acid requirement and the catalytic behavior of the acid in the reactor.

With the acid inventory control, acidity control, and analytical limits in mind, and respecting their potential effect on reported acid make-up requirements for any given operation, particularly over operating periods of less than thirty days, let us now turn to some of the specific factors affecting acid usage.

Feed Impurity Effects

The most significant impurities in the feed streams to a typical alkylation unit are ethylene, diolefins, sulfur compounds and water. Corrosion inhibitors and other chemicals used in upstream processing can also be present in some cases, and these can have harmful effects. The amount of each impurity that reaches the alkylation reactor varies considerably from refinery to refinery. If accurately determined and properly accounted for, these impurities can explain an appreciable percentage of the acid make-up reported by various operating units. The impurity data shown in Table I can be used to evaluate the merit of improved upstream process control and/or more efficient feed pretreatment methods.

Table I. Feed Impurity Effects

Compound	Sulfuric Acid Dilution Rates (Basis 98.5% Fresh, 90.0% Discard)
Ethylene	3952 Lbs. per Barrel
Butadiene	2465 Lbs. per Barrel
Sulfur Compounds	15-60 Lbs. per Pound of Sulfur
Water	4116 Lbs. per Barrel

Ethylene

If sufficient isobutane is available, most refiners today are charging a full C_3-C_4 cut to alkylation. In these operations the control of the feed deethanizer can have a noticeable effect on the amount of fresh acid needed for the alky unit, and yet the concentration of ethane-ethylene in the feed may appear to be insignificant. Ethylene reacts with strong sulfuric acid to form ethylsulfuric acid, rather than combining with isobutane to form an alkylate. The ethylsulfuric acid becomes a part of the discard acid mixture and titrates as a weak acid, thereby decreasing the calculated discard acidity and requiring the addition of more fresh acid for control. On a mole for mole basis, and assuming an acid dilution range of from 98.5 to 90.0% (wt.), the fresh acid make-up rate is 3952 pounds of acid per barrel of ethylene

in the feed. This calculated figure is consistent with rates
ranging from 2940 to 4620 pounds per barrel, which the writer has
found in common industrial use.

The alkylation acid penalty for ethylene is so large it makes
close control of the feed deethanizer absolutely mandatory for
economical overall operation. The ethane and lighter content of
the C_3-C_4 cut should be maintained nil as measured by continuous
chromatography. Even a trace amount of C_2's may represent several barrels of ethylene per day. In addition, if a slug of
ethylene is accidentally charged to the alkylation reactor it can
also have the effect of suddenly reducing the strength of the acid
catalyst to the point that alkylation ceases, olefin absorption
rapidly dilutes the acid, and chain polymerization of olefins
("acid runaways") can occur.

Diolefins

Diolefins also occur in relatively small concentrations in
olefin feed streams. Their concentration increases as cracking
severity increases, such as in fluid coking. As in the case of
ethylene, the butadienes do not appear to react with isobutane in
the presence of strong sulfuric acid. The dienes are believed to
form reaction products, most of which are acid soluble, and if
this general premise is accepted the fresh acid make-up rate for
a dilution range of 98.5 to 90.0% (wt.) can be calculated to be
2465 pounds of acid per barrel of butadiene. Industry practice is
to use rates in the range of 1890 to 4200 pounds per barrel.

Selective hydrogenation has been successfully employed commercially to convert the butadienes to mono-olefins. Current sulfuric acid prices give considerable justification for adding these
facilities in cases where severe cracking conditions are employed.

Sulfur Compounds

Before being charged to an alkylation unit, olefin streams
are normally treated for removal of sulfur compounds. The type of
treating used depends for the most part on the amount and kind of
sulfur compounds present. Quantities are normally reported in
parts per million units. According to industry sources, estimates
of the amount of sulfuric acid diluted lie in a range of 15 to 60
pounds per pound of sulfur in the feed. A very commonly used factor is 20 pounds per pound. In the case of methyl mercaptan, if
one assumes a mole for mole reaction occurs and that the reaction
products are soluble in the acid and do not titrate, then the calculated amount of acid diluted is 53.7 pounds per pound of mercaptan sulfur (dilution range 98.5 to 90.0%). This is a very severe
assumption however: commercial tests indicate about 45 pounds per
pound and this is probably more realistic. Somewhat lower dilution rates are normally assumed for H_2S and COS, typically ranging
between 15 and 18 pounds per pound of sulfur present.

In addition to the dilution effect, the presence of the sulfur reaction products in the acid catalyst promote polymerization
and other undesirable side reactions. It is therefore obvious
that operating companies should check treater performance frequently and make whatever changes are necessary to accomplish a
high degree of desulfurization of the alkylation feedstock.

Water

       Water in feedstocks, particularly that in excess of satura-
tion, is an important potential acid diluent. There are several
types of separators and driers in commercial use on alkylation
units. The simplest are coalescer devices packed with excelsior,
or other kinds of packing material. With proper maintenance,
these devices can coalesce and remove significant amounts of
finely divided and entrained water droplets in the olefin and
isobutane streams. For best results the feed streams are chilled
by heat exchange with the cold hydrocarbon effluent before they
are charged to the coalescers. In one plant test, a 55-gallon
drum of water was drawn off such a device every 12 hours. The
plant was operating at an alkylate production rate of about 2,000
barrels per day. A problem with these separators, however, is
that they develop channeling in time and their water removal efficiency declines rapidly. Unfortunately this performance deterio-
ration and need for packing replacement often goes undetected,
since there is no significant change in pressure drop, or other
data to alert the operator. The result is that fresh acid usage
steadily increases in many locations.
       A more reliable type of separator which is sometimes used is
a guard chamber packed with rock salt. These units can be built
large enough so that onstream time can be quite long. Field data
indicate essentially complete removal of free water and, in some
cases, even a portion of the soluble water. The coalesced water
is withdrawn as a brine solution. Maintenance consists of simply
restoring salt level while the unit is being by-passed for a short
time.
       The isobutane recycle from fractionation can also contain
appreciable entrained free water. If the recycle contains no olefins (from outside make-up streams, for example) this stream can
be chemically dried by contacting it with the discard acid from
the unit. This is an efficient and a relatively inexpensive procedure.
       The most efficient devices available for complete water
removal employ molecular sieves. These are widely used on hydrofluoric acid units but have been difficult to justify for sulfuric
plants. They can also be utilized to remove any sulfur compounds
remaining in the feed after pretreatment (2). If sulfuric acid
prices continue to increase, further consideration of molecular
sieves seems warranted.
       The fresh acid make-up rate for water dilution is 4116 pounds

of acid per barrel of water in the feed, again assuming an acid
spending range of 98.5 to 90.0% (wt.). Furthermore, in the view
of many technologists, increasing the water content of the acid in
the reactor has the additional undesirable effect of reducing its
effectiveness as an alkylation catalyst (3). If true, this could
result in the formation of more acid soluble reaction products and
therefore a further increase in the acid make-up.

Water also increases the corrosive nature of the acid in the
reaction section. Metal losses in the turbulent high velocity
sections of reactors can be quite rapid in units circulating high
water content acids.

Reactor Operating Conditions

The previous discussion has been concerned with impurities
in feed streams and their effects on acid make-up requirement.
All of the acid dilution rates are large and the resulting fresh
acid costs are adequate to support efficient feed fractionation
and treating facilities, frequent maintenance and close operator
observation and control of upstream operations. Assuming this is
done, the major remaining concern is the operation of the alkylation unit proper, particularly the reactor section.

In making all operating and design decisions, it is important
to keep in mind the definition of the true reaction zone. Funda-
mentally, this is the interfacial area between the immiscible
hydrocarbon and acid catalyst liquid phases in the reactor. Reactants and products flow across this boundary. The olefins in the
feed stream react instantaneously with the sulfuric acid catalyst
and combine with the relatively small amount of isobutane present
in solution in the acid catalyst to form alkylate. Alkylate
passes out through the interfacial surface reaction boundary into
the hydrocarbon phase while isobutane passes in to resaturate the
catalyst. To suppress undesirable polymerization and other reactions it is necessary to:

    a. operate the reactor and settler in such a manner
           as to produce a finely divided hydrocarbon-in-
           acid emulsion;

    b. maintain vigorous and repetitive contact between
           the hydrocarbon and acid catalyst phases and
           thereby accomplish continuous resaturation of the
           acid as the olefins react and isobutane is con-
           sumed;

    c. keep the concentration of olefin uniformly low in
           the acid catalyst phase;

    d. maintain a high concentration of isobutane in the
           hydrocarbon phase; and

    e. keep the temperature uniformly low throughout the
           reactor.

Emulsification

Perhaps the most difficult of the requirements is the emulsification and contacting. For efficient operations the emulsion characteristics must be uniform throughout the reactor. An excess of acid is preferred, since this results in a hydrocarbon-in-acid emulsion. In these mixtures the viscosity and surface tension of the continuous acid phase are effective in minimizing the tendency of the dispersed hydrocarbon droplets to coalesce and separate, acting under the influence of the very large specific gravity differential between the light and heavy phases.

The effluent hydrocarbon-acid mixture is separated either in a settling zone in the reactor or in a separate acid settler vessel. The operation of the acid settler section can have a marked effect on the overall performance of the reactor. A very successful operating technique involves operating with an acid emulsion in the acid settler and recycling the emulsion rather than a separated acid phase (4). To accomplish this the inventory of acid in the reactor settler section is lowered and the recycle rate is increased to the point that as much as 20 to 30 percent hydrocarbon is entrained in the acid as an emulsion recycle. Large improvements in acid requirements have been observed on both butylene and mixed propylene-butylene feeds when emulsion recycle has been employed.

To minimize acid make-up requirements the interfacial area of the circulating emulsion in the reactor must also be maximized. Emulsion settling rates are normally used to measure the amount of dispersed hydrocarbon phase in the emulsion. A sample is trapped in a transparent gauge glass and the time is recorded as the hydrocarbons coalesce and separate in a continuous layer on top of the heavier acid emulsion layer. Electrical conductivity measurements have also been made confirming the settling curve data. The emulsion settling curves published by Cupit, Gwyn and Jernigan (5) are typical of those observed on commercial reactors. An example of a desirable homogenous acid-continuous emulsion settling curve is illustrated in Figure 2.

Increasing emulsification has the effect of increasing the area available for mass transfer of isobutane into the acid catalyst phase. Alkylate quality improves and acid consuming side reactions decrease. Other investigators (5) have reported a large increase in the ratio of C_8 compounds to other C_6+ compounds in the product alkylate was observed when a homogeneous emulsion was produced. In the writer's experience, changing commercial reactor operation to produce the desired type of emulsion has, in some cases, had the effect of improving alkylate quality by 1.5 RON or more, and reducing acid make-up by 20 percent with other operating variables essentially unchanged.

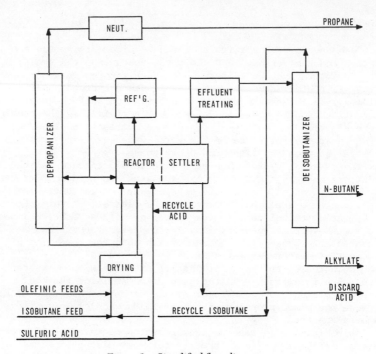

Figure 1.   Simplified flow diagram

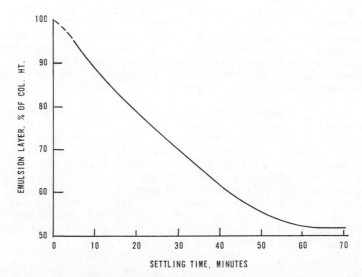

Figure 2.   Preferred emulsion settling characteristics

Contacting

The operation of the mixer used in the reactor can have a
large effect on the degree of emulsification. It controls the
rate of circulation of the emulsion within the reactor, and therefore the average number of passes the feedstocks make through the
mixing impeller during their residence time in the reactor. Most
commercial reactors have motor driven mixers, and mixer speed
cannot be varied. Some, however, are equipped with turbine and
gear drives, and some range of speed variation is available to the
operator. Impeller speeds of 620 RPM have been reported to be
very effective in improving alkylate quality in mixed feed operations (4). Improved dispersion has been credited by Knoble and
Hebert with reducing acid make-up from 1.85 to 0.659 pounds per
gallon on this commercial unit.

Use of efficient olefin feed distributors is also effective
in reducing acid consumption. The conditions at the point of
initial olefin contact with the circulating catalyst are very
important because of the extremely rapid rate of the initial reaction. Perforated sparger rings are frequently employed, and a
concentric feed nozzle is also a common feature (6).

Space Velocity

As applied to alkylation, space velocity is defined as the
gallons per hour of olefin in the feed divided by the gallons of
acid in the reactor. This term is simply a measure of the concentration of olefin in the acid phase of the reactor. As olefin
space velocity is increased, there is a greater tendency for polymerization and other undesirable reactions to occur and acid make-
up requirements increase. There is also an increase in the higher
boiling compounds in the alkylate product. Space velocity, in one
form or another, enters into almost all process performance correlations and is important to note in any plant evaluation studies.
For a given plant it is sometimes used singly in predicting plant
behavior. Figure 3 is a typical example of the combined effects
of space velocity and other variables on a butylene operation.
Note carefully however that space velocity is intimately associated with emulsification, contacting, and mixing horsepower. It
can be very misleading to compare reactors or plants of different
designs using space velocity as a primary criteria.

Isobutane Concentration

The driving force to complete the alkylation process is the
concentration of isobutane in the hydrocarbon phase in the reactor
(i.e. in the hydrocarbon effluent). This is the source of the
isobutane needed to resaturate the acid catalyst as the alkylation
reaction proceeds. The n-paraffins in the hydrocarbon phase are
inert and if they are allowed to build in concentration they can

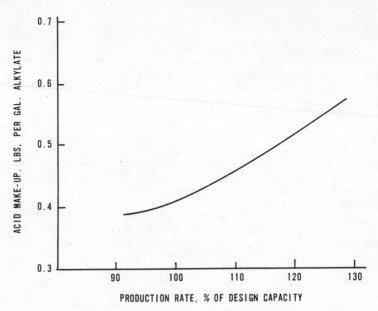

Figure 3.   Acid make-up as a function of alkylate production rate

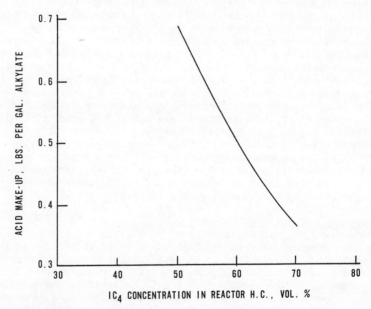

Figure 4.   Acid make-up as a function of IC₄ concentration in reactor

constitute a harmful diluent in the reaction zone. Their effect
is not always recognized, particularly by those who, for the sake
of convenience, oversimplify reactor behavior by using the external isobutane to olefin ratio in performance correlation.

To control an operating plant for minimum acid make-up, the
isobutane concentration in the reactor must be maintained at the
maximum possible level. This means operating both the depropanizer and the deisobutanizer under optimum conditions. The operator should adjust tower feed rates and operating conditions,
always using the isobutane concentration in either the total
effluent hydrocarbon or the net effluent hydrocarbon from the
reactor as his primary reference. Isobutane-to-olefin ratio,
deisobutanizer overhead purity, depropanizer recycle purity, and
refrigerant recycle purity are significant only as they relate to
reaction zone isobutane concentration.

The correlations published by Putney (7) show the effects of
isobutane concentration on octane number, and octane number on
acid consumption. These are for butylene feedstocks. By combining the data it is possible to construct the curve shown in
Figure 4 illustrating the influence of isobutane concentration on
acid make-up requirement, with all other variables constant. The
potential acid savings provide an obvious incentive for optimizing
fractionation operation in both butylene and mixed feed operations.

Propane is eliminated from the plant by charging a portion of
the stream containing the highest percentage of propane, usually
the refrigerant stream, to a depropanizer tower. In some plants
propane build-up has been reduced and isobutane concentration
increased by revamping and charging the depropanizer tower at a
higher feed rate, utilizing the full liquid handling capacity of
the tower effectively.

In the case of n-butane the operator must find the optimum
mode of operation of the deisobutanizer, or isostripper column.
For refluxed columns this involves determining the best feed
point location, and the best division between reflux and overhead
recycle to the reactor section. In some cases it has been found
that isobutane concentration in the reactor could be increased by
moving the feed point up, reducing the reflux and increasing the
recycle to the maximum that the reactor loop could handle.

Reaction Temperature

Relatively low temperatures are required in alkylation reactors using sulfuric acid catalyst. They are necessary in order to
slow down polymerization reactions and reduce the formation of
undesirable acid soluble and hydrocarbon soluble by-products.
Typically, most commercial units operate with reaction temperatures in the range of 35°F. to 65°F. Design temperatures are
usually set at 50°F. For minimum acid make-up the reactor section should be operated as cold as possible. This means operating

the refrigeration compressor for peak performance and taking maximum advantage of available cooling water. The cost of the incremental utilities required for lower temperature operation is considerably less than the cost of the incremental acid make-up. Some operators have expressed concern over the possibility of acid freezing and have restricted operation to 45°F. Others however, have reported exceptionally low acid make-up requirements at 32°F. to 35°F. and have not reported any freezing or other operating problems. Variations in the water content of the acid catalyst may explain why plant experience varies.

Fresh Acid Composition

Make-up acid for alkylation units typically contains from 98.0 to 99.5 wt. percent H_2SO_4. Recognizing the desirability of maintaining the lowest possible water content (8), some companies specify fresh acid is to be 99.0 to 99.9 % H_2SO_4 but containing no free SO_3. Others specify 98.0 to 99.0 % based on the assumption that this specification will assure there is, in fact, no free SO_3 present. In the writer's opinion, the fresh acid concentration should be specified as 99.0 to 99.5 % H_2SO_4 and containing no free SO_3. Fresh acid samples should be spot checked routinely and carefully analyzed for titratable acidity and SO_3.

Summary

The operator of a commercial sulfuric acid alkylation unit can reduce the amount of make-up sulfuric acid he must charge if he:

1. improves feed treatment for removal of sulfur compounds and inhibitors;

2. monitors feedstock composition and eliminates ethane and lighter compounds from the feed;

3. operates feed driers for maximum water removal;

4. checks discard acidity frequently and controls it at the optimum level dictated by plant experience;

5. operates the acid settler as a decanter, maintaining an emulsion in the acid settler and recycling an emulsion to the reactor;

6. maintains a hydrocarbon-in-acid emulsion in the reactor with as high an acid content as possible, consistent with settler limitations;

7. increases the speed of the reactor mixer to as high an RPM as is possible, limited by driver and settler capability;

8. operates the deisobutanizer and depropanizer as necessary to accomplish a maximum concentration of isobutane in the hydrocarbon effluent from the reactor section;

9. lowers the reaction temperature to the minimum
    consistent with the available cooling water
    and refrigeration equipment limitations; and
10. uses fresh acid of maximum strength and con-
    taining no free SO_3.

Abstract

One of the largest operating costs of a sulfuric acid alkylation unit for reacting light olefins with isobutane is the cost
of the fresh sulfuric acid make-up needed to maintain the acidity
of the catalyst at, or above, the required minimum level. Factors
affecting fresh acid requirement are classified in two categories:
feed contaminants and their effects; and the reaction conditions
that contribute to acid dilution. Estimates of acid make-up are
presented when the feed impurities are ethylene, butadiene, sulfur compounds, and water. The importance of proper sampling,
analysis, and control of the titratable acidity of the acid catalyst and potential resulting operating problems are emphasized.
Reaction conditions which affect acid make-up are: olefin space
velocity in the acid catalyst; isobutane and n-paraffin concentration in the hydrocarbon phase; the amount of surface area at
the hydrocarbon-acid catalyst interface; the degree of contacting
between the two immiscible phases; and the reaction temperature.
These conditions affect the formation of undesirable side reaction
products; such products result in increased fresh acid requirements, lower aklylate product quality, and more downstream operating problems. Methods of obtaining improved plant operation are
considered.

Literature Cited

(1) "Gas Processing Refining and Worldwide Directory" (34th
     Edition), The Petroleum Publishing Company, Tulsa, (1976-77)
(2) Collins, J. J., presented at NPRA annual meeting, San
     Antonio, Texas, (1962), Tech. 62-9
(3) Albright, Lyle F., Houle, Lawrence, Sumutka, Andrew M., and
     Eckert, Roger E. , Ind. Eng. Chem, Process Des. Develop.,
     (1972), Vol II, (No. 2).
(4) Knoble, W. S. and Hebert, F. E., Petroleum Refiner, (1959)
     Dec.
(5) Cupit, C. R., Gwyn, J. E., and Jernigan, E. C., Petro/Chem
     Engineer, (1961-1962), Dec. Jan.
(6) Kobe, Kenneth A. and McKetta, John J., Jr., "Advances in
     Petroleum Chemistry and Refining", Vol. 2, pages 315-355,
     Interscience Publishers, Inc., New York, (1959)
(7) Webb, Orlando, U. S. Patent 3,027,242 (1962)
(8) Goldsby, A. R., U. S. Patent 3,683,041 (1972)

17

SARP: Chemical Recovery of Used Sulfuric Acid Alkylation Catalyst

ARTHUR R. GOLDSBY

76 Marcourt Drive, Chappaqua, NY 10514

In the early days, in the late 1930's and early 1940's, of the alkylation of isobutane with olefins using a sulfuric acid catalyst, the acid catalyst was low in cost, the production of alkylate was limited, and the discarded or used catalyst could be used in other processes, such as naphtha and lube oil treating (1). Therefore, there was little incentive for the development of a recovery process unique to the alkylation catalyst. In addition, it was confirmed early in the development of alkylation that the current, conventional processes for the recovery or conversion of sulfuric acid containing hydrocarbons from treating processed could be used for converting the used alkylation catalyst to fresh sulfuric acid of any desired strength.

As the production of alkylate increased in the 1940's, largely because of World War II, a great deal more of the used catalyst became available for recovery. Most of it continues to be used merely as a source of sulfur, and is charged to conventional sludge conversion processes for recovery. To most chemists this did not seem to be a good solution to the problem, and it still does not. The used catalyst has a titratable acidity of about 90% H_2SO_4, and a water content of about 2-5%. The fresh charge acid to alkylation is usually white sulfuric acid of about 98.0-99.5% concentration. Currently a large amount of the used catalyst on the order of 5,000 tons per day in the United States is available for recovery, since typically about 0.5 pound of sulfuric acid is required for each gallon of alkylate produced.

In view of the foregoing, it is not surprising that many attempts have been made by many companies almost from the start of alkylation to develop a recovery process unique to the used sulfuric acid alkylation catalyst. Most of the early attempts were not intensive and continuous. Most of the work is unpublished, since it was of a preliminary nature and not very successful. An added incentive was provided for the users and licensors of the Sulfuric Acid Alkylation Process when it was discovered that hydrofluoric acid was also a good alkylation catalyst. Since hydrofluoric acid usually costs on the order of twenty times that of sulfuric acid, a recovery system by distillation was incorporated in the HF design. Hence, the figures given for acid consumption in HF alkylation are usually net figures after recovery. The figures given in H_2SO_4 alkylation are usually gross figures without recovery.

Since the hydrocarbons or conjunct polymers in the used sulfuric acid catalyst are unsaturated, attempts have been made to hydrogenate the catalyst in an effort to saturate the polymers, and thus make them insoluble in the acid. Attempts have also been made to conduct the alkylation under hydrogen pressure.

Various solvents have been used to try to extract the soluble or chemically bound conjunct polymers from the used catalyst.

Attempts have been made to separate the acid from the conjunct polymers by crystallization of the acid, and washing of the polymers from the acid crystals. This work was promising and a large pilot unit was operated. A paper on the process by Mr. S. Robert Stiles is scheduled for this Symposium (2).

Background for SARP

One line of attack, which is the subject of this paper, is to convert the sulfuric acid in the used catalyst to dialkyl sulfates, which are soluble in hydrocarbons, and then extract the dialkyl sulfates from the conjunct polymers and water. Considerable background information was available prior to the late 1950's when a study was started with the specific objective of developing such a recovery process. When it is considered that sulfuric acid had a net cost of not over about one cent per pound, or $20.00 per ton, and a net cost in alkylation of about 0.4 to 0.6 cents per gallon of alkylate, and that only 5% or less of the acid was lost in conventional recovery processes, it was realized that any new recovery process would have to have low capital and operating costs, and high yields, if it were to be competitive strictly on a savings in acid cost. Some of the background information gave hope that not only could such a process be integrated with an alkylation unit, but by using such a process, an improved alkylate of higher octane value and lower end point could be made.

Since this general line of attack is an exciting and challenging field for chemists and chemical engineers, and since such a process is of interest to most oil companies and many chemical companies for commercial operation, some of the background information will be discussed briefly.

Sulfuric acid and olefins are an intriguing pair in that vastly different results may be obtained depending on the reaction conditions, the particular acid strength and the olefin involved.

Up to about 1932 one of the most widely used analytical methods for the analysis of olefin mixtures was the Orsat

analysis. The olefin mixture in gaseous form was passed through
progressively stronger sulfuric acid. Thus, isobutylene was
removed by 62.5% acid, propylene and n-butenes by 82.5% acid,
and ethylene by 8% fuming acid.

Considerable work was done by many oil companies on the so-
called 2-stage alkylation process (3). This process had considerable appeal, since it offered a possible means of reducing
the amount of fractionation required in alkylation, and fractionation was the most expensive part of alkylation. In addition to
pilot unit work, several plant trials were made. In this process
olefin was absorbed in used or recycle alkylation acid, the inerts
and paraffin hydrocarbons eliminated in the absorption step, and
the acid-olefin mixture containing the catalyst contaminants
charged to alkylation. The elimination of inerts resulted in a
higher octane product and a reduction in the amount of fractionation required for the recycle of isobutane. However, a weakness
in the process was that it resulted in a considerably higher
acid consumption. One other weakness or complication was that
when the absorption step was carried out at a fairly high conversion of the acid to esters in the liquid phase, a considerable
amount of the neutral esters or dialkyl sulfates ended up in the
hydrocarbon liquid, rather than in the acid. Fractionation is
still an expensive part of alkylation in capital and operating
costs. However, it is not as expensive as originally, since
the introduction of effluent refrigeration (4,5) in about 1953
and isostripping in about 1956. Both of these schemes result in
a smaller deisobutanizer and a lower operating cost for fractionation to furnish recycle isobutane.

It was found in the early alkylation work that various olefin esters could be used in place of olefins to alkylate isobutane with a strong H_2SO_4 catalyst. It was also known that
diethyl sulfate was an accepted ethylating agent in the chemical
industry.

Before the days of the sophisticated mechanisms such as
those being presented in this Symposium, it was thought that
probably the olefin first formed an alkyl sulfate before reacting
with isobutane. In alkylation, reaction conditions were selected
so as to keep the concentration of the alkyl sulfates low. It
was standard procedure in batch laboratory work to have a
finishing period, that is, a period at the end of a run under
alkylation conditions except that no olefin was charged.
Especially with propylene as the only olefin, or in a high concentration in the olefin feed, it was difficult to get good
enough alkylation conditions so that the alkyl sulfate concentration would not become too high in the reaction mixture. And
even today sometimes in commercial alkylation neutral esters end
up in the alkylate leaving the reactor. This has been shown by
analysis, and in some cases by a temperature rise in the settler,
decomposition in the fractionation system, and poor lead suscep-
tibility of the alkylate. Neutral esters are not removed from

alkylate by the usual caustic washing treatment, and that is the main reason some plants use bauxite treating of alkylate. Numerous patents have been issued on various phases of an acid recovery process involving the absorption of olefins in recycle or used alkylation catalyst, extraction of the dialkyl sulfates, and charging of the dialkyl sulfate extract, after treatment with H_2SO_4, to alkylation. Reference will be made to those of particular interest in later sections of the paper. The others with which the author are familiar are included in the Literature Cited for convenience (6-16).

In the November 1965 Hydrocarbon Processing and Petroleum Refiner (17), it was indicated that the first commercial plant for such a process was being installed by Cities Service for start-up in late 1966, under license from Texaco Development Corporation. A staff article on the process in the January 2, 1967 Oil and Gas Journal, (18), was apparently based on three patents issued up to that time (19-21). A second commercial plant by Humble Oil & Refining Co. went into operation in 1969 (22).

To confirm some of the above types of information and to get some quantitative data on the absorption of propylene and buty-lenes in recycle alkylation acid, and the solubility of diiso-propyl sulfate and diisobutyl sulfates in hydrocarbons, some fundamental work, unpublished, was done by Midwest Research Institute in Kansas City, Missouri, in the early 1960's for Texaco Development Corporation.

It seemed quite clear from the alkylation process itself, and some of the attempts to extract the conjunct polymer from the used alkylation catalyst, that the polymers could not be ex-tracted from the strong sulfuric acid, such as used alkylation catalyst. However, there was no assurance that this would be the case when the acid became weak as a result of most of the acid being converted to dialkyl sulfate.

A number of different processes involving the absorption of olefin in recycle sulfuric acid alkylation catalyst, extraction of the dialkyl sulfate, treatment of the extract in some manner, and alkylation of the treated dialkyl sulfate have been con-sidered. One such process which has become known in the industry as SARP is the subject of this paper. SARP stands for Sulfuric Acid Recovery Process, which obviously has a rather generic con-notation. The process is rather specific. The writer prefers the name originally given to the process, namely, Extractylation. However, SARP has the advantage of being known and is short, so SARP will probably prevail.

The information marshalled and acquired in connection with the development of SARP has been quite helpful in getting a better understanding of H_2SO_4 alkylation itself. This is par-ticularly true for such areas as run-aways, poor lead suscep-tibility of alkylate, treating procedures for crude alkylate, the importance of low water content catalyst and what constitutes good mixing or agitation.

Experimental

From early work done for other purposes, it seemed certain that at least qualitatively olefins could be absorbed in recycle sulfuric acid alkylation catalyst, the resulting dialkyl sulfates extracted with isobutane and alkylated. Evidence as to what would happen to the conjunct polymer present in the recycle acid and the additional conjunct polymer formed in the absorption step when the acid phase of the absorption mixture was extracted with isobutane was missing. There was some concern that the conjunct polymer, or at least part of it, would also be extracted by the isobutane.

It was confirmed in the work at Midwest Research that propylene could be reacted with recycle alkylation acid to give a high yield of diisopropyl sulfate (DIPS), and that the DIPS could be extracted with hydrocarbon solvents, including isobutane. The isopropyl acid sulfate (IPS) is quite insoluble in hydrocarbons and only a small amount is extracted along with the DIPS. Since the reaction of propylene with sulfuric acid is an equilibrium reaction, some IPS is always present. It was found that some of the conjunct polymer is also extracted with the DIPS. It was anticipated that the water would stay in the acid phase or raffinate, and this was found to be the case.

It was fairly obvious that some method had to be discovered to remove the conjunct polymer from the isobutane-DIPS extract if it were to be charged to an alkylation unit. It was found that the polymer reacts quantitatively with strong sulfuric acid (19,21),such as fresh 98% acid or recycle alkylation catalyst. About an equal weight or 50% by volume of the acid in relation to the conjunct polymer is required. The acid-polymer complex forms easily and quickly merely by hand shaking in a centrifuge tube at room temperature. The complex is a heavy but fairly free flowing, viscous liquid at room temperature. It may be separated by gravity settling. It is insoluble in hydrocarbons such as isobutane, and soluble in H_2SO_4. DIPS apparently is not soluble in the complex. However, if excess acid is used, some DIPS will dissolve in the acid and acid complex. IPS is also soluble in the excess acid and acid complex.

Thus, the three step process of absorption, extraction and alkylation became a four step process of absorption, extraction, acid treatment of the extract, and alkylation.

It was concluded from the results of the work at Midwest Research and the later work on the removal of conjunct polymer from the extract with H_2SO_4 that all four steps were operable. An economic study indicated that the process would be attractive if about 90% of the acid in the used alkylation catalyst could be converted to DIPS, and about 90% of the DIPS extracted and alkylated. This amounted to an 80% recovery of the acid, and a corresponding reduction in acid consumption.

In order to obtain quantitative data on the individual steps

and overall yields, a brief pilot unit study was carried out by
Texaco Development Corporation and Stratford Engineering Corporation in Kansas City, Missouri. To expedite the study, available
Stratco Contactors were used for the absorption, acid treating
and alkylation steps. Using Texaco Development's solvent refining experience, a Rotating Disc Contactor (RDC) was used for
the extraction step.
   The SARP flow in simplified form is shown in Figure 1.
   The SARP flow as used in the pilot unit is shown in Figure 2.
and described in detail in U.S. Pat. 3,803,262 (23).
   In the following example, the feed stocks shown in Table 1
were employed in the apparatus of FIG. 2.

TABLE I

Feed Stocks

In weight percent

	Isobutane	Propane-propylene	Butane-butylene
Ethane		0.4	
Propylene		62.3	0.1
Propane	6.9	33.7	2.7
Isobutane	91.1	3.6	34.6
n-Butane	2.0		11.0
Isobutylene			15.9
Butylene-1			9.5
Butylene-2			22.9
Pentanes			3.3
Total	100.0	100.0	100.0

   10 cc. per minute of used alkylation acid from the settler
39 titrating 91.0% was charged to absorber 12 at about 25° F.
Fresh propane-propylene feed at the rate of 45.0 cc. per minute
was charged to absorber 11 at about 30° F. Unreacted propane-
propylene from absorber 11 separated in settler 23 was charged to
absorber 12. The hydrocarbon phase or unreacted propane-
propylene from absorber 12 separated in settler 14 comprised
11.8% dipropyl sulfate. The acid phase from absorber 11 separated in settler 23 comprising 80.6% dipropyl sulfate and 15.3%
propyl acid sulfate was passed at a rate of 24.2 cc. per minute
to a rotating disc contactor extractor 33 near the top. 315.0 cc.
per minute of the isobutane feed was charged to the rotating
disc contactor near the bottom. A temperature gradient was
maintained in the extractor, the bottom temperature being about
55° F and the top temperature about 65° F. Overhead from the extractor comprising 83.8% isobutane, 5.7 dipropyl sulfate, and

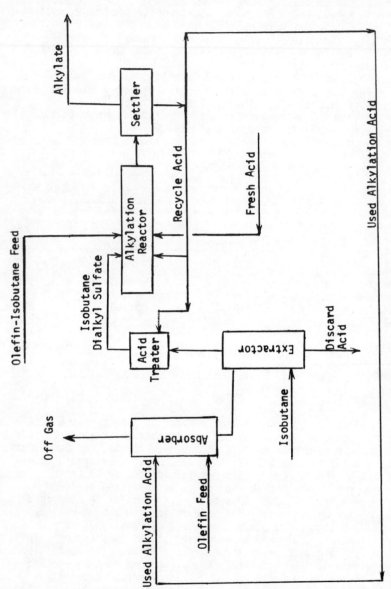

Figure 1.   Flowsheet for SARP (simplified)

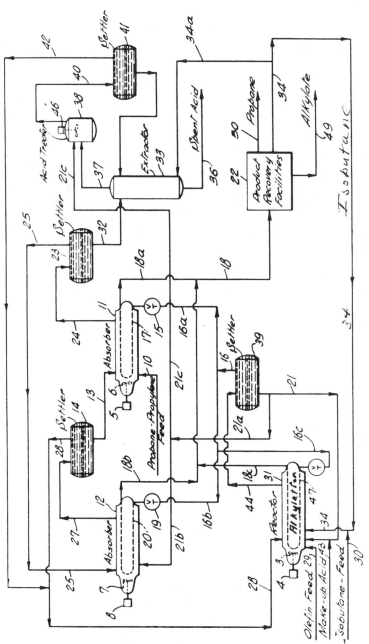

Figure 2.   Detailed flowsheet of SARP

0.3% propyl acid sulfate was passed to a 600 cc. mechanically agitated reactor at about 80° F along with 0.2 cc. per minute of used alkylation acid of 91.0% concentration. The acid treated product was passed to a settler 41 separating a hydrocarbon phase at the rate of 341.2 cc. per minute comprising 84.2% isobutane, 5.8% dipropyl sulfate and 0.1% propyl acid sulfate.

105.5 cc. per minute of the butane butylene feed in line 29, 315.4 cc. per minute of the isobutane feed in line 30, and 0.4 pound per hour of 97.5% sulfuric acid in line 43 were charged together with the hydrocarbon phases from settlers 14 and 41, recycle acid in line 21 and recycle isobutane in line 34 to a Stratco alkylation reactor 31 with efficient mixing held at 40 to 45° F. Reaction mixture was passed continously to a settler 39. The hydrocarbon phase produced at a rate of 749 cc. per minute was caustic and water washed and stablized to remove most of the butane and lighter hydrocarbon producing approximately 50 gallons per day of desired alkylate product.

Spent acid from the acid treater 38 was produced at the rate of 1.3 cc. per minute.

The spent acid from extractor 33 at the rate of 3.8 cc. per minute corresponds to an acid consumption of 0.18 pound per gallon of alkylate. A corresponding control run without acid recovery gave an acid consumption of 0.88 pound per gallon of alkylate.

Approximately 90% of the used alkylation acid was converted to dipropyl sulfate and approximately 90% of the dipropyl sulfate formed was sent back to alkylation wherein the sulfuric acid was regenerated or recovered. Overall, approximately 80% of the used alkylation acid was recovered. The reactions of sulfuric acid with propylene is an equilibrium reaction, and by the method of operation described it is not possible to convert all of the acid to extractable dialkyl sulfates; some will remain as alkyl acid sulfate. In addition, with the extraction conditions used it is not possible to extract all of the dialkyl sulfates formed.

The spent acid from extractor 33 had an analysis of approximately 2.5% water, 10% acid-polymeric oil complex (probably actually considerably higher), 10% dipropyl sulfate and 77.5% propyl acid sulfate.

The research octane of the total alkylate from the acid recovery run was 95.5 clear and 107.0 with 3.0 cc. of Tel. The motor octane was 92.6 clear and 106.8 with 3 cc. Tel. The alkylate from the control run without acid recovery had a research octane clear of 94.7 and 105.9 with 3.0 cc. of Tel.

After a little experience, the operation of the pilot unit was smooth and easy. In general the operation was stable and easier to control than in a conventional alkylation unit. In one run some water was accidentally introduced into the alkylation system. The system acid rapidly dropped to about 80% titratable acidity without causing a run-away (24), as would have been the case in conventional alkylation.

The results of the pilot unit study were about as expected
and predicted. However, the alkyl sulfates were much more stable
to time and temperature than anticipated. As a result, there was
considerable latitude for the conditions in the extraction and
acid treating steps.

A commercial unit was designed based on the pilot unit results. The design was essentially the same as in the pilot unit.
A plant was operated by Cities Service for several years. Some
years ago the entire refinery where the SARP unit was located was
discontinued. As was the case with the pilot unit, the SARP unit
was easy to operate, and in general a steadier operation was
obtained. Upsets and poor operation of units not directly a part
of the recovery section, such as the feed splitter, butane isomerization unit, and deisobutanizer were not as serious when SARP
was in operation. The general feeling was that an octane increase was obtained, as well as a demonstrated two thirds to
three fourths reduction in acid consumption. Sufficiently precise runs under stable conditions were not made in the pilot unit
or in either of two commercial plants to be certain that an increase in octane was obtained.

Up until some time after the commercial operation of SARP in
1966, it was assumed that the discard SARP acid would or could be
disposed of in the same manner as discard alkylation acid,
namely, by sending it back for credit to the acid supplier. This
was based on information from acid suppliers after samples of
acid had been sent to them. Later, it developed that although
the acid could be recovered in a sludge conversion unit, it decreased the capacity of the unit by about 50%, because of the
high hydrocarbon content of the acid. As a result, the discard
SARP acid as far as the acid supplier was concerned, had a minor
or negative value as a source of sulfur for a sludge conversion
unit.

It was found that a second SARP commercial unit of about the
same design as the first one, although considerably larger, was
also easy to operate, and in general gave little trouble. Using
an overall olefin feed of about 35% propylene and 65% butylenes,
the acid consumption was reduced to about one third. The research
octane value of the total alkylate of about 375^0 F.E.P. was about
94 clear and 105 with 3 cc. TEL. The water content on the SARP
discard acid was about 1.9% whereas without SARP it was about
4-5%, and in each case with make-up acid of 99.0-99.5%.

Originally, it was thought that quite simple and low cost
equipment could be and would be used for SARP commercially. However, more sophisticated and expensive equipment was available for
the pilot unit, so it was used. To be as sure as possible about
the scale-up, the same type of equipment was used commercially.
As a result, the cost of the recovery section (absorption, extraction and acid treatment of extract) was considerably higher
than originally projected. This became even more pronounced when
alloys had to be used for certain parts of the equipment to take

care of corrosion. With this as background, and in addition, the relatively low cost of fresh sulfuric acid and low value of discard acid, it became apparent along about 1969 that SARP could not be a success commercially just on acid savings alone, especially with fresh acid at about $20.00 per ton. SARP would have to result in other savings, such as in a higher octane product, or lower operating costs.

It should be made clear that technically from a chemical and chemical engineering viewpoint the process is a success. Each of the three steps of the recovery section works quite well, and the indications are that the operation of the alkylation section is more stable and improved because of the recovery section.

Because of the above types of factors, and the uncertainty regarding the role of Government in fuels, SARP has been dormant for several years. Alkylation also has been rather quiet. However, with the phasing out of lead in gasoline, and increased pressure for less polution from fuels, it looks as if alkylation may have yet another surge in production. And, with the rising cost of sulfuric acid, and the increasing need for high octane, clear alkylate at a reduced cost, perhaps SARP may also have a renewal of interest. Problems in SARP with suggested solutions and areas for improvement may perhaps be discussed in a later paper. A brief discussion of the four steps of SARP or Extractylation will be given.

Absorption. In the absorption step as much as possible of the olefin should be converted to dialkyl sulfate. The dialkyl sulfates are quite soluble in hydrocarbons above about 40° F, whereas the alkyl acid sulfates are not. Propylene is the preferred olefin, and n-butylenes may be used. Propylene has been used commercially. Although an exhaustive study was not made, the use of a butylene feed containing isobutylene gave poor results. Isobutylene results in a considerable loss of acid and also isobutylene. About 10-25% of the total olefin used in SARP is reacted in the absorber, assuming a net acid consumption of about 0.2-0.5%. The exact amount will depend on the net acid consumption or the amount of fresh make-up acid charged to the alkylation section.

The acid charge to the absorber is the used sulfuric acid catalyst from the alkylation settler. It usually is of about 90% titratable acidity with about 3-5% water and 3-6% conjunct polymer. The alkylation catalyst is no longer active as an alkylation catalyst when the acidity drops to about 80% due to the pick up of conjunct polymer and alkyl sulfates. However, it is still active for the absorption of propylene until substantially all of the acid is converted to alkyl sulfates. Since the formation of alkyl sulfates from olefins and sulfuric acid are equilibrium reactions, conditions should be used which will shift the equilibrium to the right, or to the dialkyl sulfate, with minimum formation of alkyl acid sulfate.

$$C_3H_6 + H_2SO_4 \underset{\leftarrow}{\rightarrow} C_3H_7HSO_4$$

$$C_3H_7HSO_4 + C_3H_6 \underset{\leftarrow}{\rightarrow} (C_3H_7)_2SO_4$$

$$2C_3H_6 + H_2SO_4 \underset{\leftarrow}{\rightarrow} (C_3H_7)_2SO_4$$

As may be seen in Figure 2, the Stratco Contactors were operated so as to get part of the advantage of countercurrent flow. The contactors give good mixing and contact of the olefin feed with the acid. Intimate mixing of the olefin feed and acid is desirable. An excess of olefin, low water content of acid, and removal of dialkyl sulfate from the acid phase of the absorption reaction mixture favor a high conversion of the acid to dialkyl sulfate. This probably can be accomplished best in a truly countercurrent operation in liquid phase, such as in a tower.

It will be noted from the flow of Figure 2 that the un-reacted olefin stream from the weaker acid absorber 11 is sent to the stronger acid absorber 12. Then the unreacted olefin stream from stronger acid absorber 12 is sent directly to alkylation. This flow under the proper conditions probably insures that no conjunct polymer, or very little, is sent to alkylation in the un-reacted olefin stream. It also enables all of the olefin not re-acted in the absorption step to be utilized in alkylation. However, the unreacted olefin stream from weaker acid absorber 11 may be substantially saturated with DIPS, and when this stream is sent to stronger acid absorber 12, it tends to retard the extent of conversion of acid and/or acid propyl sulfate to DIPS. Ignor-ing the possibility that the unreacted olefin stream from weaker acid absorber 11 may contain conjunct polymer, and just consider-ing the conversion of acid to DIPS, it would be better to send the unreacted olefin directly to alkylation.

Either liquid phase or vapor phase absorption, or a com-bination of the two may be used. Liquid phase was used commer-cially.

As in alkylation, the time required for the absorption step is determined largely by the cooling required and the efficiency of the contacting. The two contactors used in the pilot unit gave a contact or residence time of about one hour in each contactor, which is probably far in excess of what is needed.

A temperature of 35-40° F was used commercially. In some later pilot unit work better results were obtained at 50° F. However, the solubility of DIPS in hydrocarbons is much higher at 50° F than at 35° F, and this may account for a higher conversion at 50° F.

Conjunct polymerization should be kept to a minimum, as it represents a loss of olefin and acid. Retardation of conjunct polymerization is favored by a low temperature, as in alkylation, especially in that portion of the absorber with the strongest

acid, such as the upper portion of a countercurrent tower. A
short residence time and efficient mixing probably also favor
minimum conjunct polymerization.

Usually there is no difficulty in separating the acid and
hydrocarbon phases in a gravity settler, especially if the temperature of the settler is controlled. If any difficulty is encountered, it may be solved by adding isobutane to the absorption
mixture, or by cooling the absorption mixture to about 40-50° F.
If cooling is not provided in the settler, the temperature may
rise, apparently due to continued reaction, as often is the case
in the alkylation settlers in propylene alkylation.

Extraction. Solvent extraction processes are usually fairly
expensive because of the cost in recovering the solvent, such as
by heating or chilling, and usually with some loss of solvent. In
SARP we have a unique and quite favorable situation. The solvent
is a recycle stream of isobutane from the alkylation section,
which, after it is used to extract DIPS, is returned along with
the DIPS to alkylation, where the isobutane is processed as usual
in alkylation.

As will be noted in Figure 2, isobutane from the product
recovery facilities or deisobutanizer was used commercially. This
is a very large stream and is adequate in quantity.

Countercurrent operation and efficient contacting within the
extractor, a high solvent dosage, and a temperature above about
40° F favor complete extraction of the dialkyl sulfate. DIPS increases in solubility in isobutane with increasing temperature.
A temperature of about 50° F was used commercially with a Rotating
Disc Contactor (RDC). The optimum RPM of the extractor was not
determined. The volume ratio of isobutane to DIPS should be 6 or
higher. A high solvent dosage also favors a good separation of
the extract from the reaffinate in the settler. Although the DIPS
seems quite stable in isobutane solution, it is probably advantageous to operate at about 50° F rather than at about 100° F.

The aim or ideal in the extraction step is to remove all of
the available H_2SO_4 in the used alkylation catalyst as DIPS and
IPS, and to leave all of the conjunct polymer and water in the
raffinate. Just how near the ideal is realized depends on the
efficiency of the absorption step, as well as the extraction step.
It was originally anticipated that substantially all of the water
would stay in the raffinate, and no evidence or data have been
developed to indicate otherwise. In general, a low water content
and a low conversion of acid to DIPS favor retention of the polymer in the raffinate. A relatively high water content and a high
conversion of the acid to DIPS favor removal of polymer in the
extract.

The raffinate or spent acid from the extraction step comprises water, alkyl acid sulfate, dialkyl sulfate, and the conjunct polymers (probably associated with acid) formed in the
alkylation, absorption and acid treating steps. The extract comprises the isobutane solvent, dialkyl sulfate, a relatively small

amount of alkyl acid sulfate, and part of the conjunct polymer.
The composition of the SARP reject acid or extractor bottoms
will vary depending on a lot of factors, but for most of the work
with less than optimum conditions in the absorption, extraction
and acid treating steps, a fairly typical analysis is shown in
Table 2, below. For orientation purposes an analysis of used
alkylation acid is also shown.

TABLE II

SARP Reject Acid

Weight Percent

	Alkylation Acid	SARP Acid
Oil (conjunct polymer)	6.0	15.0
Water	3.0	4.6
Sulfuric acid	91.0	28.0
Isopropyl acid sulfate	0.0	44.8
Diisopropyl sulfate	0.0	7.6
Total	100.0	100.0
Sulfur	29.7	20.7
Hydrocarbon in acid-oil complex and alkyl sulfates	6.0	31.9
Equivalent H_2SO_4	91.0	63.5

There is some question as to whether there is actually any
free sulfuric acid in the SARP acid. Probably in most cases
there isn't and preferably there should not be any. In most cases
it is believed that approximately an equal weight of acid and
conjunct polymer are loosely tied up as a complex. Under more
nearly optimum conditions in all four steps of the process, and
after equilibrium or steady state conditions have been reached,
the water content probably would be nearer 2%, or less, and the
amount of alkyl sulfates would be considerably lower. The ideal
or ultimate would be just conjunct polymer and water. It should
be noted that the hydrocarbon content of the SARP acid is much
higher than that of alkylation acid.

Acid Treatment. Under most conditions, some of the conjunct
polymer is removed by the isobutane in the extraction step. Just
how much and whether it will be removed depends on the specific
composition of the absorption mixture and the extraction conditions. In the pilot unit, relatively weak make-up acid of 97.5%
was charged to the alkylation step, so the recycle acid charged
to the absorber was relatively high in water. If fresh make-up
acid of 99.5% concentration, the usual or better drying of alkylation streams, and a strictly countercurrent absorber were

used, it might very well not be necessary to acid treat the
isobutane-DIPS extract.

Commercially the extract was treated with used alkylation
catalyst to remove conjunct polymer, since it was anticipated that
quite likely under the conditions used some conjunct polymer would
be in the extract. It was shown conclusively in a pilot unit that
when the extract was not treated, the fresh make-up acid or acid
consumption in alkylation with SARP was no lower than without SARP.

Fortunately, used or discard alkylation catalyst is satisfactory for the acid treatment, and only a small amount is required. Usually about 0.05 pound per gallon of alkylate was arbitrarily used commercially. Since operating conditions in commercial operation vary from day to day, and analytical tests are
often slow and infrequent, an excess of acid was usually used. As
indicated earlier, an excess of acid causes solution of DIPS in
the acid. Thus, to recover the DIPS in the excess acid, the raffinate or acid phase from the acid treatment step was sent to the
extractor.

In refineries or areas in which two or more alkylation units
are operated, one SARP unit large enough to process the used alkylation acid from all of the alkylation units could be installed.
The recovered acid in the form of isobutane-DIPS extract could be
sent to the various alkylation units as needed or desired. Such a
larger SARP unit would not only have a lower cost per unit of capacity, but would also have additional advantages if further processing of the SARP reject acid were carried out prior to conventional recovery.

Alkylation. In general the conditions for alkylation are, or
may be, the same with or without SARP. A good discussion of the
conditions and variables in alkylation is given in the paper by
Mr. Orlando Webb (25) scheduled for this Symposium. Only those
conditions which have some unique or peculiar relation to SARP
will be discussed herein.

About three fourths of the make-up acid charged to alkylation
is in the form of alkyl sulfates, and only a minor portion or
about one fourth of the usual 99.5% concentration. The catalyst
is, or may be, somewhat different with SARP, especially as to its
water and conjunct polymer content. Without SARP the catalyst
usually contains about 2.5-5.0% water. With SARP the water content should usually be in the 0.5-2.0% range. A low water content
of the catalyst is believed to be desirable (26). If less drying
of feed stocks in SARP were practiced, the water content of the
catalyst might be as high as without SARP.

A large excess of isobutane is used, e.g., as much as 60-80%
by volume of the hydrocarbons in the alkylation reaction mixture.
Consequently, a large quantity of isobutane must be recovered and
recycled. It is also available for SARP. The isobutane concentration is one of the most important variables, as far as
quality of product and acid consumption are concerned. A high
concentration of isobutane tends to promote alkylation and

suppress conjunct polymerization. The isobutane concentration can be increased by adding more isobutane, or by reducing the amount of inerts, such as propane and n-butane.

In addition to the particular design of the alkylation reactor and the amount of horsepower going into the mixing, the mixing and alkylation results are usually improved by having 50% or more by volume of acid catalyst in the emulsion so that the acid is in the continuous phase. Recycle of emulsion containing 20-35% hydrocarbon rather than flat acid from the settler to the reactor also helps the mixing and gives improved results, as would be expected. This is especially true at high through put or space velocity*, e.g., 0.5 or higher, or with high propylene content feeds. It will be recognized that recycle of emulsion also is in effect a way of returning isobutane to the alkylation reactor without resorting to fractionation.

Lower temperatures down to about 30^0 F favor lower acid consumption and higher quality product. Most commercial units are operated in the 40-50^0 F, with an increased acid consumption. This isn't so serious if the used catalyst from alkylation is sent to the SARP recovery section. In some cases, the size of the reactor is determined by the amount of cooling surface required for the design temperature. With SARP not as much cooling is required in the alkylation reactor, since part of the heat of the alkylation reaction is evolved in the absorption section. It has been estimated that 50% or more of the heat of alkylation is in the absorption step.

Efficient mixing of the hydrocarbon and acid is desirable to keep the acid catalyst as nearly saturated as possible with isobutane, and in as high a concentration as possible. The solubility of isobutane in the acid catalyst increases with decreased water content (26) and increased polymer content. Thus, it should be possible to have a higher concentration of isobutane in the SARP alkylation catalyst for a given titratable acidity, and it should be possible to maintain a higher concentration of isobutane in the catalyst.

Based on a study of commercial plants some years ago, it was concluded that often the mixing was quite poor in that the emulsion was not uniform. There was usually a small amount of good emulsion and a lot of poor emulsion. A major portion of the reaction mixture was loose, fast breaking emulsion, and even large amounts of unemulsified hydrocarbon were present. The design and operation in many cases are still about the same today, and hence the mixing and emulsion probably are too. Although proof is not available, it is believed the use of the isobutane-DIPS stream as part of the hydrocarbon feed in SARP helps improve the mixing. The propylene represented by the DIPS amounts to

*Gallons per hour of olefin in feed divided by gallons of acid in alkylation reactor.

10-25% of the total olefin feed to alkylation. The rest of the olefin feed may be propylene, butylenes, amylenes or polymers. The acid represented by the DIPS amounts to about 65-75% of the total make-up acid.

The alkylation reactors used in the pilot unit operation and also the commercial operation were Stratco Contactors. However, the writer knows of no reason, and no evidence is available, to indicate that any other type reactor, such as the cascade and pump and time tank, would not be satisfactory. It is believed that they would give at least as good results with SARP as without SARP for the same operating conditions and olefin feed stock, and of course with a much lower acid consumption.

Summary and Conclusions

A process termed SARP has been developed for the recovery of the used or discarded catalyst from the alkylation of isobutane with olefins using strong sulfuric acid as the catalyst. The process was used commercially for several years by two companies, but currently is not in commercial operation.

The used catalyst is reacted with propylene to convert the acid to diisopropyl sulfate. The diisopropyl sulfate is extracted with isobutane, the extract treated with a small amount of used catalyst, leaving a weak acid containing the conjunct polymer and water. The isobutane extract free of conjunct polymer and water is charged to alkylation, along with additional olefin and fresh make-up acid. A discussion of the reaction conditions required and the variables involved in the four steps of the process are given.

With SARP the net acid consumption is reduced by about 65 to 75%, or to about 0.2 of a pound of sulfuric acid per gallon of alkylate. The alkylate product is of about the same quality, or perhaps somewhat better, as alkylate produced without SARP. A 375° F.E.P. total alkylate had a clear octane of about 94 and 105 with 3.0 cc. of TEL.

Literature Cited

1. Goldsby, A. R., U.S. Pat. 2,267,458 (Dec. 23, 1941).
2. Stiles, R. S., This Book, Chapter 18, 1977.
3. Goldsby, A. R., U.S. Pat. 2,420,369 (May 13, 1947).
4. Goldsby, A. R. and VanGundy, J. C., U.S. Pat. Re 22,146 (July 28, 1942).
5. Goldsby, A. R. and Putney, D. H., Pet. Ref., Sept. 1955.
6. Goldsby, A. R., U.S. Pat. 3,422,164 (Jan. 14, 1969).
7. Goldsby, A. R., U.S. Pat. 3,428,705 (Feb. 18, 1969).
8. Goldsby, A. R., U.S. Pat. 3,448,168 (June 3, 1969).
9. Goldsby, A. R., U.S. Pat. 3,691,252 (Sept. 12, 1972).
10. Goldsby, A. R., U.S. Pat. 3,742,081 (June 26, 1973).
11. Massa, H. E., U.S. Pat. 3,442,972 (May 6, 1969).

12. Goldsby, A. R. and Pevere, E. F., U.S. Pat. Re. 22,510 (July 4, 1944).
13. Goldsby, A. R., Pevere, E. F., Clarke, L. A. and Hatch, G. C., U.S. Pat. 2,348,467 (May 9, 1944).
14. Goldsby, A. R. and Clarke, L. A., U.S. Pat. 3,000,991 (Sept. 19, 1961).
15. Goldsby, A. R. and Gross, H. H., U.S. Pat. 3,083,247 (March 26, 1963).
16. DeJong, U.S. Pat. 2,381,041 (1945).
17. Hydrocarbon Processing and Petroleum Refiner 14, (1965).
18. Oil and Gas Journal, Jan. 2, 48 (1967).
19. Goldsby, A. R., U.S. Pat. 3,227,774 (Jan. 4, 1966).
20. Goldsby, A. R., U.S. Pat. 3,227,775 (Jan. 4, 1966).
21. Goldsby, A. R., U.S. Pat. 3,234,301 (Feb. 9, 1966).
22. Oil and Gas Journal, Jan. 2, 54 (1967).
23. Goldsby, A.R., U.S. Pat. 3,803,262 (April 9, 1974).
24. Goldsby, A. R., U.S. Pat. 3,683,041 (Aug. 8, 1972).
25. Webb, O., This Book, Chapter 16, 1977.
26. Albright, L. F., Houle, H., Sumutka, A. M. and Eckert, R. E., Ind. Eng. Chem. Proc. Dev. 11, 446, No. 3 (1972).

18

Recovery of Sulfuric Acid Alkylating Catalyst by Crystallization

S. ROBERT STILES

Westinair Associates, P. O. Box 2091, Morristown, NJ 07960

Alkylation of isobutane with olefin is accomplished by the carbonium ion mechanism when the sulfuric acid catalyst purity is maintained at a high level — generally above 90-92% titratable acidity. At lower concentrations olefin polymerization to the dimer is accelerated and alkylation to tri-methyl-pentanes retarded. Alkylate product quality is effected.

The free uncombined sulfuric acid+ester+water composition of a typical alkylation acid, present in a commercial alkylation reactor, is illustrated by the check (✓) point data plotted in the three component diagram of Figure 1. Water content ranges between 2-3% and ester diluents between 5-9%. Normally a continuous purge stream of reactor acid is withdrawn, processed through an acid plant for rejection of hydrocarbon esters, and the regenerated acid at 98.5-99.5% H_2SO_4 purity returned to the reactor to control the catalyst acidity within this range.

Alkylation catalyst activity can be maintained more economically by concentrating these esters to 30-40% content level in a small purge stream, prior to being processed through the acid regeneration system. Savings in acid regeneration and alkylate processing costs that result are illustrated by Figure 2.

To accomplish these objectives recycle alky-acid is chilled, to freeze and extract 100% H_2SO_4 crystals from the purge acid, according to procedures patented by Stiles, Skelly and Felter (1-6). The sulfuric acid is returned to the alkylation reactor and the quantity of ester concentrate purge is reduced to approximately 1/6th of its normal quantity. Operating data obtained from a commercial installation is reported by Ø point composition data plotted in Figure 1.

Discussion

During alkylation a small quantity of reactants and impurities remain trapped in the acid phase as acid soluble esters. The term ester, as employed here, refers to any compound formed by the replacement of at least one of the acid-hydrogen by a hydrocarbon

302

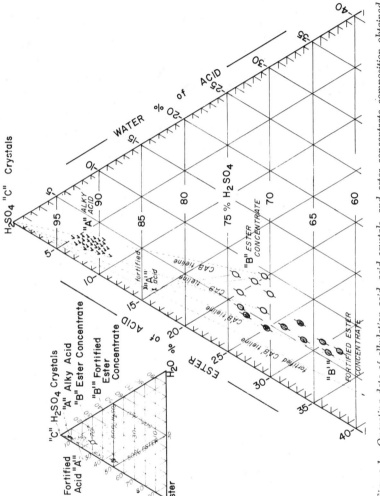

Figure 1. Operating data of alkylation acid, acid crystals, and ester concentrate composition obtained in an alkylation unit with acid crystallization

radical. Hydrocarbon radicals may also contain sulfur, oxygen and/or nitrogen often found in feed contaminants, such as amines, carbonyl sulfide, carbamates, thio-carbamates and mercaptides.

Being polar, esters are preferentially soluble in the acid phase and accumulate to dilute or contaminate the acid; and being a high molecular weight diluent, they have a temperature depressant effect on the freezing point of the acid liquid phase.

Generally these esters exist only in acids at low water content, and readily dissociate or hydrolyze when the water concentration is increased. Mono-esters with the HC-radical replacing only one hydrogen illustrated by:

$$C_4H_8 \ + \ H_2SO_4 \ \longrightarrow \ (C_4H_9)HSO_4$$

dissociate readily at relatively low temperatures, whereas the dimer-ester illustrated by the equation:

$$C_4H_8 \ + \ (C_4H_9)HSO_4 \ \longrightarrow \ (C_4H_9)_2SO_4$$

requires higher temperatures and longer time for dissociation to occur. Trimer-esters, often called trialkyl esters, are formed from higher molecular weight radicals such as:

$$C_4H_8 \ + \ (C_4H_9)_2SO_4 \ \longrightarrow \ (C_8H_{17})SO_4(C_4H_9)$$

These trimer-esters are extremely slow to dissociate and require elevated temperatures at boiling-acid conditions to complete the reactions. Tars are produced with release of sulfur dioxide.

Differences in hydrolysis characteristics are used in analytical test procedures to identify the easily dissociated esters by polymer and titratable acidity, and the less reactive by total acidity procedures. Total hydrocarbon content is determined by carbon analysis, and uncombined H_2SO_4 by the aniline sulfate procedure.

Hydrolysis has been used in some acid regeneration systems but the accumulation of excessive amounts of trimer-esters and tars have an adverse effect on alkylate quality.

Regeneration by combustion to totally decompose the acid and esters to CO_2, SO_2 and H_2O is generally preferred. This decomposition is performed in an acid plant where the resulting SO_2 is then catalytically converted to SO_3 and regenerated to the 98.5-99.5% H_2SO_4 that is returned to the reactor system.

Alkylation and Ester Reaction Mechanism

It is well known (7,8,9) that on start-up with fresh acid, the sulfuric acid must be "conditioned" by adding the acid rapidly to circulating isobutane containing less than 2% olefins to generate the carbonium-ion. Mixing acid with olefin-free isobutane will not generate this ion. However if the acid is added too slowly, to be overwhelmed by the olefin, production of "red-oil" results. It is also known that olefin feed can be stopped at any time without causing any problems, but interruption of isobutane recycle

flow — without immediate cut-off of olefin feed flow — will deteriorate the alkylate quality which, if carried on for any period
of time, will result in a disastrous "acid run-away". A reactor
system can be kept on "stand-by" conditions for several days or
weeks, as long as isobutane recirculation is maintained. Alkylate
is slowly but continuously released from the acid phase into the
isobutane liquid during such operation and the catalyst "acidity"
increases.

Alkylation, ester formation, polymerization, cracking, isomerization and other refinery operations depend on the carbonium-ion.
Whitmore (10) suggests that the carbonium-ion is formed by the addition of hydrogen ion, from an acid, to an olefin double bond:

$$H_3C\text{-}\overset{\overset{\displaystyle CH_3}{|}}{C}\text{=}CH_2 \; + \; H^+ \; + \; X^- \; \rightleftharpoons \; H_3C\text{-}\overset{\overset{\displaystyle CH_3}{|}}{\underset{+}{C}}\text{-}CH_3 \; + \; X^-$$

which then performs the basic refinery reaction:

1. Addition of negative ion:

$$R\text{-}\overset{+}{C}\text{-}C \; + \; X^- \; \rightleftharpoons \; R\text{-}\overset{\underset{\displaystyle X}{|}}{C}\text{-}C$$

2. Elimination of proton to produce olefin:

$$C\text{-}\overset{+}{C}\text{-}C \; \rightleftharpoons \; C\text{-}C\text{=}C \; + \; H^+$$

3. Migration of proton (order of stability: tertiary - secondary -
primary):

$$C\text{-}\overset{\overset{\displaystyle C}{|}}{C}\text{-}\overset{+}{C}\text{-}C \; \rightleftharpoons \; C\text{-}\overset{\overset{\displaystyle C}{|}}{\underset{+}{C}}\text{-}C\text{-}C$$

4. Migration of methyl group:

$$C\text{-}\overset{\overset{\displaystyle C}{|}}{C}\text{-}\overset{+}{C}\text{-}C\text{-}C \; \rightleftharpoons \; C\text{-}\overset{+}{C}\text{-}\overset{\overset{\displaystyle C}{|}}{C}\text{-}C\text{-}C$$

5. Addition of olefin; reverse reaction occurs by scission, two
carbon atoms removed from charge:

$$C\text{-}\overset{+}{C}\text{-}C \; + \; C\text{=}C\text{-}C \; \rightleftharpoons \; C\text{-}\overset{\overset{\displaystyle C}{|}}{C}\text{-}C\text{-}\overset{+}{C}\text{-}C$$

6. Hydrogen transfer reaction with tertiary hydrogen:

$$C\text{-}\overset{\overset{\displaystyle C}{|}}{\underset{\underset{\displaystyle C}{|}}{C}} \; + \; C\text{-}\overset{+}{C}\text{-}C \; \longrightarrow \; C\text{-}\overset{\overset{\displaystyle C}{|}}{\underset{\underset{\displaystyle C}{|}}{C}}{+} \; + \; C\text{-}C\text{-}C$$

Any non-reversible reaction that terminates the equilibria by
forming dimer or trimer esters dilute the acid and hinder the reaction of the carbonium ion.

Ideal reaction conditions of high isobutane concentration, and
isobutane diffusion into the acid phase at a rate greater than the
olefin diffusion rate, result in high quality alkylate product.
Ester formation, exclusive of those formed from feed contaminants,

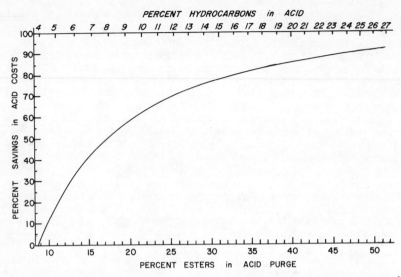

Figure 2. *Savings in alkylation acid processing costs resulting from operation of ester concentration system*

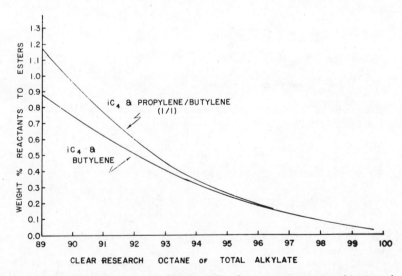

Figure 3. *Ester production from olefin and isobutane reactants as a function of reaction conditions related to alkylate quality as measured by Research Octane Number*

is nil. As reaction conditions deteriorate and alkylate quality
decreases, the rate of dimer and trimer ester formation increases.
Ester production, relative to alkylate quality, is shown by Figure
3 & 4. Generally ester production from feed contaminants is more
difficult to predict. Data obtained from commercial operations and
plotted in Figure 4, show the combined effect of ester production
from these two sources.

Yield of trimer or trialkyl esters oscillates as data of Figure 5 indicates. They are difficult to analyze and do not show in
the acidity tests. However their effect on acid make-up requirements is appreciable. Variations in "free" H_2SO_4 content of acid
are plotted as dotted lines in the top of this figure and show oscillations that are typical of most commercial operations. An
ester concentrator that can extract these esters as they are generated to remove them from the reactor system, will accomplish a
valuable service. Smoother operation producing better quality alkylate will result.

Ester Concentrator - Process Description

In this process as shown by Figure 6, a portion of the Alkylation Reactor acid recycle, that is separated from the isobutane-
rich alky-reactor effluent, is chilled by direct contact with countercurrent flowing isobutane liquid in an agitated chiller vessal
or crystallizer. Crystals of H_2SO_4 are formed. Referring to data
points in Figure 1, we have the alky-acid of composition "A" in the
area of the check (✔) point data, chilled to form crystals of composition "C" that are essentially 100% H_2SO_4. As these crystals
form, the free H_2SO_4 content of the liquid is reduced and its ester
plus water content increases along tie-line "CAB". This mixture of
crystals in alky-acid liquid flows downward thru colder upflowing
isobutane liquid and more crystals are formed. The freezing point
of the liquid is lowered as its ester plus water concentration increases, until at point "B" the freeze-point temperature equals the
temperature of the incoming chilled isobutane liquid and no further
crystallization occurs.

Crystallization is performed in isobutane liquid to remove
heat of fusion, but more important - to maintain a high isobutane
concentration, so that-as the ester and water concentration is increased, isobutane will continue to react with the mono ester.
The alkylate released from the acid phase, will be absorbed by the
isobutane liquid and removed from the crystallizer. Without this
removal of alkylate, crystal size growth control is difficult. A
portion of the alkylation reactor auto-refrigerant recycle from
the bottom of the depropanizer tower is used for this purpose.
This stream is free of olefins and moisture and is chilled by exchange with evaporating propane before entering the bottom of the
crystallizer.

A portion of this chilled isobutane liquid flows upward to
chill the downflowing acid, remove the heat of fusion and convey
the alkylate back to the alkylation reactor. The remainder conveys
the acid crystals plus ester concentrate mixture to the centrifugal

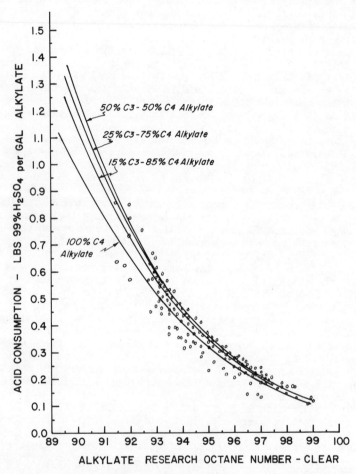

Figure 4.   Sulfuric acid consumption rate related to alkylate production and quality

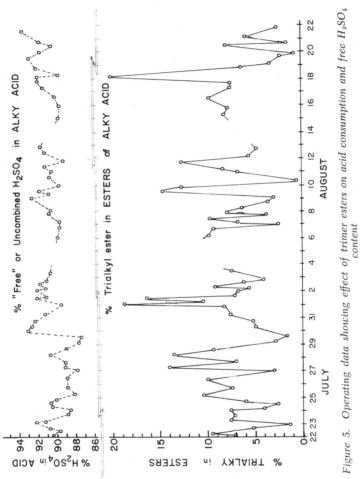

Figure 5. Operating data showing effect of trimer esters on acid consumption and free H_2SO_4 content

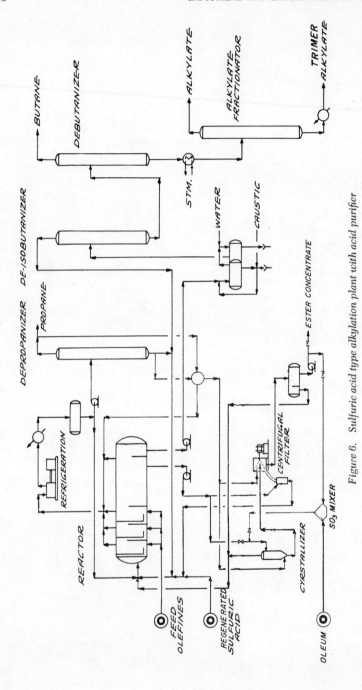

Figure 6.   Sulfuric acid type alkylation plant with acid purifier

filter. Here the acid crystals "C" are separated from ester concentrate "B" as the slurry enters the apex of the spinning conical
screening centrifuge. Crystals larger than the screen openings are
retained on the screen, while the ester concentrate and isobutane
liquid, containing smaller crystals, pass thru the screen and flow
into the decanter vessel. Isobutane liquid separate from the ester
concentrate and is returned to the alkylation reactor.

Crystals "C" slide across the surface of the conical screen
propelled by the angular force resulting from centrifugal force and
the increasing diameter of the spinning conical screen. The angle
of the cone is selected to allow crystal slippage while subjecting
the crystals to centrifugal force to remove adherent mother liquor.
Continuous flushing of isobutane is maintained to assist in this
liquid-solid separation. As the crystals reach the outer-diameter
of the spinning cone, they are discharged into acid recycle flowing
thru the crystal receiver and are returned to the alky reactor.

Ester concentrate "B" that decants from the isobutane, is sent
to acid regeneration facilities for removal of hydrocarbons and
water. The regenerated 98.5-99.5% H_2SO_4 is returned to the reactor.
A portion of the ester concentrator is continuously recirculated
from the decanter to the crystallizer inlet and mixed with incoming
alky-acid, to provide "seed" crystals of the small crystals that
had passed thru the screen.

Freezing point temperature of ester concentrate is lowered by
depressant effect of both esters and water. Higher concentration
of esters "B'" can be obtained at a given crystallization temperature, if the water content is reduced. Closed ∮ data points of
Figure 1 show results of operations where oleum was mixed with alky
acid, reacted with water:

$$SO_3 + H_2O \longrightarrow H_2SO_4$$

so that the resulting concentrate, obtained after subsequent crystallization and separation of H_2SO_4 crystals, had a higher ester
content. Complete reaction must be accomplished before chilling
since oleum freezes at crystallization temperatures, see Figure 7.

Water reaction with oleum maybe accomplished with incoming
acid or with recycle acid-ester concentrate as shown in Figure 6.
Warmer crystallization temperatures can be used to obtain a given
ester concentrate level, if the system is operated on "block-flow"
or intermittent procedure. In this procedure ester concentrate
collected from alky acid is accumulated in the decanter for a time
without discharging any to the acid plant. Flow of alky-acid is
then stopped and the accumulated inventory run-off via the oleum
contactor with H_2SO_4 crystals returned to the reactor in normal
manner. For example ester concentrate of composition "B" obtained
from alky acid "A" is fortified to composition "A'" and crystals "C"
separated, to result in ester concentrate "B'" that is then sent to
acid regeneration. A smaller quantity of concentrate results and
processing costs are reduced.

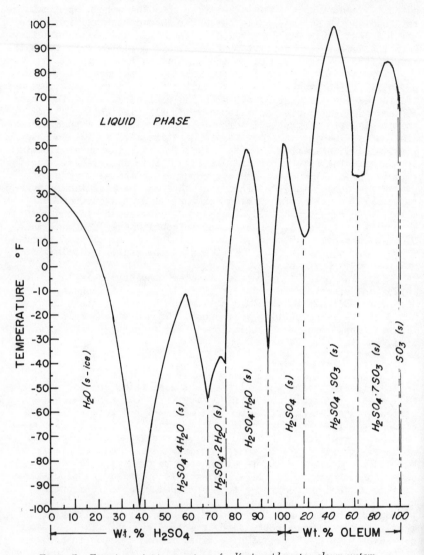

Figure 7.   Freezing point temperature of sulfuric acid–water–oleum system

CONCLUSION

Today durable laminated 200 mesh screens are available that can operate in conical screening centrifuges with excellent service factor to retain crystals as small as 75 microns. The centrifuge available in the earlier plant required a more rugged 40 mesh sieve that could only retain crystals larger than 400 microns. Greatly improved yield of crystals per pass result. This factor combined with a crystallizer design that controls alkylate release more efficiently, greatly improves the performance and economics of the process.

Considerable reduction in acid requirements and alkylation production cost are obtained by ester concentration, as shown by Figure 2. The operation of the earlier plant demonstrated the process operability of the system, but disclosed mechanical and operational limitations of equipment available that made the process uneconomical at that time. Today energy costs have changed and sulfuric acid is no longer worth $20/ton. This process, installed in an alkylation unit with the equipment available today, can be an integral part of the alkylation system to substantially reduce alkylate production costs.

LITERATURE CITED

(1) Felter, R.H., US Patent 2,593,128 (1952)
(2) Skelly, J.F. and Stiles, S.R., US Patent 2,716,592 (1955)
(3) Stiles, S.R., US Patent 2,831,043 (1958)
(4) Stiles, S.R., US Patent 2,862,791 (1958)
(5) Skelly, J.F. and Stiles, S.R. US Patent 2,863,724 (1958)
(6) Stiles, S.R., US Patent 2,903,339 (1959)
(7) Stiles, S.R., World Petroleum, Annual Refinery Review (1956)
(8) Durrett, L.R., Taylor, L.M., Wantland, C.F., Duoretzky, I., J.Am.Chem.Soc. $\underline{84}$ (1962)
(9) Cupit, C.R., Gwyn, J.E., Jernigan, E.,PetroChem.Engr., 203 Dec (1961), 207 Jan. (1962)
(10) Whitmore, F.,Chem.Eng.News, $\underline{26}$, 668 (1948): Ind.Engr.Chem. $\underline{26}$, 94 (1934)

19

Alkylation: Its Possible Impact on the Sulfuric Acid Industry

WARD A. GRAHAM

Stratford/Graham Engineering Corp., 4520 Madison Avenue, Kansas City, MO 64111

I.   Present world alkylation capacity and the proportion thereof that is based on sulfuric acid, broken down by region (North America, W. Europe, etc.). Approximate annual consumption of sulfuric acid. Proportion of this that is recovered and regenerated.

Alkylation is here defined as the process of alkylating isobutane with light olefins, which is the major petroleum refinery process for producing both aviation gasoline and motor car gasoline blending iso-octanes. Combining published data (1) with our own job records, the estimated alkylation capacity of the free world refining industry is 1,098,000 BPSD. The two principal catalysts used are sulfuric acid and hydrofluoric acid; the above worldwide capacity is approximately 55% sulfuric and 45% hydrofluoric, as tabulated in Table I.

TABLE I

	H_2SO_4	HF
North America	554,440	436,925
Europe	7,200	30,200
Other	43,460	25,775
TOTAL	605,100	492,900

Acid consumed in the process is also in need of definition. Hydrofluoric acid is literally consumed; it reacts to form heavy polymers and ultimately is pumped from the system to a combustion chamber or to a neutralization stage where the products of neutralization are hauled off to a dump or used as landfill. Sulfuric acid is consumed in a similar manner but only as a small fraction of that charged to the process. The primary effect upon sulfuric acid is that of dilution, rather than actual consumption. The fresh sulfuric acid catalyst is typically 98.0-

314

99. 5% titratable acidity. When the acid enters the alkylation
system it is gradually diluted both with water and hydrocarbon
polymers and esters to a spending strength of about 90% titratable acidity. Although the alkylation reactions will occur at
lower acidities, the usual 90% cut-off strength is dictated by the
corrosivity of carbon steel with weaker acid. In a National
Petroleum Refining Association Question and Answer session in
Philadelphia(2), we reported laboratory data showing successful
alkylation at as low as 82% titratable acidity where the water
diluent was low and the greatest diluent was hydrocarbon polymers. Some water is necessary, but only a few per cent. This
confirms previous laboratory work at Purdue University(3).
Economics are complicated but constitute the final answer in
determining the optimum acid spending strength for a given
plant. This will be discussed in more detail below.
     Keeping in mind the sulfuric acid dilution rate and the
actual consumption rate, a typical acid dilution rate for alkylation is 0. 5 lbs. acid/U.S. gallon of alkylate produced. Thus, a
plant producing 1,500 BPSD of alkylate would require a fresh
acid feed rate of about 16 short tons of acid per day which would
be diluted to about 90% before regeneration. It is estimated
that only 2-5% of this acid is actually consumed by side reactions, the remainder is pumped to storage and then to regeneration. The efficiency of regeneration varies between 98. 0 and
99. 5%(4). Thus, if the acid dilution rate is assumed at 0. 5 lbs./
gallon and 5% is assumed to be consumed in the alkylation and
regeneration steps, the worldwide acid consumption in sulfuric
acid alkylation and regeneration is [0. 05 (0. 5 x 605, 100 x 42/
2000)] or 318 short tons/day. However, the acid regeneration
plant capacity must be 20 times this, or 6, 360 tons/day to provide the total alkylation acid requirement. This is, of course,
in addition to the actual acid production capacity for other uses.

II.   Likely trends in alkylation capacity in view of gasoline
lead regulations, demand patterns, etc.
     Likely trends in alkylation capacity are almost impossi-
ble to predict. The entry of federal and local governments into
the control of more and more aspects of industry, especially
the petroleum refining industry, is alarming and renders
decision-making on technical and economic bases virtually
meaningless. The pell-mell rush to "control" has prompted
grave misgivings in every sector of society. It is stated in one
article(5) that ". . . policies may be adopted by power politics
that could bring on unknown, possibly catastrophic results".

Other headlines(6) include: "Feelings went first from annoy-
ance to anger, and now to rage...", "Government bungling
helped bring the U. S. energy crisis", "There is an energy
policy stalemate in Washington ... and it stems from the fact
that ... critical oil and gas issues ... are bogged down in
partisan politics". In this atmosphere, prediction of trends in
the alkylation process must be based first upon most probable
government decisions and then upon technical/economic considerations.

Many excellent articles have addressed various facets
of the technical/economic aspects of alkylation in recent years.
Data shown in Table II were abstracted from an article published in 1971(7) ; these data indicate that reducing gasoline lead
levels from 3.0 to 0.0 cc TEL/U.S. gal., holding a constant
M. O. N. of 86.0, could increase alkylation charge rates from
47 to 68% in refineries with capacities ranging from 60,000 to
240,000 BPSD, assuming unlimited butane availability. Other
refinery process charge rates are also changed; the reader is
referred to the original article for complete details. From
Table II, it can be seen that present-day ratio of alkylation to
crude of about 6% could increase to about 9% in the smaller
refineries and about 10% in the larger.

TABLE II

Refinery Size, BPSD	60,000		120,000		240,000	
Gasoline Properties						
M. O. N.	86.0	86.0	86.0	86.0	86.0	86.0
R. O. N.	94.0	96.7	94.0	97.3	94.0	96.3
Lead level, cc. TEL/						
U. S. Gal.	3.0	0.0	3.0	0.0	3.0	0.0
Gasoline Pool,						
% Alkylate	13.4	19.7	13.4	19.7	13.4	22.4
Alkylation Unit Charge,						
MBPSD	4.5	6.6	9.0	13.2	17.9	30.0
% Increased at						
0 cc TEL		47		47		68

In 1974, it was reported(8) that "for meeting the challenge of unleaded gasoline, U.S. refiners have built up their octane making capability to the highest level in history. A rough index of
this is the combined catalytic reforming and alkylation capacity
as a percent of crude capacity. That number has risen from
about 25% ten years ago to 30% in 1974". It is understood that

most of this increase is in reforming. However, reformate has a lower motor octane rating and alkylate remains the best blending stock for improving motor octane.

In this article (8), it is stated, "A typical blend to make an unleaded gasoline of 93 RON and 85 MON could contain the following:

	Vol. %
Reformate (100 RON)	31
Alkylate	15
Cat-cracked gasoline	32
Light hydrocrackate naphtha	15
Butane	7".

The unleaded gasoline of 93 RON 85 MON was predicted as a probable requirement of 1976-1980 cars with catalytic converters. The original unleaded requirement of 91 RON was predicted (7) to be inadequate for long term use in the 1974-1975 model cars in the U. S. Initial preliminary data (9) indicate that a higher percentage of these cars will be "not satisfied" with 91 RON, but the difference is believed to be small.

Obviously, the gasoline supply picture is subject to constant change. The European Parliament has reportedly (10) "urged the Common Market Commission to postpone its plans for reducing the lead content in gasoline. . . ". In October, 1976, the U. S. EPA altered the lead phasedown schedule.

If the lead is phased out or reduced, the new manganese antiknock compound (11) could become widely used and, in this event, the gasoline blending schedules would be revised again. The response of alkylate to the manganese compound is similar to its response to lead, so it appears alkylate would remain as a prime octane blending ingredient. The percentage of alkylate in the finished gasoline would undoubtedly vary. However, this product is now being investigated to determine if it can satisfactorily replace lead. These are but two of many examples of new influences which will enter into future decisions concerning alkylation.

In a report (12) dated August 11, 1976, one firm estimated that new octane-producing facilities would be required for 1978 and 1979, as follows:

Catalytic Reforming 540,000 BPD; Alkylation 298,000 BPD; Isomerization 332,000 BPD; and Cracking 899,000 BPD.

III.   Discussion of the basic principles of current sulfuric
acid and HF alkylation technology and of any likely developments
which could lead to a change in the present balance between the
two.

The basic principles of sulfuric and hydrofluoric acid
catalyzed alkylation reactions have been described in many
different articles and books, some of which are tabulated in the
bibliography (13-23). The complexity of the reaction is such
that many details could not be isolated until the advent of sophisticated analytical equipment and techniques. The fact that
commercial refinery alkylation units almost always receive
feeds of varying rate and/or composition makes the analysis of
such plants' performance very difficult. Even with a constant
feed, the number of olefinic compounds usually present in the
feed promotes different reactions and side-reactions, the products of which generally end up in the alkylate product. Laboratory studies of alkylation of isobutane with individual pure
olefins have provided significant data on reaction rates and
yields as influenced by the common reaction variables (24).

The basic chemical reaction involved is believed to be
that of carbonium ion production by the addition of a proton to
an olefin where the proton is supplied by the protonic acid
catalyst, in this case either sulfuric or hydrofluoric acid (20).
A further statement in this same article explains the important
addition made by modern analytical equipment: "... the rapid
development of spectroscopical methods of analysis, of which
nuclear magnetic resonance spectroscopy was the most important, has enabled Deno (25) and many others to show that
carbonium ions can now be directly observed in solution. This
has made it possible to determine the structure of such ions.
Deno states that the HSO_4^- salts of substituted cyclopentenylca-
tions form the sludge in commercial alkylation acid. He did not
encounter any cations possessing less than 10 carbon atoms. "

Although the basic chemical reactions have been thought,
for many years, to be similar with either sulfuric or hydro -
fluoric acid catalyst, extensive work with sulfuric acid described by Albright et al (24) has demonstrated that this is probably
not the case. The need for more detailed work with hydrofluoric acid is cited. The following summarizes their report:
"The reaction mechanism for hydrogen fluoride alkylation also
seems to be radically different than that for sulfuric acid alkylation. The alkylation mechanism which has been widely accepted
in the past seems to be somewhat more satisfactory for hydrogen fluoride alkylation even though it is not for sulfuric acid

alkylations. " The announcement of this Alkylation Symposium
contained the following: "Some important new developments
have recently clarified to a significant extent the mechanism of
the alkylation of both isobutane and aromatics. "

One possible advantage of the HF process in propylene/
butylene alkylation is the production of isobutylene from isobutane effected by the hydride-ion transfer to propylene. Isobutylene is the C_4 olefinic isomer which produces a significantly
higher octane alkylate with HF. This shift, however, converts
as much as 22% of the propylene to propane and is a debit.
Some normal butane is also produced from butylenes but this is
estimated at only 4-6%. The higher octane isobutylene alkylate
and a claimed yield increase must be contrasted with normal
paraffin production from olefins and a higher isobutane requirement. The typical mixed $C_3 = / C_4 =$ feed can be made to produce
a high octane alkylate with either acid catalyst by the optimization of other variables. The highest alkylate octane numbers
reported are produced with sulfuric acid catalyst, alkylating
with a typical cat cracker butylene olefin.

The cost of the acids, the relative hazards of each, the
actual alkylate yield and quality all must be considered when
selecting the alkylation process to use. The installed cost of
the plant and operating costs also must be considered. Cost
savings can be realized if octane quality is lowered.

Supplemental processes which can be operated in conjunction with alkylation and/or sulfuric acid production can influence the overall economics. Examples are (1) the integration of normal butane-to-isobutane isomerization with alkylation,
utilizing common fractionation equipment and (2), utilizing 65%
sulfuric acid extraction of isobutylene or isoamylene from olefins fed to alkylation, justified by monetary return on sale of the
high purity iso-olefin as a petrochemical feedstock, which reduces quantity of alkylate produced and reduces isobutane required while producing still higher quality alkylate with sulfuric
acid catalyst.

At the present time, the trend seems to be toward reduced alkylate production and reduction or elimination of propylene as a feedstock. This is based upon currently reduced
gasoline demands and high value of propylene as a petrochemical. In any event, if the percentage of butylenes in the feed increases, sulfuric acid alkylation will assume a more clear-cut
advantage on the basis of barrel-octane superiority.

Numerous factors have always required consideration in
attempting to evaluate which alkylation process to use. New

reaction equipment has been patented (27, 28) which reportedly
promotes intimate mixing with reduced power input; some
claims are made for alkylation of isopentane with isopentene
using either sulfuric or hydrofluoric acid as the catalyst. Reduced plant and operating costs would obviously improve the
relative position of alkylation in the refinery's processing
scheme. A recent verbal report from one refiner indicated HF
acid consumption over a two-year period was extremely low,
perhaps only one-tenth of that normally reported. New developments in sulfuric acid effluent refrigerated alkylation indicate
reductions in operating costs can be made. Another verbal report indicates still higher octane alkylate is being produced in a
pilot plant than reported at any time in the past.

Expansion of existing plants is an interesting potential. The
effluent refrigeration alkylation process jointly offered by Texaco Development Corporation and Stratford Engineering Corporation features the evaporation of the reactor effluent to cool
and remove the reaction heat from the reactor. This flashed
vapor, rich in isobutane, is compressed and, after removal of
propane and other light ends, is returned to the reactor; this
step can reduce the usual C_4 fractionation tower size as much
as 50-70%. If an existing alkylation plant, either HF or H_2SO_4,
is not effluent refrigerated, this means the deisobutanizer
tower is about twice the size required with effluent refrigeration. Such a plant can be expanded in alkylate capacity two to
three times by converting to effluent refrigeration. Metallurgi-
cally, HF alkylation plants can be converted to H_2SO_4, but
H_2SO_4 cannot be converted to HF. Although effluent refrigeration has not, to date, been applied to HF alkylation, numerous
patents have been issued for such use. (29)

One disturbing factor is revealed in a Loss Information
Bulletin (30) published by the Oil Insurance Association, en-
titled "Alkylation Units". This report includes the following:
"During the period from 1961 to 1972 the OIA experienced 17
alkylation unit losses and has knowledge of at least one additional loss. Of this total of 18 losses, 5 occurred in H_2SO_4
units and 13 in HF units. This indicates more than a two to
one edge of losses in HF units over H_2SO_4 units. Another
significant feature is the fact that no losses have been recorded
in H_2SO_4 units since 1967, while HF unit losses first occurred
in 1965 and have increased in successive years in both magnitude and number.

These 18 occurrances in aggregate represent total
property damage losses exceeding $6,800,000. This does not

include loss of earnings resulting from these events. Business Interruption coverage was not provided for all these instances, but in the three cases where it was involved, the lost earnings represented an additional total of approximately $1, 400, 000.

Losses in HF units account for about $5, 600, 000 of the aggregate property damage total, and of this amount nearly $4, 000, 000 is accountable to the past two and a half years. "

Thus, 82% of the losses in the period 1961-1972 occurred in HF plants. A further set of loss reports (31), covering the period from 1973 through November 1976 indicates this has continued and, if anything, has increased. Therefore, for the past 15 years, sulfuric acid alkylation has been demonstrated to be a much safer process.

So the competition will apparently continue, each refinery alkylation plant and catalyst selection being evaluated on an individual basis.

IV.   Developments in sulfuric acid recovery and regeneration methods.

The regeneration of spent sulfuric acid from alkylation has historically (32) been that of decomposition in a combustion chamber to reduce the acid to essentially SO_2 and H_2O, oxidizing the SO_2 to SO_3 catalytically, and absorbing the SO_3 in weak acid to produce fresh acid and even oleums. This process is widely used. The typical sulfuric acid plant must be designed to include the capability for handling spent alkylation acid. A plant making acid from molten sulfur is the most economical and easiest to operate because of the relative purity of the SO_2 produced. If spent alkylation acid is burned, carbon oxides and water vapor are also produced and this complicates the plant. The water vapor must be removed prior to the catalytic oxidation step. The gas balance to catalytic oxidation must be carefully adjusted to accommodate the carbon oxides while maintaining the surplus oxygen/sulfur dioxide ratios required for efficient oxidation of the SO_2 to SO_3. (Although spent alkylation acid has been used in fertilizer manufacture, this practice is phasing out because of the odors emitted from such plants and also because of the grayish color imparted to the final fertilizer product.)

Independent chemical companies own and operate most of the sulfuric acid plants. Plant capacities are generally in the range of 500-1500 tons fresh sulfuric acid/day. However, a number of acid plants are installed within, or immediately adjacent to, the petroleum refinery area; most of these acid

plants are owned outright by the refining company. Capacities
of these plants range from 50 to as much as 600 tons/day.

The economic importance of the refinery-owned sulfuric
acid plant is increasing. The immediate benefit is that of reduced acid costs, both handling costs and production costs.
Economic data were published in 1972 in an article by the
author (33) and showed that net direct manufacturing costs of a
refinery-owned acid plant were about 1/4 to 1/2 of that of the
acid market price. The industry practice in the past has been
to allow a credit for the spent alkylation acid returned for regeneration; today that practice is changing in some areas and an
additional charge is being made for handling the spent acid.
This has happened because of pollution control laws which require the acid plant operators to turn down their production rate
and/or to install expensive stack gas cleanup systems to avoid
atmospheric pollution.

Thus, a new acid plant within the refinery, built to meet
today's standards for conversion efficiency and non-polluting
emissions, has an even greater potential economic benefit to
the refiner. The fact that a major West Coast (U.S.A.) refiner
built (in 1972-1973) and now operates an acid plant within the
stringent Los Angeles pollution-control area, that the plant
produces fresh acid from H_2S and also regenerates spent alkylation acid, attests to this. The capacity of this plant is approximately 275-300 tons/day.

Other acid plants are under serious consideration by refiners at this time. Plants as small as 15 tons/day have been
quoted in the fourth quarter of 1976. Once the first small plant
has been built and operated successfully, it is the writer's considered opinion that many of these plants will be built. This
will definitely influence the ratio of sulfuric/hydrofluoric alkylation plants; many small HF alkylation plants were built simply
because sulfuric acid was unavailable.

An additional economic benefit to the refiner is based
upon the government-imposed requirement to desulfurize almost all the products leaving the refinery. This has resulted in
the installation of sulfur-recovery plants in a high percentage
of the U.S.A. refineries and will undoubtedly follow in most of
the world. Hydrotreating is the primary process for removing
sulfur. Hydrogen sulfide is the primary product. Plants to
convert hydrogen sulfide to sulfur in quantities as small as
4 tons/day are in operation within petroleum refineries; plants
as large as 375 tons/day have also been reported (34). (It is
estimated that sulfur from petroleum will account for 10% of the

free world supply by 1980.) (35)

Once the first sulfur plant has been required, a second
standby plant is usually required to provide absolute capability
for handling all sulfur-containing streams in the event of the
first plant shutdown. If sulfur oxide emissions are limited by
law, then the refiner is faced with shutting down the refinery
if the first sulfur plant is down and a second plant is not available. Hydrogen sulfide can no longer be disposed of by burning
where the sulfur oxides are discharged to the atmosphere.

This is where the sulfuric acid plant merits additional
consideration. If the sulfuric acid plant is economically favorable on the initial basis of supplying and regenerating alkylation
catalyst, the economics become even more favorable if the acid
plant is designed with enough capacity to serve as the "standby
sulfur plant". Instead of converting hydrogen sulfide to sulfur,
it can be converted to fresh sulfuric acid. H_2S is ideal fuel in
an alkylation acid regenerating plant and, in sufficient quantities
can eliminate the need for supplemental fuel in the combustion
chamber. One ton of H_2S produces almost one ton of sulfur but
that same ton of H_2S can produce almost 3 tons of sulfuric acid.
Thus, the sulfuric acid plant within the refinery can provide
alkylation acid catalyst at lowest prices and can permit the
refiner to market his by-product sulfur optionally as sulfuric
acid if the monetary return is greater.

A further processing advantage is available to the refiner if a sulfuric acid plant has been built for handling both
alkylation acid and all the refinery H_2S. When the market conditions dictate converting H_2S to sulfur instead of sulfuric acid,
the idle capacity within the acid plant can be utilized by pumping
the acid through alkylation at a faster rate, raising the final
acid spending strength which produces a higher octane alkylate
(especially with butylenes alkylate). The extra utilities cost for
regenerating more acid must be offset by the increased value
of the higher octane alkylate. In most cases, this economic advantage will be significant.

The newest improvement in further reducing the sulfur
dioxide in the stack gas from a contact sulfuric acid plant was
described in 1974 (36). To the best of our knowledge the plant
described is the cleanest acid plant, i.e., non-polluting, in
the world today.

A Soviet process reported in 1971 was claimed to "slash
acid-making costs". This was based upon process modifications and a new catalyst. (The "technology is being offered by

Newton Chambers Engineering Limited, Sheffield, England"...)
(37).

Other patents have been issued claiming increased efficiency in acid production. One claims "an important advantage",
achieved by using reaction heat to preheat recycled tail gas for
final conversion to SO_3 (38).

An article published in Europe in 1972 described a "pressure process for making H_2SO_4". The conversion of SO_2 to
SO_3 was cited as 99.85% and the investment cost as about 10%
less than that of a conventional plant.(39)

One patented process (40) was introduced in the mid-'60s
to reduce the amount of sulfuric acid required by alkylation; it
was called the Sulfuric Acid Recovery Process (SARP) and was
jointly licensed by Texaco Development Corporation and Stratford Engineering Corporation. Chemically, SARP proved all
claims made for it. Utilized only with propylene/butylene
alkylation the acid requirement was reduced as much as 70%;
actual acid dilution rates were lower than 0.2# acid/gallon
alkylate. However, the spent acid from SARP was different
and could not be regenerated at the same rate as regular spent
alkylation acid. This caused the chemical companies to increase the charges for regenerating the SARP spent acid to a
point where there was no economic incentive to operate SARP.
The two commercial SARP installations are not in use at the
present time although new possibilities for SARP have arisen
just in the past few months.

V.   Taking into account II, III and IV, likely future trend in
sulfuric acid consumption and regeneration.

As stated earlier, predictions of future trends in alkylation and acid use are too dependent upon political decisions to be
even attempted on a practical basis. If political considerations
are ignored, then an estimate of a maximum alkylation capacity
and maximum acid requirement can be attempted, although such
a prediction must be very general and subject to challenge from
many sources.

Assuming the predicted worldwide petroleum refining
increases 45% by the year 1990 (41), that gasoline is 40% of the
refinery output and that tetraethyl lead is reduced to zero in
60% of the gasoline, which results in a 40% increase in alkylation, the present day alkylation capacity of 1,098,000 BPSD
will increase by a factor of 1.45 x 1.4 to 2,229,000 BPSD, or an
increase of 1,131,000 BPSD. If the present ratio of 55% sulfur-
ic/45% hydrofluoric is maintained, then 622,000 BPSD of this

increase will be sulfuric acid catalyzed. If we further assume most of this is butylene alkylate, with a typical acid dilution rate of 0.4#/gallon alkylate, the actual new acid regeneration plant capacity required will be 5225 tons/day. If new developments in sulfuric acid alkylation resulted in its use in all the new alkylation, new acid plant capacity would be about 10,000 tons/day.

Other uses for sulfuric acid in the Free World have been reported as 82,000,000 short tons in 1970, with a predicted increase of 25% through 1975 and a further increase of 25% by 1980. This would indicate new acid production requirements of about 28,000,000 tons/year 1976-1980, or approximately 79,000 tons/day. Thus, the new capacity for alkylation acid would be in the range of 7-12% of the total.

Literature Cited

1.   "Worldwide Directory, Refining and Gas Processing, 1974-1975", (32nd Edition), Oil and Gas Journal.

2.   1975 NPRA Q & A Session on Refining and Petrochemical Technology, Pg. 4.

3.   Albright, L. F. et al, "Alkylation of Isobutane with Butenes: Effect of Sulfuric Acid Compositions", Ind. Eng. Chem., Process Des. Develop. (1972) Vol. 11, (No. 3), pp. 446-450.

4.   Donovan, J. R. & Stuber, P. J., Chemical Engineering, (Nov. 3, 1970), pp. 47-49.

5.   Thompson, R. G. and Lievano, R. J., Hydrocarbon Processing, (October, 1975) pg. 73

6.   Vervalin, C. H., Hydrocarbon Processing, (November, 1975), pg. 9.

7.   Dunmyer, Jr., J. C., et al, The Oil and Gas Journal, (May 17, 1971, pp. 132-150.

8.   Aalund, L. R., The Oil and Gas Journal, (August 26, 1974), pp. 41-44.

9.   "Octane Requirements of 1975 Model Year Automobiles Fueled with Unleaded Gasoline", U. S. Environmental Protection Agency, Report 75-28JLB, (August, 1975).

10.   Unzelman, G. H., The Oil and Gas Journal, (November 17, 1975), pp. 49-57.

11.   The Oil Daily, No. 6, 024, (November 20, 1975), pg. 6.

12.   Bonner & Moore Associates, Inc. "Technological Feasibility of Reducing Lead Anti-Knock Additives in Gasoline 1976-1980", (August 11, 1976).

13.   Mrstik, A. V., Smith, K. A., and Pinkerton, R. D.,
       "Commercial Alkylation of Isobutane", Progress in
       Petroleum Technology, American Chemical Society,
       ACS #5, (August 7, 1951).

14.   Schmerling, Louis, "Alkylation of Saturated Hydro-
       carbons", The Chemistry of Petroleum Hydrocarbons,
       pp. 363-408, Reinhold. 1955.

15.   Iverson, J. O. & Schmerling, L., "Alkylation of Para-
       ffins", Advances in Petroleum Chemistry and Refining,
       Interscience, (1958), pp. 336-383.

16.   Payne, R. E., "Alkylation - What you Should Know About
       This Process", Petroleum Refiner, Vol. 37, (No. 9),
       (Sept. 1958), pp. 316-329.

17.   Putney, D. H., "Sulfuric Acid Alkylation of Paraffins",
       Advances in Petroleum Chemistry and Refining, Inter-
       science, (1959), pp. 315-355.

18.   Hofmann, J. E. and Schriesheim, A., "Ionic Reactions
       Occurring During Sulfuric Acid Catalyzed Alkylation",
       Journal American Chemical Society, XXXIV, (March
       20, 1962), pp. 953-961.

19.   Cupit, C. R., Gwyn, J. E. & Jernigan, E. C., "Catalytic
       Alkylation", Petro/Chem Engineer, (December, 1961),
       pp. 42-55, & (January, 1962), pp. 49-59.

20.   Buiter, P., Van't Spikjer, P., Van Zoonen, D., "Ad-
       vances in Alkylation", P. D. No. 17, 7th World Petro-
       leum Congress, (1965).

21.   Jernigan, E. C., Gwyn, J. E. & Claridge, "Optimizing
       Alkylation Processes", Chemical Engineering Progress,
       Vol. 61, (No. 11), (November, 1965), pp. 94-98.

22.   Albright, L. F., "Comparisons of Alkylation Processes",
       Chemical Engineering, (October 10, 1966) No. 7, pp. 209-
       215.

23.   McGovern, L. J., "Developments in Commercial Alkyla-
       tion of Isobutane the Past 25 Years,", presented at the
       164th National American Chemical Society meeting,
       (August, 1972).

24.   Li, K. W., Eckert, R. E., & Albright, L. F., "Alkyla-
       tion of Isobutane with Light Olefins Using Sulfuric Acid",
       I, II and III, presented at the American Chemical Society
       Meeting, (Sept. 7-12, 1969).

25.   Deno, N. C., Chemical Engineering News, (May 10, 1964),
       42 (40), pp. 88-100.

26.   Private correspondence, (November 24, 1975).

27.   Clonts, K. E. , "Liquid-liquid Mass Transfer Process
       and Apparatus", U. S. Patent No. 3, 758, 404, (Sept. 11,
       1973).
28.   Clonts, K. E. , "Alkylation Utilizing Fibers in a Conduit
       Reactor", U. S. Patent No, 3, 839, 487, (Oct. 1, 1974).
29.   U. S. Patent Nos. 2, 906, 796 (1959); 2, 949, 494 (1960);
       2, 977, 397 (1961); and 3, 925, 501 (Dec. 9, 1975).
30.   Oil Insurance Association report (November, 1972).
31.   Private Communication.
32.   Duecker, W. W. & West, J. R. , "The Manufacture of
       Sulfuric Acid", (1959), Reinhold.
33.   Graham, W. A. , "Alkylation Integrates Acid Plant",
       Hydrocarbon Processing, (August, 1972).
34.   Anon. , "Why Recover Sulfur from H$_2$S?", The Oil and
       Gas Journal, (Oct. 28, 1968), pp. 88-101.
35.   Buckingham, P. A. & Homan, H. R. , "Sulfur and the
       Energy Industry", Hydrocarbon Processing, (August,
       1971), pp. 121-125.
36.   Collins, J. J. et al, "The Pura Siv S Process For Re-
       moving Acid Plant Tail Gas", Chemical Engineering
       Progress, Vol. 70, (No. 6), (June, 1974), pp. 58-62.
37.   Lawrie, N. , "Soviet Process Slashes Sulphuric-Acid-
       Making Costs", Chemical Engineering (Mar. 8, 1971).
38.   Jaeger, W. , "Process for the production of Sulfur Tri-
       oxide by the Cold Gas Process", U. S. Patent No.
       3, 647, 360 (Mar. 7, 1972).
39.   Vidon, B. , Chemical & Process Engineering, (July,
       1972), pp. 34-35.
40.   U. S. Patents, Nos. 3, 234, 301; 3, 227, 774; 3, 227, 775;
       3, 442, 972; 3, 534, 118; 3, 544, 653; and 3, 665, 050 (1966-
       1970).
41.   Anon. , "Oil Still The Key Fuel", The Oil and Gas
       Journal, (Nov. 10, 1975), pp. 159-178.

20

The IFP Dimersol Process for Dimerization of Propylene into Isohexenes

An Attractive Alternate to Propylene Alkylation

JOHN ANDREWS and PIERRE BONNIFAY

Institut Francais du Petrol, North American Office, 450 Park Avenue, New York, NY 10022

The 1974/75 slump in gasoline demand is over, the 1976 demand is expected to top 6% while the 1977 demand will return to a more normal 3 to 5%. The question of lead phase-down appears relieved with an EPA postponement, but as a result of the steadily increasing number of vehicles that use unleaded gasoline (about 10% increase annually on the total number of automobiles in the United States) by 1980 some 62% of the total gasoline pool will have to meet no-lead gasoline specification. (1)

This is what refiners are faced with in their planning studies. An obvious answer to this problem of supplying not only more gasoline but increasing percentages of no-lead grade, is expanded crude capacity. However this leads to high investment costs for crude units and downstream processing units and just now there seems to be some reluctance in the refining industry to large scale expansions. Refiners are looking for other means to increase their gasoline producing capacity while optimizing their ability to produce increasing volumes of no-lead gasoline.

One solution to the problem is couched in raising fluid catalytic cracking capacity and conversion.

Fluid Catalytic Gasoline (FCC) full boiling range gasoline has a respectible RON of about 92 and an R+M/2 of around 87. Since current no-lead grade specifications include 87 R+M/2, FCC gasoline can be utilized directly in the no-lead pool.

The C3 and C4 streams from the FCC product distribution can be converted into high octane blending components. One favorite method of doing this is by means of alkylation of the C3/C4 olefins with isobutane.

Alkylate is produced from C3/C4 FCC cuts or sometimes from the C4 cut alone. These products have good

RON numbers which vary depending upon the olefin feed
composition.

	RON	R+M/2
C3 Alkylate	91	90.5
C4 Alkylate	96	95
C3/C4 Alkylate	93.5	92.5

Alkylates are very desirable no-lead ingredients
and obviously the C4 alkylate is preferred because of
its substantially higher octane quality. Moreover,
the manufacture of C4 alkylate requires about 15% less
isobutane than C3 alkylate.

However, isobutane is becoming more expensive as
its supply decreases:
- Both gas processing throughput and natural gas-
  liquids production for the past three years signify
  that U.S. production of gas liquids is in a downward trend. Specifically, the production of
  isobutane from this source has decreased almost 8%
  from 1974 to 1975. (2)
- Isobutane from hydrocracking plants continues to be
  available, but construction of new hydrocrackers is
  very expensive and therefore an increased supply
  from this source is, short term at least, doubtful.
- Even when energy costs were appreciably lower than
  now, isomerization of butane to isobutane was no
  real bargain because of the large energy requirement
  for fractional separation of isobutane and butane.
  Today the situation is aggravated by high energy
  cost.
- Therefore in many cases economics will not permit
  the refiner to make available sufficient isobutane
  to alkylate his entire C3/C4 FCC stream and --
  there is demand for a new or modified approach which
  will rethink the entire problem of C3/C4 utilization.

Reviewing these thoughts and applying them to the
refinery situation -- one scenario which has found
favor is the following:

FCC Unit. Utilizing new catalysts and recently
developed technology including reactor modifications,
raise raw oil charge rate and/or conversion.

Gas Concentration. If a good FCC revamp is possible, it follows that the gas concentration system
will need to be extensively revamped.

Alkylate and Dimate. The degree of revamp necessary or the need to construct a new alky unit will
depend not only upon the existing operation but by the
isobutane the refiner will be able to bring to the
alkylation unit.

For example, in the event a refiner has been
alkylating a mixed C3/C4 stream and to expand the alkylation capacity would strain his isobutane supply
sources, the refiner can install a C3/C4 splitter and
alkylate only the C4's.

The advantages are numerous:
1. Revamping of the alkylation may not be necessary
2. Alkylate octane number will improve
3. Isobutane efficiency will increase (less iC4
   required for C4 alkylation than C3 alkylation).

However, what will the refiner do with the C3's,
and how about his need for larger quantities of no-
lead gasoline?

The IFP answer to this question not only cuts
new construction costs, reduces operating costs, reduces the refiners dependence on isobutane, but also
increases the RON and the (R+M)/2 of the gasoline
produced from the C3/C4 olefins.

The answer is the IFP DIMERSOL PROCESS to produce
isohexenes, which we will refer to as dimate, and the
purpose of this paper is to demonstrate how the Process accomplishes this.

Let us procede as follows:

Comparative Economics At A Glance

Figure I offers a means of comparing the econo-
mics of charging a FCC C3 stream to a new Dimersol
Process Unit for dimate production or a new Alkylation
Process Unit for alkylate production.

The series of slanted parallel lines represent
variable isobutane prices which when related to the
appropriate propylene price give the resultant cost
of the alkylate product identified as gasoline in
¢/gal. The cents per gallon values for propylene relate to the price for the FCC C3 stream with propane
at the same price as propylene. Included in the
correlation are the alkylation operating costs shown
in Table IV at 0% return on investment.

The single broken line relates propylene price
to dimate product cost identified as gasoline in ¢/gal.
Included are the Dimersol operating costs shown in
Table III & IV at 0% return on investment.

This superimposed plot then makes it possible to
determine the most economic method of dispatching

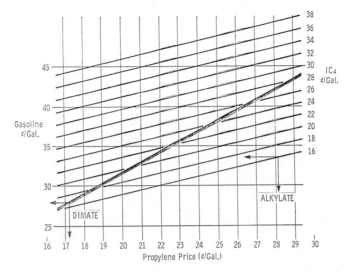

Figure 1. Comparative economics for charging propylene to dimersol or alkylation

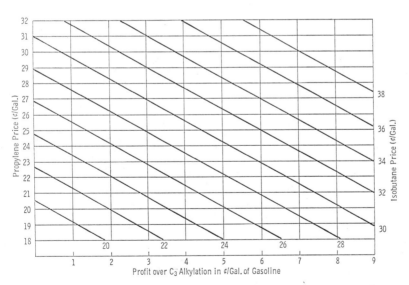

Figure 2. Comparative economics incentive for charging propylene stream to dimersol

refinery propylene into gasoline depending upon propylene and isobutane costs and including respective process operating costs.

For example, if a refiner's isobutane costs 25¢/gal, the case is made for Dimersol operation (over alkylation) for any value of propylene up to 25.7¢/gal.

Any set of conditions located over the broken diagonal line would favor Dimersol operations while the lower isobutane prices (below the line) would favor the C3= alkylation route.

Still another method of utilizing this chart is under the conditions which say that alkylate operation sets the value on the FCC C3 stream. For example: if a refiner pays 25¢/gal for isobutane and sells C3 alkylate into a 35¢/gal gasoline market, his propylene costs out at about 18.7¢/gal. This same 18.7¢/gal propylene would yield gasoline at 30¢/gal in the Dimersol Process thus realizing a 5¢/gal profit. Figure 3 expresses the same data in another way: in terms of Dimersol profit over C3 Alkylation in ¢/gal of product.

Take A Closer Look

If from this rather quick economic surveillance dimate looks favorable, then one should examine:
1. the dimate product qualities comparing them with those of C3 alkylate
2. the comparison of the more detailed economics including investment and operating costs.

Table I compares properties of Dimersol Dimate and C3 alkylate. In studying this table one sees a basic difference between dimate and C3 alkylate volatility manifested in distillation and RVP. In gasoline blending this becomes important when one is blending with reformate. To illustrate, we will show a series of bar charts giving the RON's of consecutive individual 10 volume % increments of full boiling range gasolines.

Starting with a typical reformate in Figure 3, note the severe drop-off in the front end octane. When this same reformate is blended with 15% C3 Alkylate as shown in Figure 4, the change in octane is evenly distributed over practically the entire gasoline boiling range.(3) However, when 15% dimate is blended with the identical reformate, the octane effect is concentrated in the 10-30 vol % range where the octane slump was most pronounced. The dimate's low boiling range coupled with a 97 RON, serve to fill this gap

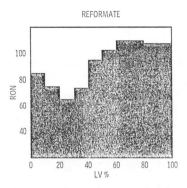

Figure 3. Research octane distribution over gasoline boiling range

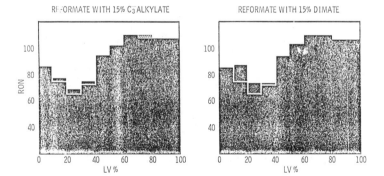

Figure 4. Research octane distribution over gasoline boiling range

TABLE   I

PROPERTIES OF

DIMERSOL DIMATE AND C3 ALKYLATE PRODUCT

	DIMERSOL DIMATE	C3 ALKYLATION ALKYLATE
PRODUCT QUALITY		
SP. GR.	0.69	0.70
R.V.P.	6.5	2.0
RON CLEAR	97	91.0
RON + MON/2	89.5	90.5
ASTM BOILING RANGE		
IBP	133	150
10	136	195
50	140	210
90	160	250
EP	370	400

FOR DIMERSOL GASOLINE BLENDING VALUES - SEE NOTE 6.

not only in octane level but in volatility as well. A balanced octane/volatility gasoline makes any automobile perform better in any season.(4)

Table II shows the yields and product distribution in a Dimersol Unit charging 3803 BPSD of mixed C3's containing 2700 BPSD propylene. The conversion level on propylene is 95%.

Table III shows the detailed direct operating costs based upon the following base data:

Steam per 1,000 lb
High pressure $3.50
Medium pressure $2.50
Power, per KW $0.012
Water cooling, per 1,000 gal $0.03
Operators, per shift
Operation   1.1
Supervision   0.2

In Table IV these same economics are compared against C3 alkylation costs for a given availability of propylene and referenced against a barrel of product from each respective process. The base data is the same as for Table III.

Process Description

Figure 5 illustrates the simplicity of the Dimersol Process. Dried feedstock, which may vary widely in propylene composition, is charged to the reactor where soluble catalyst is added in low concentration. Close reactor temperature control is obtained by circulation of the reaction mix through a cooler which may be of the air or water type. The reaction occurs at essentially ambient temperature, allowing for the exothermic heat of reaction, and at a pressure sufficiently high to maintain a liquid phase.

Upon leaving the reactor, the effluent enters a drum where ammonia injection neutralizes the catalyst, forming salts: subsequent water injection dissolves the salts removing them from the system. The scale of this washing operation is small because the catalyst concentration is low-- with only about 15 gpm of process water required for 1000 BPSD of isohexenes product. If required, a subsequent small treating step can be supplied with the unit to remove anions from the wash water.

The depropanizer operates to separate the isohexenes from C3 LPG overhead.

TABLE   II

IFP DIMERSOL PROCESS

CONVERSION   =   95% C3

FEED	VOL %	BPSD	LBS/HR
PROPYLENE	71	2700	20,550
PROPANE	29	1103	8,170
	100	3803	28,720

PRODUCTS

LPG

	VOL %	BPSD	LBS/HR
PROPYLENE	4.2	135	1,030
PROPANE	34.6	1103	8,170
DIMATE	61.2	1952	19,520
TOTAL	100	3190	28,720

PROPERTIES OF DIMATE

	WT %	
ISOHEXENES	92.0	RON = 97.0
NONENES	6.5	(RON+MON)/2 = 89.5
HEAVIER	1.5	

TABLE   III

IFP DIMERSOL PROCESS

DIRECT OPERATING COSTS

1952 BPSD DIMATE	$/BBL	
LP STEAM	0.26	
COOLING WATER	0.04	
POWER	0.01	
CATALYST	0.40	
TOTAL UTILITIES & CATALYST		0.71
LABOR	0.11	
SUPERVISION	0.03	
MAINTENANCE	0.03	
		0.17
TOTAL DIRECT OPERATING COSTS		0.88/BBL

TABLE IV

IFP DIMERSOL VS ALKYLATION ECONOMICS

TOTAL OPERATING COSTS
($/BBL PRODUCT)

	DIMERSOL	ALKYLATION
BPSD PROPYLENE CHARGE	2700	2700
BPSD ISOBUTANE CHARGE	0	3645
BPSD PRODUCT	1952	4750
ESTIMATED BATTERY LIMITS ERECTED COST $MM	1.8	8.0
UTILITIES & CHEMICALS	0.71	1.38
OPERATING LABOR, SUPERVISION AND MAINTENANCE	0.17	0.11
DIRECT OPERATING COSTS	0.88	1.49
OVERHEAD & CAPITAL COSTS	0.48	0.67
TOTAL OPERATING COSTS	1.36	2.16
¢/GAL @ 0% R.O.I.	3.2	5.1
¢/GAL @ 20% R.O.I.	4.5	7.5

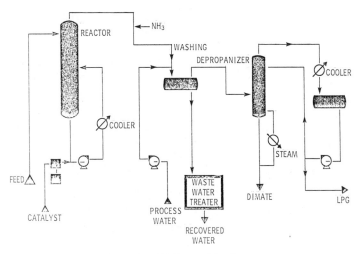

Figure 5. IFP dimersol for dimate

Practical Application

The process operates at practically ambient temperature and at sufficient pressure to maintain a liquid phase. Material of construction is essentially all carbon steel.

Though the economics used in this paper are based upon a 100% new plant, the IFP Dimersol Process can be adapted to existing equipment because of its simplicity and modest operating requirements. For example, the reactor section for small to medium size units may utilize an existing LPG bullet-type vessel or an existing polymerization unit can be converted into the Dimersol Process.

Conclusions

One has to be impressed with the economics of the Dimersol Process and the simplicity of the design which leads not only to ease of operation but to fast construction as well. The construction time schedule for one unit which is now under design foresees operation by the end of 1977.

The Dimersol Process offers refiners a unique method of quickly solving problems associated with the production of a gasoline pool radically different in composition from that produced just 2 or 3 years ago.

LITERATURE CITED

1. "Octane Crunch Threatens Petrochemicals," Chemical
   Week, (Sept. 15, 1976)

2. Congram, Gary E. "U.S. Gas-Processing Throughput
   Dips While Ethane Recovery Rises 5.7%," The Oil &
   Gas Journal, (July 5, 1976)

3. Albright, Lyle; K.W. Li, Eckert, Roger E., "Alkyla-
   tion of Isobutane with Light Olefins Using Sulfuric
   Acid" Industrial & Engineering Chemistry, (1970)
   Vol. 9, Alkylation.
   Data source used to construct Figure 4-1.

4. Barbier, J.C.; Douillet, D.; Franch, J.C.; Raimbault, C.; Bonnifay, P; Cha,B.; Andrews, J. "Gasoline Pool: What it Needs," Hydrocarbon Processing,
   (May, 1975)

5. Barbier, J.C.; Douillet, D.; Franch, J.C;
   Raimbault, C.; Bonnifay, P.; Cha, B.; Andrews, J.
   "Propylene Dimer; Fuel or Chemical" Hydrocarbon
   Processing, (April, 1976)

6. Chauvin, Y.; Gaillard, J.F.; Quang, D.V.; Andrews
   J.A. " The IFP Dimersol Process for the Dimeriza-
   tion of C3 & C4 Olefinic Cuts: NPRA Convention
   Center, San Antonio, Texas, (1973)

21

Monsanto's Ethylbenzene Process

A. C. MACFARLANE

Monsanto Chemical Intermediates Company, Texas City, TX 77590

Ethylbenzene is commercially produced almost entirely as an
intermediate for the manufacture of styrene. Since only a
limited amount can be made by the superfractionation of C_8
petroleum aromatics, most ethylbenzene is produced by the alkylation of benzene with ethylene. The alkylation reaction can
occur either in the vapor phase or the liquid phase. A number
of proven processes exist. The liquid phase processes using
aluminum chloride catalysts are currently the most widely used.
The purpose of this paper is to describe a new and improved
version of this latter process which has been commercialized.

Friedel-Crafts Chemistry

The oldest method of alkylation with ethylene is the liquid
phase reaction using anhydrous aluminum chloride as the catalyst.
This reaction is a form of the classic Friedel-Crafts reaction
and was discovered in 1879 by Balsohn. Most Lewis and Bronsted
acids are known to be active for olefin alkylations. Alkylation
by H_2SO_4 and H_3PO_4 was first shown by Ipatieff, et al, in 1936
who extended the reaction to isoparaffins. For the liquid phase
alkylation of benzene with ethylene, however, aluminum chloride
is preferred over the other acids, although a co-catalyst or
promoter is usually needed to obtain efficient alkylation. $AlCl_3$
when dissolved in benzene containing some HCl forms a complex
which can be simply described as:

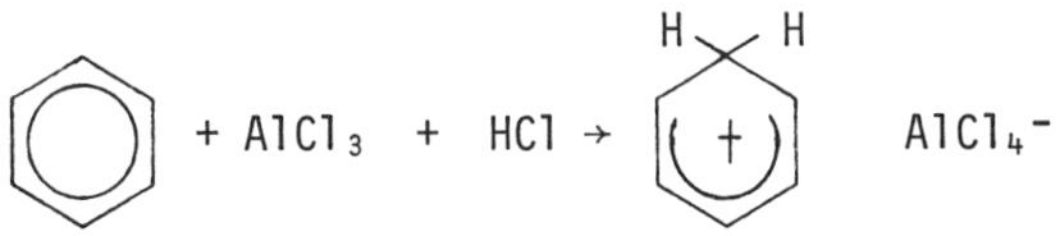

$$\text{C}_6\text{H}_6 + AlCl_3 + HCl \rightarrow [\text{C}_6\text{H}_7]^+ \ AlCl_4^-$$

341

This complex then reacts with ethylene to give ethylbenzene.

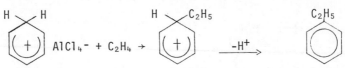

The alkylation reaction is complicated by the occurrence of minor side reactions such as cracking, polymerization, hydrogen transfer, etc. However, of major importance is the formation of polyalkylated products. The first alkyl group formed activates the aromatic nucleus so that the second alkylation proceeds more readily than the first and so on at least until steric hindrance intervenes, although hexaethylbenzene is quite readily formed. This results in a reaction product containing a mixture of mono, di, tri, and higher ethylbenzenes together with unreacted benzene. The ratio of ethylene to benzene in the reactor feed can, of course, be chosen to maximize the formation of mono-ethylbenzene; but the other products cannot be eliminated. Fortunately, the reaction is reversible, e.g., diethylbenzene will react with benzene under the influence of $AlCl_3$ to form monoethylbenzene.

$$C_6H_5(C_2H_5)_2 \;+\; C_6H_6 \rightleftharpoons 2\; C_6H_5 \cdot C_2H_5$$

This transalkylation reaction permits virtually all the ethylene and benzene fed to the reaction system to appear eventually as the monoethylbenzene product; the reaction was first demonstrated in 1894.([1]) The equilibrium amounts of the various products are shown in Table I.

The alkylation reaction can be performed under two rather different conditions. If, for example, in a well-stirred laboratory semi-batch reactor a given amount of ethylene is added rapidly to the benzene under given reaction conditions such that the ethylene is completely absorbed, it will be found that the product formed immediately after all ethylene is added is relatively rich in higher polyethylbenzenes and relatively poor in the desired monoethylbenzene. However, if the same amount of ethylene is added relatively slowly to another batch of benzene, the reaction product will be relatively rich in monoethylbenzene and poor in the undesired polyethylbenzenes. The rapid initial absorption and reaction of ethylene forms polyethylbenzenes unselectively, and the slow liquid phase reaction of the polyethylbenzenes with unreacted benzene results in an approach towards thermodynamic equilibrium. That is, although the amounts of higher polyethylbenzene is negligible under thermodynamic control, they can be considerably under kinetic control.

TABLE I

THERMODYNAMIC EQUILIBRIA OF BENZENE AND ETHYLBENZENES LIQUID PHASE[4]

Molar Ratio E/B[1]	Weight Percentage						
	Benzene	EB[2]	Di-EB	Tri-EB	Tetra-EB	Penta-EB	Hexa-EB
0.2	75.6	22.9	1.5	0.02	tr[3]	tr	tr
0.4	56.0	38.1	5.7	0.20	tr	tr	tr
0.6	40.6	46.8	11.9	0.72	0.02	tr	tr
0.8	28.6	50.1	19.4	1.77	0.09	tr	tr
1.0	19.6	49.3	27.3	3.57	0.25	tr	tr

1. Molar ratio of ethyl groups to aromatic rings.
2. Ethylbenzene.
3. Trace, less than 0.01 wt %.
4. Calculated from free energy data.

Commercial Alkylation Process

Emmet Reid and his co-workers at Johns Hopkins demonstrated in the 1920's the practicality of using ethylene for ethylbenzene preparation with the liquid $AlCl_3$ catalyst. It was, however, Dow Chemical in the United States who developed the Friedel-Crafts reaction with $AlCl_3$ into a continuous ethylbenzene manufacturing process in 1937. BASF were also similarly active in Germany in the 1930's. Dow showed the importance of HCl as a promoter and the necessity of maintaining strictly anhydrous conditions. The Dow process operated at about 95°C and at a small positive pressure. In 1942, with the cutoff of natural rubber supplies from the Pacific area, a rapid increase in styrene capacity was required. Monsanto operated a plant at Texas City for the government that used their own version of Friedel-Crafts chemistry for the ethylbenzene step. Carbide also installed their own $AlCl_3$ process at Institute, West Virginia. Dow Chemical expanded their original ethylbenzene production capacity by building in Los Angeles, California, and Freeport, Texas.

The various ethylbenzene processes, although having individual differences, all seemed to have been based on similar principles. Invariably present were three phases--ethylene gas, aromatic liquid, and a liquid catalyst complex. Reaction took place in the catalyst complex, and equilibrium was established between the catalyst complex and the organic phase. The liquid reaction product was then cooled and the two liquid layers separated. The lower catalyst complex layer was recycled to the reaction system. $AlCl_3$ was lost from the system in two ways, by solubility in the organic layer and by withdrawal of a slipstream of catalyst complex to allow addition of fresh catalyst. The catalyst complex was then hydrolyzed separately to produce an aqueous $AlCl_3$ waste solution and an organic layer which was added back to the system. The organic reaction product from the alkylation step was washed to remove dissolved $AlCl_3$ and HCl. The washed alkylate was separated into its components in a series of three distillation columns. This then is the well proven $AlCl_3$ process that is presently widely used in very large plants around the world.

Disadvantages of Commercial Process

Styrene, and thus ethylbenzene, are now commodity chemicals. Economics have forced manufacturers to construct modern, highly engineered production units with capacities of 1,000 million pounds/year and higher. Thus, the basic chemistry and the manufacturing techniques have to be closely scrutinized in the light of present day raw material shortages, energy costs, product purity demands, and environmental considerations. In these billion pound/year plants, relatively small improvements can often lead to important money savings provided they are reliable

and well demonstrated. In plants of this size, not very much
uncertainty can be permitted.

   In some versions of the ethylbenzene manufacturing process,
Figure 1, a relatively large volume of catalyst complex, after
cooling and separation from the alkylate, is recycled through the
alkylator to moderate the temperate rise of the exothermic reaction by acting as a heat sink.(2,3,4) The catalyst complex,
however, tends to absorb preferentially the higher polyethylbenzenes. These are then subjected to a highly reactive environment at an elevated temperature for a fairly long average
residence time. It is, thus, not surprising that such a system
tends to make considerable polymers, high boilers, and tars.
This not only leads to a significant loss in yield but also to a
higher than necessary usage of aluminum chloride catalyst, since
the tars tie up the catalyst irreversibly and so render it
inactive. Thus, even more of the catalyst complex has to be
withdrawn from the system. The presence of increased amounts of
impurities also tends to produce a relatively impure ethylbenzene product with many of these impurities carrying all the
way through to the product styrene. This catalyst complex stream
is also very corrosive and requires high alloy materials of
construction in piping and equipment handling the complex.

Improved Alkylation Process

   A few years ago Monsanto at Texas City, Texas, decided to
re-examine its position in ethylbenzene manufacture. Other processes were available or were becoming available and, as we have
seen, the $AlCl_3$ chemistry was quite ancient. An examination of
the other processes revealed that they too possessed certain
serious shortcomings. One of many potential problems with a new
process is the risk involved in being the first to build a
1.7 billion lb/year EB plant using a not fully proven technology.
1.7 billion times almost anything is a lot of money to be made or
lost so we had to be sure that any change was to the most economical process available. We concluded that rather than abandon
the old $AlCl_3$ technology we ought first to give it a fair chance
and try to drastically improve it. Thus, we decided to take a
new look at the basic Friedel-Crafts chemistry. As a result of
this, we have developed and commercialized a process using the
basic $AlCl_3$ catalyst but with chemistry in an area completely
different from that previously used. This new homogeneous process
eliminates or greatly reduces most of the problems associated with
the two-phase system.

   The Monsanto Texas City laboratory discovered that ethylene
would react completely and virtually instantaneously with benzene
containing only a small amount of dissolved $AlCl_3$. The $AlCl_3$ is
used only once and there is no recycle. In this homogeneous
alkylation system, care has to be taken to minimize the formation
of the higher ethylbenzenes. These compounds are more basic than

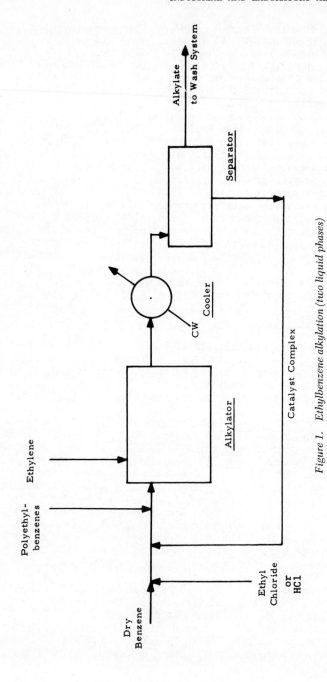

Figure 1. Ethylbenzene alkylation (two liquid phases)

benzene, in a Lewis sense. In fact, the tetraethylbenzenes are
probably basic enough to tie up the small amount of aluminum
chloride catalyst present as a relatively stable salt, thus
stopping the alkylation.

Doering, et al,(5) in 1958 had shown that catalysts in
reactions of the Friedel-Crafts type rapidly become inactivated
due to the presence of higher alkylated benzenes. For example,
when working with methylbenzenes, they showed that a heptamethyl-
benzenium ion even more basic than penta and hexamethylbenzene
could be formed. This heptamethyl ion was so basic that it was
even extractable by aqueous hydrochloric acid!

Since only a small amount of aluminum chloride is used in
this homogeneous alkylation process, more care has to be taken
to control the method and rate of addition of ethylene to the
benzene. The alkylation reaction vessel is designed to accommodate simultaneously both the very rapid ethylene-benzene reaction
and the relatively slow polyethylbenzene transalkylation reactions. By careful design of the reactor and control of operating
conditions, the formation of higher polyethylbenzenes can be
minimized.

The absence of a catalyst complex phase has certain advantages. The alkylation temperature is no longer limited to the
100°C range. Much higher temperatures can now be used without
excessive yield losses, tar formation, etc., and with only a
moderate AlCl$_3$ usage in order to maintain high catalytic activity.
The use of higher temperatures in alkylation also permits recovery of the heat of reaction in a waste heat boiler, generating
steam at a useful pressure. This steam has a significant effect
on the economics. The higher temperatures also permit decreased
use of aluminum chloride. Additionally, it also leads to a less
corrosive environment.

The transalkylation and isomerization reactions can be
satisfactorily explained by the Streitwieser mechanism(6). This
mechanism proposes a 1,1-diphenylethane-type intermediate. For
example, para-diethylbenzene. (Figure 3) Such an intermolecular
mechanism is consistent with the experimental data and does not
require the assumption of a sequence of intramolecular 1,2 shifts.
The decay of the polyethylbenzenes towards equilibrium is consecutive and not concurrent. The catalyst seems to be associated with
the most basic center; and when it reaches steady-state, the
catalyst transfers to the next most basic one. There is also a
concurrent intramolecular isomerization such as 1,2,4 triethylbenzene going to 1,3,5 triethylbenzene. There is hence a
movement towards isomer equilibrium as well as product equilibrium.

Monsanto's reactor layout, in simplified form, is shown in
Figure 2. Dry benzene, ethylene, catalyst, and promoter are fed
continuously to the reactor. The alkylator effluent is mixed with
polyethylbenzenes, mainly diethylbenzene recycled from the subsequent recovery system. The transalkylator allows enough residence
time for the product to approach equilibrium. The aluminum

348   INDUSTRIAL AND LABORATORY ALKYLATIONS

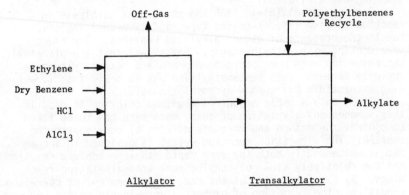

Figure 2.   Alkylator reactor scheme (single liquid phase)

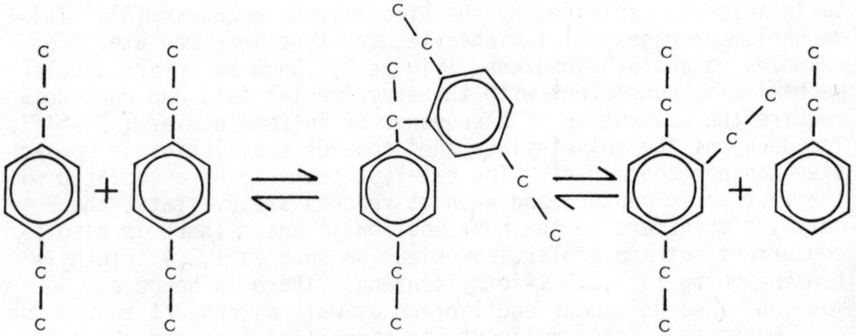

Figure 3.   Disproportionation of p-diethylbenzene

chloride catalyst is fed to the alkylator in very small amounts
together with the feed materials. The alkylated hydrocarbon
product is no longer saturated with $AlCl_3$. The $AlCl_3$ fed is used
once through only.

The material leaving the transalkylator is adiabatically
flashed by simple pressure reduction (Figure 4). The flashed
organics and HCl are recovered and recycled to the alkylator
while off-gas is flared or burned. With high purity ethylene,
the off-gas stream is very small since all the ethylene is reacted. However, with low purity ethylene, the off-gas becomes
significant and has appreciable fuel value. Recovery and recycle
of HCl allows control of this operating variable over a wide
range while still maintaining very low usage. The major vessels
in the reaction system are constructed of brick-lined carbon
steel. The use of higher alloys is required only for relatively
short runs of pipe and for the alkylator waste-heat boiler.

Purification of Crude Ethylbenzene

The purification of the liquid alkylate from the reaction
system is done in a conventional fashion (Figure 5). Before
feeding the crude ethylbenzene to distillation, the aluminum
chloride and the residual HCl must be completely removed. Water
washing accomplishes most of this task; final traces are removed
by a caustic soda treatment. This process is completely reliable, is free from emulsion formation, and there is no downstream
fouling or corrosion to worry about.

The washed alkylate is fed to a series of three distillation
columns where benzene, ethylbenzene, and diethylbenzene-triethyl -
benzene mixtures are removed as overhead products. The benzene
is recycled to the benzene drying column before feeding again to
the alkylator. The diethylbenzene mixture is recycled to the
transalkylator. The bottoms from the last column is what we call
"flux oil." This consists mainly of diphenylethanes. The amount
of this material is good measure of the overall process yield.
The yield loss in this process, depending on the alkylation
system design and control, is between 0.6 and 0.9 lb flux oil
per 100 lb ethylbenzene. Steam is generated by waste-heat
boilers. These together with the alkylation reactor waste-heat
boiler account for 90 percent of the energy required for the
entire ethylbenzene unit.

Catalyst Manufacture

The aluminum chloride catalyst for the alkylation process
may be purchased material or the aluminum chloride may be
prepared in situ. For large plants preparation in situ from
cheap powdered aluminum and anhydrous HCl is preferred; however,
ethyl chloride is just as effective as HCl although more
expensive. A small supplementary agitated vessel is used to

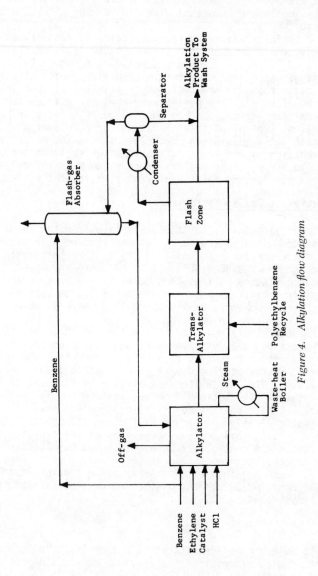

Figure 4.   Alkylation flow diagram

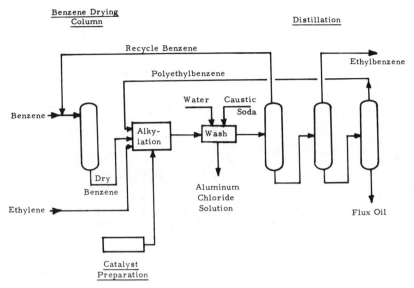

Figure 5.   *Ethylbenzene flow diagram*

AlCl$_3$ + C$_2$H$_5$Cl + C$_6$H$_5$·C$_2$H$_5$ → Catalyst Complex + Alkylate

(or HCl)

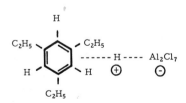

Figure 6.   *Catalyst complex preparation*

prepare the catalyst in the form of the heavy liquid catalyst complex of fixed composition $[C_2H_5)_3C_6H_4^+ \cdot Al_2Cl_7^-]$ (Figure 6). This is a most convenient way to handle, pump, and meter the addition of the $AlCl_3$ catalyst to the alkylator, where it promptly dissolves in the benzene mixture giving a completely homogeneous solution.

Spent Catalyst Disposal

The aluminum chloride catalyst used in the alkylation is finally rejected from the system as an acidic aqueous solution. This by-product $AlCl_3$ is usually best used for waste water treatment. Sale or use of this material is often as an alum or copperas substitute in water clarification. It is also used in U.S. municipal sewage treatment plants, where a valuable side effect is phosphate removal. In Japan it is converted to poly-aluminum chloride (PAC) and used for water treatment. Other uses for this aqueous aluminum chloride stream exist.

The Monsanto Texas City plant chooses to use an evaporative treatment for aqueous $AlCl_3$ effluent from the wash system.(7) This concentrates the $AlCl_3$ and recovers the HCl as muriatic acid. This latter material is, of course, very useful in the regeneration of zeolite water treating beds. An additional advantage of the evaporative treatment is the freedom of the concentrated by-product $AlCl_3$ from possible organic contamination. The increased markets for aqueous aluminum chloride coupled with the decreased $AlCl_3$ requirements in the homogeneous alkylation process are such that disposal of this by-product is not foreseen as a problem.

Kinetic Modeling

This homogeneous version of the classic Friedel-Crafts $AlCl_3$ chemistry has another interesting by-product; that is, the development of an accurate kinetic model of the reaction. Earlier attempts to model the reaction with the heavy catalyst complex phase present were unsatisfactory. The information used in building the present model, and it might be added for designing the plant, was gained from a one-liter laboratory autoclave opera-ting under semi-batch conditions and also from a fully automated bench unit using reactors with diameters of 1 1/2 inches.

The modeling work went hand in hand with the new insight gained into the chemical reactions taking place in the system. Two important premises made were first, that the alkylation was an irreversible reaction with transalkylation a separate and independent reaction, and second, that the reaction rate of molec-ular ethylene in the liquid was not rate controlling as long as the concentrations of the higher polyethylbenzenes were low. This allowed the kinetic constants for the transalkylation model to be evaluated from transalkylation data; when the combined alkylation-transalkylation model was fitted to alkylation data, only the relative reaction rates of the benzene, ethylbenzene and polyethylbenzenes had to be evaluated in terms of the process variables. The resulting kinetic model extrapolates correctly to extreme values of reaction conditions and also accurately fits the area of practical interest now used in the operating plant.

Choice of Operating Conditions

The major variables controlling the alkylation and transalkylation reactions are as follows:

TABLE II

$AlCl_3$ Catalyst Concentration
HCl Promoter Concentration
Temperature
Residence Time
E/B Molar Ratio (Ethyl Groups/Benzene Rings)
Pressure

The detailed mathematical model of the chemical reactions has revealed the presence of certain relatively sharp optima for these reaction variables. If, for example, it is desired to reduce the $AlCl_3$ usage to a minimum value, then the kinetic model can be used to show the effect of each of the variables on the $AlCl_3$ catalyst concentration. To make the comparison effective, it should be made at the same quality of alkylate. This, for example, can be conveniently expressed as the concentration of tetraethylbenzene in the product. It will be recalled that in the homogeneous system the amounts of higher polyethylbenzenes must be severely restricted because of their relatively basic nature. Figure 7 shows a plot of percent tetraethylbenzene versus temperature at various $AlCl_3$ usages. The graph shows that at given percentage the $AlCl_3$ requirements can be reduced by increases in temperature. For simplicity of presentation, the other variables have been held constant. Similarly, Figure 8 shows the effect of liquid residence time on the $AlCl_3$ requirements.

The graphs showing the effects of the other variables on $AlCl_3$ usage all show the same type of curves with fairly pronounced breakpoints. It is, of course, most advantageous to operate at or close to the breakpoints. The homogeneous alkylation process allows a surprisingly wide range of operation. Changes in one or more of the variables can be used to minimize another of the variables. Thus, the reaction conditions can be tailored to optimize the economics for a particular set of local conditions. For example, a plant can be designed to optimize total cost considering capital, raw materials, catalyst, and

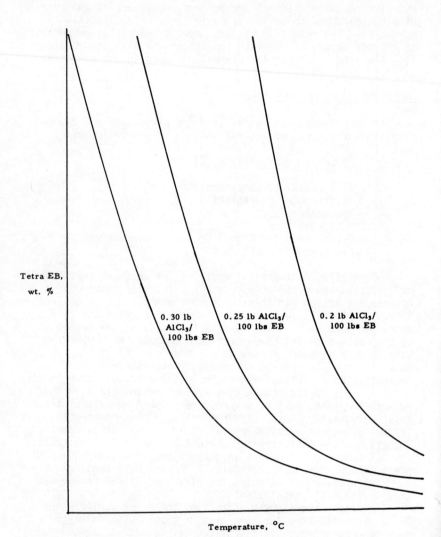

Figure 7. Alkylator—effect of temperature

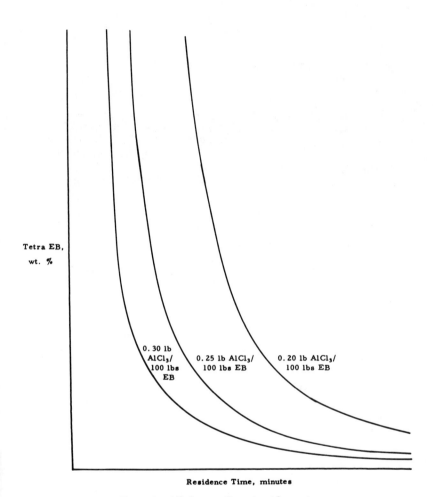

Figure 8.   *Alkylator—effect of residence time*

utilities. Once the plant is built it can be operated for
minimum cost with operating conditions changed as required by
changing energy and raw material costs.
   The homogeneous alkylation process is also adaptable to the
use of dilute ethylene streams. Here the alkylation pressure
has to be somewhat higher, but the processing is quite similar.
A modern styrene facility is currently under construction in
Australia using this dilute C_2H_4 homogeneous alkylation process.

Conclusions

   It is concluded that elimination of the separate catalyst
complex phase in the $AlCl_3$ alkylation process adds significantly
to its attractiveness. In addition to the ease with which a
liquid phase catalyst can overcome any poisoning and get back
on stream, the new homogeneous process can operate at higher
temperatures and recover the heat of reaction to generate steam.
It can also operate in a less corrosive environment while producing an ethylbenzene product of exceptional purity and can
reduce the amount of aluminum chloride required several fold.
   The payoff in any development such as this is in ultimate
plant operation. This process was conceived in the laboratory
and scaled up with a good understanding of the basic chemistry.
The new alkylation process has been in operation at Monsanto's
Texas City plant since February 1974. Typical operating data
are shown in Table III. It is believed to be the world's
largest single-train alkylator (Figure 9). It has a demonstrated
capacity of 30% to 115% of design and has operated extremely
smoothly since startup without process or mechanical problems.
Operations have been characterized by low maintenance costs and
low catalyst usage. We are well pleased with this process and
believe it to be the best proven alkylation available. Five
companies have recently selected this patented process (8) for
installation around the world; including Pemex at their new
complex at La Cangrejera.

Acknowledgement

   The author wishes to express his appreciation to his co-
workers on this project, particularly to F. Applegath and
L. E. Dupree, for their suggestions and assistance with this
paper.

Abstract

   The majority of the ethylbenzene manufactured in the world
is produced by the classical Friedel-Crafts reaction involving
the ethylation of benzene in the presence of aluminum chloride
as a catalyst. It has been customary for the aluminum chloride
to be present in fairly large amounts as a separate heavy

Figure 9. Single train alkylator used to produce 1.7 billion pounds of ethylbenzene per year

TABLE III

TYPICAL OPERATING DATA
ETHYLBENZENE FROM BENZENE AND ETHYLENE

Feed Requirements, lb/lb Ethylbenzene (100% Basis)

Ethylene	0.267
Benzene	0.742

Co-Products, lb/lb Ethylbenzene

Steam	0.99
Aluminum Chloride Sol., 25%	0.008

Catalysts and Chemicals, lb/lb Ethylbenzene

Aluminum Chloride	0.0019
Anhydrous HCl (a)	0.0006
Caustic Soda (100% Basis)	0.00065

(a) May substitute ethyl chloride
on a mole for mole basis.

Utilities, per lb Ethylbenzene

Steam (200 psig)	0.037
Fuel (LHV), Btu	832
Cooling Water (16°F Rise), lbs	23
Process and Boiler Feed Water, lbs	1.11
Electricity, kwh	0.013

catalyst complex phase which is subsequently separated from the alkylated organics layer and continuously recycled to the alkylator. This catalyst complex phase has been found to be not only unnecessary but actually harmful in obtaining maximum yields, purity, and minimum catalyst usage. This paper describes the single-train Monsanto ethylbenzene unit installed at Texas City having a capacity of 1.7 billion pounds a year.

Literature Cited

(1)   Radziewanoski, C., Ber., $\underline{27}$, 3235 (1894).
(2)   Hornibrook, J. N., Chem. and Ind., May 19, 1962, p. 872.
(3)   Donaldson, J. W., "ETHYLENE," Chapter 11 (edited by
      S. A. Miller) Benn London 1969.
(4)   Albright, L. F., Process for Major Addition Type Plastics
      and Their Monomers, McGraw-Hill, 1974.
(5)   Doering, W. Von E., et. al., Tetrahedron, 1958, Vol. 4,
      p. 178.
(6)   Streitweisser, A., and Reif, L. J., Am. Chem. Soc., 1960,
      82, 5003.
(7)   Campbell, D. N. This Book, Next Chapter, 1977.
(8)   U.S. Patent 3,848,012 (Nov. 12, 1974) to Monsanto Company.
      U.S. Patent 3,899,545 (Aug. 12, 1975) to Lummus Company.

22

The Use of By-Product AlCl₃ from an Ethylbenzene Plant to Treat Potable or Process Water

The Use of By-Product AlCl$_3$ from an Ethylbenzene

Plant to Treat Potable or Process Water

DAN N. CAMPBELL

Monsanto Chemical Intermediates Co., Texas City, TX 77590

Most of the ethylbenzene used in the manufacture of styrene is produced by the classical Friedel-Crafts alkylation. The process involves the reaction of benzene and ethylene in the presence of an anhydrous aluminum chloride catalyst promoted by hydrochloric acid. An alkylation mixture containing mono-, di-, tri-, and higher ethylbenzenes together with unreacted benzene is formed in this reaction which is separated by conventional distillation.

Prior to feeding the alkylate to distillation, the aluminum chloride and the hydrochloric acid are removed and recovered. The recovery of the spent alkylation catalyst involves two or three washing steps. The alkylate is first washed (hydrolyzed) with water to remove the dissolved AlCl$_3$ and then with dilute caustic in a second step to get rid of very small amounts of residual acid. A third water wash may be used as a final clean-up if necessary.

A by-product aqueous stream is generated in the first washing step that contains dissolved HCl, alkylate and AlCl$_3$. The uses for this by-product depend upon its purity. It has been used directly in industrial waste water treating and it has been shown to effectively remove phosphate from municipal sewage.[1] In Japan it is converted to polyaluminum chloride, a highly desirable coagulant for waste water.

An additional use of this material is afforded through the use of an evaporator system to purify the aqueous AlCl$_3$. Such a system provides a means of recovering the HCl as muriatic acid, concentrating the AlCl$_3$ and removing most of the soluble hydrocarbon contamination by water azeotroping.

By-product $AlCl_3$ purified in this manner has been
used as a substitute for alum in the paper industry.
There is still another economically attractive
use for this material. It can be moved into the more
lucrative and larger potable water treating markets
as a replacement for either alum or copperas ($FeSO_4$).
Heretofore, this has not been possible because of
hydrocarbon contaminants. This paper describes a
process that has been developed to upgrade the
$AlCl_3$ solution into a product that is suitable for
this use. The quality of the purified $AlCl_3$ and
the results obtained when it was used as a coagulant
in a commercial clarifier are presented.

PURIFICATION OF $AlCl_3$ CONCENTRATE

Even though the concentrated $AlCl_3$ by-product
has been flash-distilled, it still contains some
soluble organics (from the alkylate) at the ppm
level. These must be removed before it can be
used in treating potable water because of possible
harmful physiological effects. Moreover, some
preliminary laboratory work has indicated that a
27% $AlCl_3$ solution containing as little as 1 ppm
benzene can impart odor to the treated water.
The data suggest that the nose can detect benzene
in water at a concentration of 10 parts per trillion.
There are two obvious ways of removing the
remaining hydrocarbon contamination. Multi-stage
stripping could be used but the most economical
and simplest method is to add a carbon bed adsorption step to the process as indicated in Figure 1.
Carbon bed removal can be justified only in conjunction with the evaporator system where the hydrocarbon
contamination is very low. The concentration of
soluble organic components in a system where the
evaporator is not used is simply too high and the bed
is exhausted too quickly.
A fiber-plastic column, 1 foot in diameter,
filled with 12 feet of Calgon Corporation's CPG
activated carbon was used to prepare hydrocarbon-
free $AlCl_3$ for a test. This column was used simply
because it was available and a much larger diameter
column with less height would be designed for a
commercial unit. A relatively fast flowrate of
1 gal/min was used in the adsorption column.

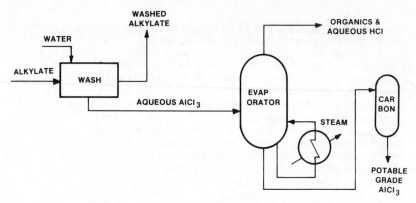

Figure 1.   Commercial purification of AlCl₃ solution

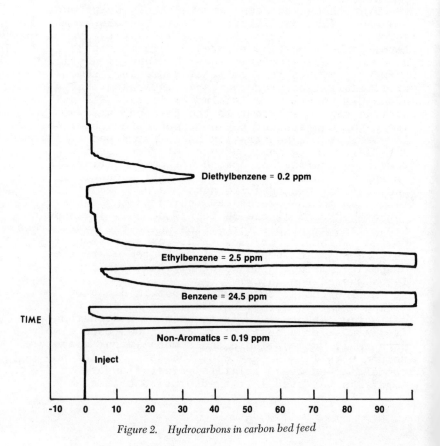

Figure 2.   Hydrocarbons in carbon bed feed

Analytical results for the AlCl₃ solution before
and after carbon treatment are shown in Table I.

Table I

ORGANICS IN BY-PRODUCT 27% AlCl₃ SOLUTION

Contaminant	Feed, ppm	Carbon Bed Effluent, ppm
C_1-C_4 Non-Aromatics	0.19	0.0242
Benzene	24.5	<0.0001*
Ethylbenzene	2.5	<0.0002*
Diethylbenzene	0.2	<0.0005*
Polynuclear Aromatics	-	<0.003*
Organic Chlorides	-	<0.003*

*None Detected

Polynuclear aromatics in the effluent were
determined by hexane extraction and ultraviolet
absorption of the concentrated extract. Benzene
and the other low boiling organics were analyzed
by vapor phase (N_2) partitioning and gas chromatography. Suspended solids which are not shown in
this data were in the 100-200 mg/l range. A
cartridge filter will reduce the solids to the
50 mg/l usually specified by producers of activated
carbon very easily.
     As indicated in Table I, only a very small
amount of methane through butane (C_1-C_4 Non-
Aromatics) remained in the carbon bed effluent.
And as to be expected the carbon had a high affinity
for the aromatics. The gas chromatograph strip
chart in Figure 2 illustrates the sensitivity of the
gas chromatograph method for the hydrocarbons.
Benzene and ethylbenzene were off scale in the feed
whereas there was no evidence of even a trace of
benzene in the carbon treated AlCl₃.

In some plants the aluminum chloride catalyst is made in-situ from scrap aluminum and anhydrous HCl. Table II shows the level of metallic components present in the AlCl₃ used in this work.

Table II

TRACE METALS IN AQUEOUS AlCl₃ SOLUTION

Component	ppm
Copper	469
Iron	92
Mercury	0.00007
Manganese	12
Magnesium	569
Zinc	725
Others	<5 ea.

Copper, magnesium and zinc at the 1-2% level are normal components of scrap aircraft aluminum alloys.

CLARIFIER OPERATING VARIABLES

Coagulation effectiveness of the AlCl₃ was studied in a 164,000-gallon Infilco Accelator (clarifier). This commercial unit is a combined clarification/softening operation using 30-35 ppm copperas, a by-product of the metal pickling industry, as the coagulant. And in addition, lime is injected into the clarifier sludge blanket to adjust the pH to 10, resulting in precipitation of the $Ca(HCO_3)_2$ temporary hardness as $CaCO_3$. Copperas has been the preferred coagulant for the turbid and hard plant raw water that is obtained from the Brazos River. The effluent from this unit is used for both process and potable water.

OPTIMUM AlCl₃

Turbidity and residual aluminum were the main points of concern in substituting the AlCl₃ for copperas. Excess aluminum carryover could foul the ion exchange resins in the plant demineralizers and cause AlPO₄ or aluminum silicate scale in downstream cooling equipment. Optimum coagulant treatment level was established first; and as shown in Figure 3, 10-15 ppm was most effective. Turbidity and residual aluminum were at a minimum at these levels.

Most of the total residual aluminum shown was floc carryover rather than dissolved aluminum. This is significant because floc carryover is removed by an anthracite coal bed filter prior to entering the potable water system. In Figure 3 the "dissolved" aluminum was taken as that remaining in the supernatant liquid after the samples had been standing for 4-5 days. Aluminum analyses were done by atomic adsorption using a heated graphite atomizer to get reliable low (<1 ppm) data. As indicated in this graph, the dissolved aluminum was near that obtained with a copperas-treated potable water.

Clarifier effluent turbidity levels were higher than normal [3.5 vs 3.0 nephelometric turbidity units (NTU) for copperas] because the unit was operating with the sludge level in the mixing chamber too low. Figure 4 illustrates the effect of sludge level on turbidity. The points on this curve were taken at 2-hour intervals and are scattered, but regression analysis shows a good correlation. The curve indicates that the optimum percent sludge had not been reached. Sludge concentrations of 12-15 vol % (measured after a 5-minute settling time) are normal in this unit with copperas.

Effect of pH

Adjustment of the pH in the clarifier with lime produced surprising results considering the amphoteric nature of aluminum. At pH 8, as shown in Figure 5, both total (floc carryover + soluble) and dissolved aluminum were very low (turbidity also) but as the pH was increased to 9, they increased sharply. From pH 9-10, total and dissolved aluminum decreased again. This was not expected because with NaOH aluminum is more, not less, soluble at pH 10. The data in Figure 5 are in replicate and the same phenomenon has been observed in laboratory jar test also. What causes this has not been resolved at this time. It could be a result of changes in the charges at the colloid double layer or

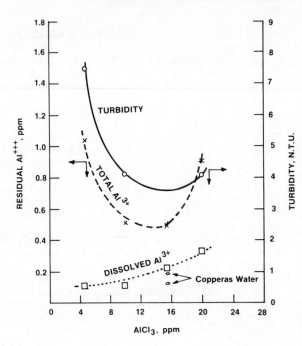

Figure 3.   Optimum AlCl₃ concentration (res. time = 82 min, pH = 10)

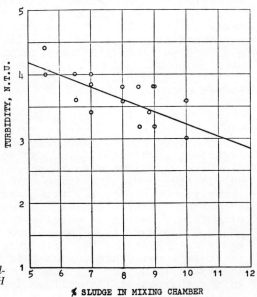

Figure 4.   Effect of sludge volume (15 ppm AlCl₃, 82 min, pH 10)

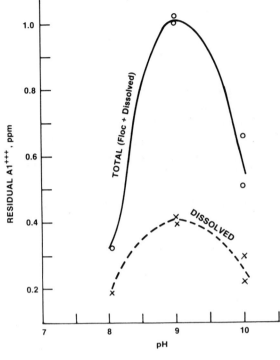

Figure 5. Residual Al³⁺ vs. pH (82 min, 15 ppm)

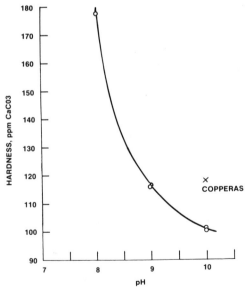

Figure 6. Hardness vs. pH (82 min, 15 ppm AlCl₃)

$Ca(AlO_2)_2$ may have a lower solubility at pH 10.
Nevertheless, the lower aluminum at pH 10 is a
significant advantage since as with copperas, cold
lime softening and clarification can be done in
one step. Figure 6 shows the reduction achieved
in the Brazos River water hardness as a function
of pH. Approximately 18 ppm less hardness was
obtained with the $AlCl_3$ than with copperas.
Sulfuric acid, a natural impurity of copperas,
causes an increase in permanent hardness in lime
softening operations.

POTABILITY, ETC.

   Results of drinking water analyses by an
independent laboratory using standard methods of
the Environmental Protection Agency (EPA) are listed
in Table III for the $AlCl_3$ treated and the raw
Brazos River waters.
   In all cases, the aluminum chloride water met
EPA criteria and it is potable. The EPA has been
contacted about this work, and it was the opinion
of one of their representatives that there is no
reason why $AlCl_3$ would not be an acceptable
coagulant in treating potable water. He expressed
an interest in getting the work published so that
the information would be available to all.
   Additional analyses are given in Table IV to
illustrate further the treated water quality. Copper
and zinc analyses are included to show that these
ions are carried down with the floc as would be
expected. These two were the highest heavy metal
contaminates of the by-product $AlCl_3$ (see Table II).
Zinc present in the raw river water was also removed.
   Brazos River water contains a relatively high
concentration of total organic carbon (TOC) in the
summer. Some of this carbon is removed in the floc
as indicated by both the CCE (Carbon Chloroform
Extract) and TOC.
   Chloride ion concentration in a closed cooling
water system may be concentrated as much as seven
times and it can contribute to corrosion of stainless
steel equipment. Brazos River water contains as
much as 200 ppm during dry periods. Thus, it is
doubtful that the increased chloride (7 ppm) would
cause measurable additional corrosion in any case.

TABLE III

<u>DRINKING WATER STANDARDS-PLANT CLARIFIER TEST WITH AlCl₃</u>
July 16, 1976

CONTAMINANT	EPA PROPOSED MAXIMUM, ppm	AlCl₃ COAGULATION, ppm	RAW BRAZOS RIVER, ppm
Arsenic	0.05	<0.01	<0.01
Barium	1.0	<0.5	<1.0
Cadmium	1.01	<0.01	<0.01
Chromium	0.05	<0.01	<0.01
Fluoride	1.4-2.4	0.4	0.2
Lead	0.05	<0.05	<0.05
Mercury	0.002	<0.001	<0.001
Nitrate	10	6.65	7.3
Selenium	0.01	<0.01	<0.01
Silver	0.05	<0.01	<0.05
Turbidity, (NTU)	1	1*	>10

---

*Potable water is filtered

Table IV

ADDITIONAL WATER QUALITY ANALYSES PLANT CLARIFIER
TEST WITH AlCl₃

CONTAMINANT	MAXIMUM, ppm	AlCl₃ COAGULATION, ppm	RAW BRAZOS RIVER, ppm
Cyanide		< 0.1	< 0.1
Organics, CCE		0.5	2.43
TOC	-	81	87
TDS*	500	381	413
Chloride	250	79	72
Aluminum	1	0.3	0.15
Copper	1	< 0.01	< 0.01
Zinc	5	< 0.01	4.9

---

*Total dissolved solids.

ECONOMICS

The significance of the development is that a
low-value by-product from a styrene plant has been
upgraded to a very useful chemical. A one-billion-
pound-per-year styrene plant using the most modern
technology will produce approximately 2 million
pounds of by-product aluminum chloride. This is
enough to treat the water required annually by
a City of approximately 250,000 population. This
estimate is based upon the amount of coagulant
needed to clarify the high hardness and turbidity
of waters in Texas' Brazos River.

Total costs of the evaporator system and the
carbon bed to upgrade the by-product have been
estimated to be about $0.05/lb of contained $AlCl_3$.
Recent potable grade copperas coagulant has been
quoted at $0.10/lb delivered in the Houston area
and thus, the recovered $AlCl_3$ would be more than
competitive. This use in potable water treating
also will provide an expanded and more stable
market for the aqueous $AlCl_3$.

CONCLUSIONS

In conclusion, the aqueous $AlCl_3$ from an
ethylbenzene plant can be upgraded to a by-product
suitable for use in treating both process and
potable water. The process is economical, easy
to operate, and all of the necessary engineering
technology is readily available.

REFERENCES CITED

1. Connell, C. H., University of Texas Medical
   Branch, Galveston, Texas for the Environmental
   Protection Agency, "Phosphorus Removal and
   Disposal From Municipal Waste water," Project
   No. 17010DYBO2171.

23

Alkylation of Benzene with Propylene over a Crystalline Alumina Silicate

E. F. HARPER
P. O. Box 836, Bristow, OK 74010

D. Y. KO
E. I. du Pont de Nemours, Wilmington, DE 19899

H. K. LESE, E. T. SABOURIN, and R. C. WILLIAMSON
Gulf Research & Development Co., P. O. Drawer 2038, Pittsburgh, PA 15230

There are many examples of the alkylation of aromatics with
olefins to produce alkylbenzene in textbooks, the open literature,
and in numerous patents. This reaction is catalyzed by both proton and Lewis acids in a homogeneous phase and in heterogeneous
phases. The latter systems are characterized by both proton
(H_3PO_4) and Lewis acids (BF_3) on supports and the amorphous and
crystalline alumina silicates. And, the reaction has been
studied extensively. However, up until the start of this investigation (1969) there had not been a systematic investigation of
the kinetic parameters nor an adequate catalyst aging study on
the alkylation of benzene with propylene over a crystalline
alumina silicate.

The reaction is known to proceed through activation of the
olefin by the catalyst (a rare earth exchanged Y Zeolite in this
work) and this activated olefin then reacts with benzene and
alkylbenzene. However, since the concentration of activated
olefin is not known, the total concentration of olefin is used
in obtaining kinetic parameters according to the following
reactions.

$$Benzene + Propylene \xrightarrow{k_1} Cumene \quad (1)$$

$$Cumene + Propylene \xrightarrow{k_2} Di\text{-}isopropylbenzene \quad (2)$$

$$Di\text{-}isopropylbenzenes + Benzene \underset{k_4}{\overset{k_3}{\rightleftharpoons}} 2\ Cumene \quad (3)$$

Other polyalkylbenzenes are also formed by the sequential reaction
of di-isopropylbenzenes with propylene to form tri-isopropyl-
benzenes which can further react. However, at a molar ratio of
benzene to propylene of 6 or higher these compounds are not
important.

Development of A Kinetic Model

The above overall reaction scheme was used to represent the process. The kinetic models were developed to describe these reactions as shown. No attempt was made to study the basic reaction mechanisms. The following point rate models were formulated and tested:

$$r_1 = k_1 C_B C_P \tag{4a}$$

$$r_2 = k_2 C_C C_P \tag{4b}$$

$$r_3 = k_3 C_D C_B \tag{4c}$$

$$r_4 = k_4 C_C^2 \tag{4d}$$

$$k_4 = k_3/K \tag{4e}$$

The subscripted r's are the four reaction rates representing the reactions shown above, and the units are moles per hour per unit weight of catalyst. The k's are the rate parameters. The subscripted C's are the concentration of benzene, B; propylene, P; cumene, C; and di-isopropylbenzenes, D. Equation (4e) defines the equilibrium constant, K, for reaction (3). The concentration units used are moles of component per unit mass of total reactor fluid. Over the range of conditions used in this work, the reactor fluid was a liquid phase maintained by sufficient pressure at all conditions of temperature and composition.

These basic rate models were incorporated into a differential mass balance in a tubular, plug-flow reaction. This gives a set of coupled, non-linear differential equations which, when inte-grated, will provide a simulation model. This model corresponds to the integral reactor data provided by experimentation. A material balance is written for each of the four components in our system:

$$\frac{dC_B}{d\tau} = -r_1 - r_3 \tag{5}$$

$$\frac{dC_P}{d\tau} = -r_1 - r_2 \tag{6}$$

$$\frac{dC_C}{d\tau} = r_1 - r_2 + 2r_3 \tag{7}$$

$$\frac{dC_D}{d\tau} = -r_3 + r_2 + r_4 \tag{8}$$

The initial conditions (concentrations) are given for each
component. The independent variable, τ, is the total liquid
weight hourly space time, weight catalyst-hour per weight total
liquid feed. This form of the differential equations was used
in the simulation and model building. These were convenient for
representing the concentration profiles in the same form in
which the data were taken. An energy balance was not required
at this stage, because the reactors were operated isothermally
for each set of runs. Equations (5)-(8) were mechanized and
solved on an anolog computer (EAI 680). The parameters were
adjusted and determined for each set of composition-space time
data. The parameters obtained should characterize the reaction
rates, the activation energies, the chemical equilibrium, and
its temperature dependence.

The main point was apparent what kinds of experimental data were
required. For a given reactor inlet composition, and a given
operating temperature, reactor data were collected at various
space times. This provided a composition profile for each
component from which the parameters could be found by simulation.
Sequentially, other temperatures and initial compositions were
selected and additional data were obtained. The experimental
data and simulated curves are shown in Figures 1, 2, and 3.

Using data of these types and a postulated model, it was
possible to simulate the observed conditions, determine the
parameters, and pool the results. These pooled parameter values
were combined and reduced to one simple model representing all
the data, and thus the process under study. The kinetic model
forms were also inspected for reasonableness and consistency.

At the higher temperature levels, it was noticed that some
unreacted propylene was present at a space time of 0.0125, or 80
space velocity. At higher space time, there was no unreacted
propylene. This meant that the primary alkylation reactions
were extremely fast--implying very high catalytic activity for
alkylation. The major portion of the reactor space was used for
transalkylation, giving an improvement in selectivity. This was
seen by noting the rapid decay of the di-isopropylbenzenes
content after they reached their maximum. It was also evident
that equilibrium was closely approached at the highest space
times.

The values of the rate and equilibrium constants at two
temperature levels, frequency factors, the activation energies,
and the heat of the transalkylation reaction for the model
finally obtained are listed in Table I. It is seen that the
equilibrium constant increases with decreasing temperature,
indicating the transalkylation is an exothermic reaction. The
heat of reaction for transalkylation is about 6.5 kcal per mole.
This means that the cumene content at equilibrium is favored by
decreasing the temperature, although the effect is not particularly great.

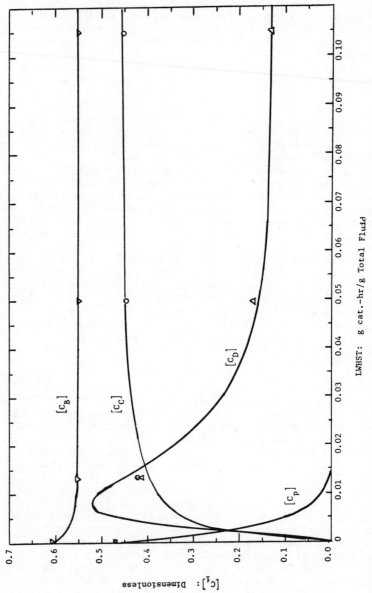

Figure 1. Consecutive reactions sequence model. $Bz/pp = 10.25$, $T = 232°C$. $[c_i] = c_i/c_{imax}$. c_i, c_{imax}: moles $i/100$ g total fluid. $c_{Bmax} = 2$, $c_{pmax} = 0.25$, $c_{cmax} = 0.25$, $c_{Dmax} = 0.02$.

LWHST: g cat.-hr/g Total Fluid

$[c_i]$: Dimensionless

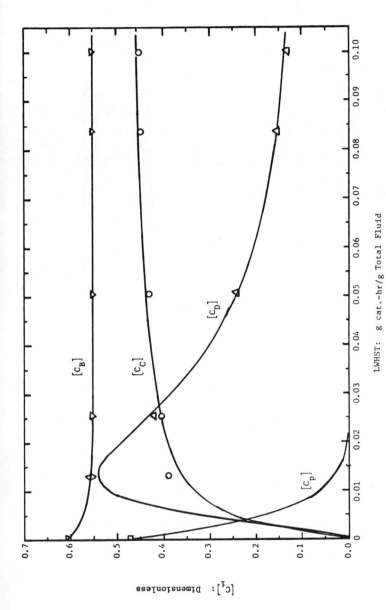

LWHST: g cat.-hr/g Total Fluid

$[c_i]$:   Dimensionless

Figure 2.   Consecutive reactions sequence model. Bz/pp = 10.2, T = 214°C. $[C_i] = C_i/C_{imax}$. $C_i C_{imax}$: moles/100 g total fluid. $C_{Bmax} = 2$, $C_{pmax} = 0.25$, $C_{Cmax} = 0.25$, $C_{Dmax} = 0.02$.

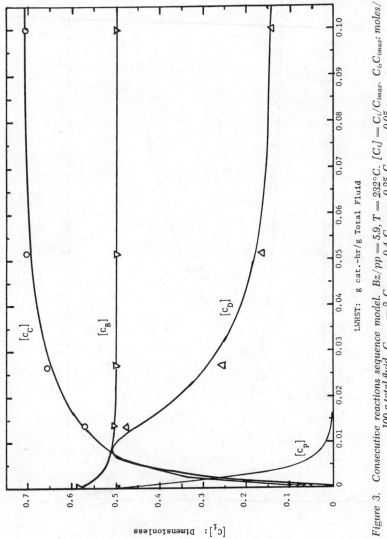

Figure 3. Consecutive reactions sequence model. $Bz/pp = 5.9$, $T = 232°C$. $[C_i] = C_i/C_{imax}$. $C_{i}C_{imax}$: moles/ 100 g total fluid. $C_{Bmax} = 2$, $C_{pmax} = 0.4$, $C_{Cmax} = 0.25$, $C_{Dmax} = 0.05$.

Table I

REACTION RATE AND EQUILIBRIUM PARAMETERS

Rate Constants: $k_i = k_{io} e^{-E_i/RT}$, (g Total Fluid)^2/mole-g cat-hr

Equilibrium Constants: $K = K_o e^{-\Delta H/RT}$, Dimensionless

Temp, °F	$k_1 \times 10^{-5}$	$k_2 \times 10^{-5}$	$k_3 \times 10^{-4}$	$k_4 \times 10^{-4}$	K
450 (232°C)	0.2485	0.8890	0.5805	0.1340	4.34
417 (214°C)	0.1578	0.5580	0.3250	0.0590	5.51

k_i	k_{io}	E_i (Kcal/g mole)
1	5.45×10^9	12.32
2	2.628×10^{10}	12.64
3	3.80×10^{10}	15.75
4	5.83×10^{12}	22.27

K^1	Ko	ΔH (Kcal/g mole)
	6.52×10^{-3}	−6.52

----

^{1} $Ko = \dfrac{k_{30}}{k_{40}}$, $-\Delta H = E_4 - E_3$

With the kinetics models available, one can now study the behavior and the performance of various types of reactor systems, such as those used for commercial operation. However, before that can be done, the thermophysical properties of the components at the reaction condition have to be determined.

Thermophysical Properties

This is a liquid-phase catalytic reaction system and the reaction conditions are very close to the critical conditions of the reactants propylene and benzene. The values of the thermophysical properties (e.g., heat of formation and heat capacity) are generally not available at the reaction conditions and are difficult to evaluate accurately. We evaluated how well the thermophysical properties were estimated by simulating a commercial cumene reactor, and comparing the adiabatic temperature rise of the simulation with that of the observed data available.

Most published data of the heat of formation are for the components at the ideal gas state. Heats of formation at the liquid state can be evaluated by subtracting heats of vaporization at the temperature of interest from heats of formation at the gas state (ΔH°_{fg}). For the present system, all ΔH°_{fg} values are available except that of the di-isopropylbenzenes. The values of ΔH°_{fg} of di-isopropylbenzenes were obtained by group contribution methods and averaged to present the average isomer distribution observed in the reaction.

The heat capacities used for all liquid components were estimated from available engineering correlations. These values were reduced to equation form. The models and coefficients and the heat of formation at the liquid state are listed in Table II.

Energy Balance. It was decided that a simple, fixed-bed, adiabatic reactor would be required in this process. Since the reactions involved release considerable heat, this influences the local temperature, which in turn influences the reaction rates. An energy balance, or heat balance, having the following general form, was added along with the mass balance in all subsequent simulations:

$$\frac{dT}{d\tau} = \frac{\overline{M}}{\overline{C}_p} \left(\sum_{i=1}^{n} R_i \, \Delta H_{fi} \right) \qquad (9)$$

$\overline{M}$ = average molecular weight of the total fluid

$\overline{C}_p$ = average heat capacity, cal/g mole, F

R_i = net rate of production of the i-th component, g-mole/hr-g cat.

ΔH_{fi} = heat of formation of i in liquid state, kcal/g-mole.

Table II

THERMOPHYSICAL PROPERTIES AT LIQUID STATE

Heat of Formation: ΔH_{fl} = Kcal/mole

Heat Capacity: $C_p = \alpha + \beta T$, C_p = cal/mole F, T = F

Compound	ΔH_{fl} (330–480°F)	α	β
Propylene	-8.431	8.118	0.01424
Benzene	12.652	15.511	0.02194
Cumene	-10.503	25.454	0.034
Dicumene	-32.87*	36.315	0.045

*Estimated by group contribution methods.

This energy balance when coupled with material balances, Equations (5) through (8), generated point temperatures over the range of space times considered.

   Digital Computer Simulation Program. A digital reactor simulation program was used to integrate the steady state differential mass balances and the energy balance for a plug-flow reactor. Numerical solutions for the resulting set of coupled, non-linear differential equations were obtained by using the fourth-order Runge-Kutta method. The entire set of differential equations was integrated simultaneously along the reactor length to obtain the composition and temperature profiles. The main program was general in nature and the reaction rate equations and thermodynamic properties were introduced in a subroutine. This program developed a print-out containing composition and temperature information along the space-time integration path. In addition, yields, selectivities, and conversions were also generated. In this way, all details for the reactor simulation were made available at many specified space-time conditions.

   Adiabatic Reactor Simulations. Using the chemical reaction system model and this digital program, we conducted many simulations. Since all of our simulations were conducted to approach an equilibrium condition, adiabatic temperature rise information was generated directly. A summary of these results is shown in Figure 4. This temperature rise varied from about 30 to 61°C over the range of variables studied. A major variable studied was the cumene selectivity based on propylene. Figures 5, 6, and 7 show generated results relating selectivity to three process variables. At a given inlet composition to the reactor, the selectivity is influenced differently by temperature. The magnitude and direction depends on the space time employed. The optimum selectivity at high space times was increased with reduced inlet temperature. Whereas, intermediate selectivities at low space times were increased by an increase in inlet temperature over a space time range from about 0.04 to 0.1.

Effects of Operation Variables
on the Performance of the Reactor

   Three operation variables, benzene-to-propylene ratio (R_B), temperature (T), and space time (τ) are of prime concern to a reactor designer. Since alkylation reactions are so fast, the controlling performance of the reactor is determined by the reaction selectivity. At a constant average temperature, examination of the reaction paths and kinetics models indicated that selectivity increased with increasing benzene-to-propylene ratio regardless of the space time applied. The temperature effect was slightly more complicated. Since most of a process reactor

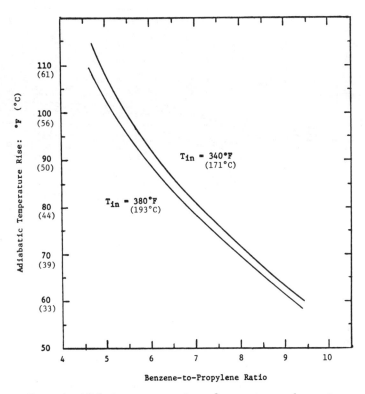

Figure 4. Adiabatic temperature rise vs. benzene-to-propylene ratio. Parameter = reactor inlet temperature.

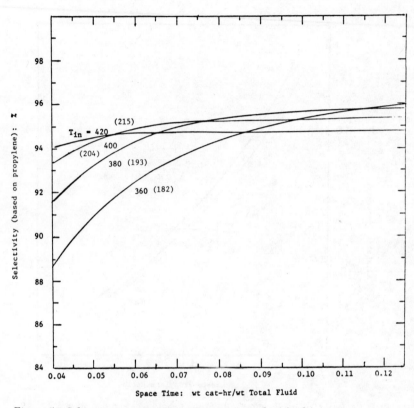

Figure 5.   Selectivity vs. space time. Benzene/propylene (mole) = 10; parameter = reactor inlet temperature (T_{in}), °F (°C).

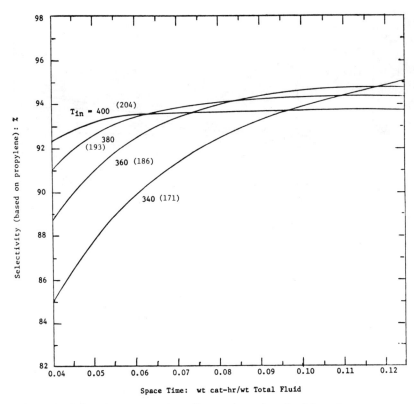

Figure 6. Selectivity vs. space time. Benzene/propylene (mole) = 8; parameter = reactor inlet temperature (T_{in}), °F (°C).

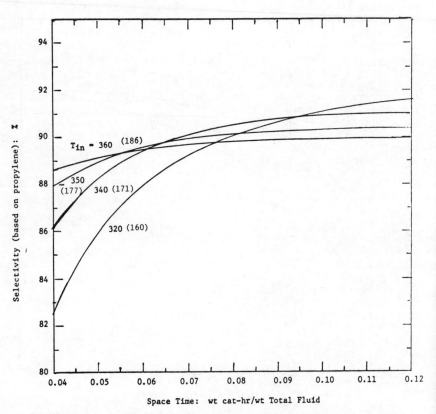

Figure 7. Selectivity vs. space time. Benzene/propylene (mole) = 5; parameter = re-actor inlet temperature (T_in), °F (°C).

will be operated in the transalkylation reaction region, and the
latter reaction is a reversible, exothermic reaction, the selectivity is favored by lowering the temperature when the reactor
is operated close to transalkylation equilibrium. However, at
low space time, the selectivity is determined by how quickly
cumene can be formed. Thus, kinetics instead of equilibrium
would be the dominating factor for selectivity in the low space
time region. In other words, when the kinetic temperature
effect is evaluated, space time must be considered simultaneously.
The following adiabatic reactor analysis will elaborate on and
substantiate the above discussion.

In an adiabatic operation, inlet temperature rather than
reactor average temperature is one of the prime operating
variables, because it can be easily manipulated. Table III
shows the effect of the reactor inlet temperature on cumene
selectivity. At a low space time ($\tau = 0.05$), when reaction rate
is still the determining factor, higher inlet temperature gives
higher selectivity, because of the higher reaction rate. However,
at a higher space time ($\tau = 0.1$) where equilibrium is virtually
reached, a higher inlet temperature results in a lower selectivity, because of the temperature effect on transalkylation equilibrium.

Table IV shows the effect of benzene–to–propylene ratio in
the feed on selectivity at two space velocities. It is very
apparent that when the ratio is increased, the selectivity
becomes higher.' The reasons for this are three-fold. Kinetics
of the two alkylation reactions favor production of cumene if
the benzene-to-propylene is higher. The equilibrium concentration
of cumene is higher, because of higher benzene concentration.
At the same inlet temperature, a higher benzene concentration
results in lower outlet temperatures which in turn increase the
equilibrium constant of the transalkylation reaction. It is
interesting to note that the increase in selectivity is more
significant at a higher space time because of the more pronounced
effect of the equilibrium.

Optimal Adiabatic Reactor Design. The selectivity curves
generated in Figures 5, 6, and 7 show a relationship between the
benzene/propylene molar ratio, inlet temperature, and space time
for an adiabatic reactor. One would like to have a formula,
chart, or graph that would easily predict optimal yields as a
function of these variables. Figure 8 does this, and is really
another way of presenting the envelopes defined in Figures 5–7.
Figure 8 also allows one to select proper operating conditions
as the catalyst ages.

Catalyst Aging Characteristics and Product Purity

All of the work done to establish the kinetic parameters
was conducted on the same catalyst bed used for a long-term

Table III

SELECTIVITY VERSUS INLET TEMPERATURE

Benzene/Propylene = 8.27 (Mole), π = 514.7 psia

Run	1	2	3	4	5	6
Inlet Temperature, °F (°C)	380 (193)	390 (199)	400 (204)	380 (193)	390 (199)	400 (204)
Outlet Temperature, °F (°C)	448 (231)	457 (236)	466 (241)	448 (231)	457 (236)	467 (241)
LWHSV, wt Liq/wt Cat-hr	20.83	20.83	20.83	10.0	10.0	10.0
Selectivity Based on Benzene	96.15	96.38	96.45	97.16	96.99	96.78
Selectivity Based on Propylene	92.69	93.16	93.34	94.58	94.30	93.98

Table IV

SELECTIVITY VERSUS BENZENE-TO-PROPYLENE RATIO

Run	1	2	3	4	5	6
Bz/pp (Mole)	6	8.27	10	6	8.27	10
Inlet Temperature, °F (°C)	390 (199)	390 (199)	390 (199)	390 (199)	390 (199)	390 (199)
Outlet Temperature, °F (°C)	476 (247)	457 (236)	446 (230)	477 (247)	458 (236)	447 (230)
LWHSV wt Liq/wt Cat-hr	20.83	20.83	20.83	10.0	10.0	10.0
Selectivity Based on Benzene	95.29	96.38	96.75	95.50	96.99	97.55
Selectivity Based on Propylene	91.10	93.16	93.85	91.48	94.30	95.56

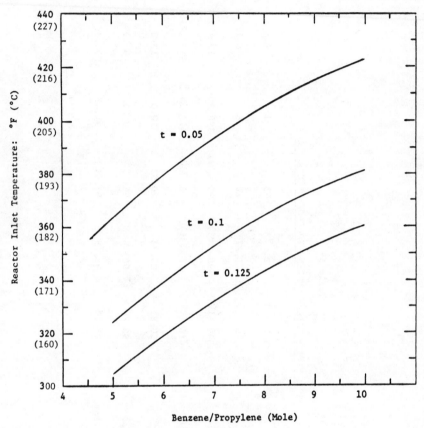

Figure 8.   Inlet temperature vs. benzene/propylene for maximum selectivity. Parameter = space time, t.

isothermal (232°C) aging study. This study lasted for 9 months
at an average LWHSV = 10 with the catalyst having produced 900 g
cumeme/g-catalyst. The study was then terminated with no evidence
of the catalyst having aged with the exception of a slight
dimunition in selectivity over this time period. A typical
weight percent analysis on the product was the following:
cumene, 16.53%; di-isopropylbenzenes, 0.69%; ethylbenzene,
0.043%; a trace of n-propylbenzenes; and benzene, 82.53%. These
data give an efficiency to cumene of 93.7% based on propylene
and when coupled with the aging data would apparently justify
the catalyst being used in a cumene unit. However, as we will
soon see, what happens in an isothermal reactor does not necessarily occur in an adiabatic reactor.

A larger pilot plant unit was built and properly insulated
to assure essentially no heat loss. High feed rates were used
to simulate what happens in a unit area of a commercial reactor
and to accelerate a very slow rate of catalyst aging to one we
could measure over a period of 2-3 weeks. The first experiment
was designed to approximate the isothermal reactor. However,
the conditions actually were 180°C inlet temperature, 221°C
outlet temperature, and 8/1 molar ratio benzene/propylene at a
linear hourly space velocity of 13. This experiment started out
as expected with the alkylation reaction occurring in the first
2-3 inches of the catalyst bed and the reactor was adiabatic.
However, after about 3 hours, it was obvious the catalyst was
aging quite rapidly as evidenced by the alkylation zone moving
up the reactor. After 20 hours, 12 inches (90 grams) of the
catalyst had been completely deactivated and the experiment was
terminated. The rate of aging was 0.57 inches/hour. This rapid
aging was quite unexpected since the feed was pumped over the
catalyst only 1.3 times faster than during the isothermal run
and the catalyst was cooler in the alkylation zone which should
have been beneficial according to the teachings of Wise.(1)
A more careful examination of the feed rate data showed that,
while it was true the benzene/propylene feed was pumped over the
entire catalyst bed 1.3 times faster than during the isothermal
run (LWHSV adiabatic = 1.3 LWHSV isothermal), the feed rate
through the alkylation zone was much higher. A convenient way
of expressing this rate was to determine how many grams of total
feed passed through a unit (cm^2) of catalyst surface area per
hour. This rate was termed as superficial velocity and had
g/cm^2-hr dimensions. For the isothermal reactor the superficial
velocity was 180 g/cm^2-hr and in the adiabatic run it was
1400 g/cm^2-hr or superficial velocity ratio of about 8. This
could account for the aging rate differences (less than 10 g
catalyst in 9 months for the isothermal reactor versus 90 g
catalyst in 20 hours adiabatic) if the aging rate difference was
a tetramic function of this ratio.

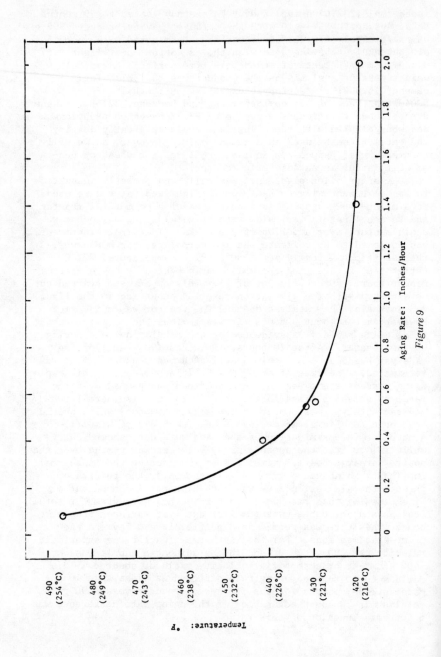

Figure 9

This hardly seemed a reasonable rational of the fast aging rate, but neither did one involving lower temperature. A series of experiments were done to ascertain which of these if either was the dependent variable. Figure 9 shows that temperature was a very unexpected and heretofore unknown inverse function of the aging rate. All of the temperatures shown in Figure 9 are those of the hot spots leading edge. Feeds entered about 59°C lower. The effect of superficial velocity was shown to be directly proportional to the concentration of propylene but was only slightly higher than unimolecular. The linear aging rate then has the following form.

$$\text{Linear Aging Rate} = \frac{A \cdot f(\text{propylene})}{ft} \text{ inches/hour}$$

However, this rate is dependent on the cross sectional area of the reactor. A more accurate relationship would be one showing the volume of catalyst aged per unit time.

$$\text{Volume Aging Rule} = \text{Linear Aging Rate x Area}$$

The data in Figure 9 should then be multiplied by 0.8152 to convert to in^3/hr or by 13.3586 to convert to cm^3/hr if these data are to be compared to that obtained in other reactors.

While the aging characteristics of the catalyst were being determined, GLC analyses of the product showed that larger quantities of ethylbenzene and n-propylbenzene were produced at the higher temperatures. These products arise from a cracking and an isomerization mechanism, respectively. Both of these products would then have temperature and catalyst residence time as a dependent variable. A study of both these products as a function of temperature and LWHSV was then done by using a small adiabatic alkylation reactor operated at high space velocity followed by a larger isothermal or transalkylation reactor. This is essentially what an adiabatic reactor is, but the temperature in the isothermal zone is dependent on the adiabatic heat generated. The results of this study are shown in Table V. The data show that lower temperatures and higher flow rates give a cleaner product with higher efficiencies. However, at temperature lower than 221°C and LWHSV's higher than 5, efficiencies are lowered due to the equilibrium reaction 3.

Experimental

The catalyst used in most of the work reported here was Union Carbide's SK-500. The catalyst is a Y faujasite with the following composition: $(La^{+++})_{8.8}(NH_4^+)_{21.1}(Na^+)_{8.3}[(AlO_2)_{55.7}(SiO_2)_{136.3}] \cdot ZH_2O$. The catalyst was activated by heating to 550°C for 90 minutes with a N_2 stream to drive off NH_3 and H_2O.

Table V

ALKYLATION CONDITIONS: 8/1 B/P, LHSV = 23 h^{-1}, 4.2 MPa
(600 PSIG), 252°C (485°F)

Transalkylation LHSV			2.3	4.5
Transalkylation Temperature				
221°C (430°F)	% NPB		0.025	0.023
	% EB		0.037	0.042
	% NPB/Cumene		0.18	0.14
	Benzene Selectivity		96.1	95.4
227°C (440°F)	% NPB		0.037	
	% EB		0.061	
	% NPB/Cumene		0.27	
	Benzene Selectivity		96.1	
232°C (450°F)	% NPB		0.051	0.038
	% EB		0.086	0.085
	% NPB/Cumene		0.38	0.23
	Benzene Selectivity		95.6	95.2
252°C (485°F)	% NPB		0.166	0.112
	% EB		0.351	0.280
	% NPB/Cumene		1.10	0.70
	Benzene Selectivity		92.2	92.9

NPB = n-Propylbenzene

EB = Ethylbenzene

Gulf's commercial grade benzene (99[+]% benzene) and alkylation grade propylene (73–75% propylene and 27–25% propane) were
used in all the work.

The kinetic studies were done in a 1/2-inch pipe jacketed
reactor with a 1/8-inch thermowell down the middle. The first
7 inches of the reactor was filled with quartz chips and used as
a preheater. Ten grams of 10–20 mesh SK-500 was then mixed with
twice its volume of quartz and placed on top of the preheat
section. The feed was pre-mixed in a pressurized tank and then
pumped upflow through the reactor. The temperature was controlled
by pumping Dow Therm through the jacket with heat being supplied
externally. This resulted in the reactor being isothermal.

The remainder of the experimental results were obtained in
a 1-inch pipe reactor with a 1/4-inch thermowell which was well
lagged with 2.5 inches of asbestos and equipped with heat compen-
sation to insure an adiabatic reactor. The lower 12 inches of
the reactor was used as a preheat section and filled with glass
beads. The catalyst section was filled to 72 inches with a
1/16-inch extrudate which weighed 565 grams. Benzene and propylene were pumped into a pressurized autoclave which serves as a
partial preheater and mixer and then upflow through the reactor.
The desired temperature was obtained by adjusting the preheat
section to a temperature about 43°C below the desired one and
the exotherm then raised it to the desired level. The pressure
on the reactor was maintained at 3.47 MPa (500 psig, 35.4 kg/cm^2)
at 216–227°C and 4.32 MPa (700 psig, 49.4 kg/cm^2) at 252°C.

Off-gas analyses were done by mass spectrometry and reactor
effluent samples were analyzed by glc. Most of the glc work was
done with an 8-foot 1/4-inch OD column containing 10% SE-30 on
acid washed Chromsorb W. However, the separation of n-propylbenzene from cumene had to be done with a dual 3/16-inch copper
column consisting of a 12-foot section having 10% Bentone 34 and
10% Dow Corning silicone gum 550 on 60–80 mesh acid washed
Chromsorb W and a 6-foot section containing 20% Apiezon L on
60–80 mesh acid washed Chromsorb P.

Discussion of the Results

The kinetic data, while new, presented nothing unusual.
The data were accurate in that they did predict product distribution from an adiabatic reactor.

The inverse temperature relationship on the rate of aging
was unknown prior to this investigation. Two explanations of
this are offered without substantiating data. The first is that
there are two distinctly different reactions occurring on the
catalyst. One of which is a reaction leading to large bulky
molecules which are slow or unable to escape the 13 Å windows
of the catalyst and is associated with a low energy of activation.
These bulky molecules then dehydrogenate to form coke or polynuclear aromatics. The other reaction, alkylation, has a much

higher energy of activation. As the temperature is increased, this becomes essentially the only reaction to occur. The second rational might really be another way of describing the first. This says that at low temperatures the reaction of propylene with benzene is reaction rate controlled and is associated with high concentrations of propylene around the active catalyst sites. At higher temperatures the reactions become mass transfer controlled (reaction rate very fast) and this leads to very low concentrations of propylene at the active sites.

The enormous differences in aging rates between the iso-thermal and adiabatic runs can be easily understood if one again looks at Figure 9. In an adiabatic reactor the feed must enter at a temperature considerably lower than that generated by the heat of reaction. This feed then initially passes through the catalyst at temperatures which produce very fast aging. In the isothermal reactor, the entire catalyst must have been hot enough to produce essentially no deactivation.

One might then think that if the feed was brought into an adiabatic reactor at around 232°C, little or no aging would occur. This was observed to be true, but the high temperature (271-277°C) generated led to excessive by-product formation.

Process Design

Figure 10 shows the rough design for this process. The alkylation or adiabatic reactor might be split into two smaller parallel reactors. One of which would be used for regenerating catalyst while the other was used for alkylation. The alkylation reactor would be operated at rather high space velocities (15-25 LWHSV), 7 or 8 molar ratio benzene/propylene, and 210-215°C inlet temperature - 250-255°C outlet temperature. The product would then be heat exchanged to reduce the temperature in the transalkylation or isothermal reactor to 220°C. This reactor would be operated at an LWHSV = 5 and should never age since it would never see any propylene. By recycling the di-isopropyl-benzene back to the reactor, the efficiency of the process can be increased from 93% (based on propylene) to 97-98%.

Conclusions

Of all the data which can be obtained from an isothermal alkylation reactor, only the kinetic and thermodynamic data can be applied to an adiabatic reactor. If the process or reaction is divided into alkylation and transalkylation sections, then one can select conditions for the process to give high catalyst life times and produce a pure product at high efficiencies.

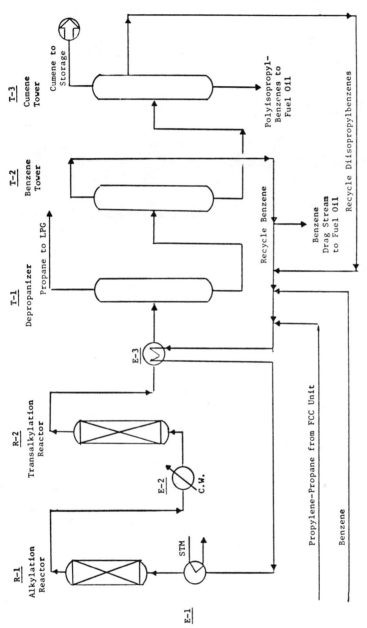

Figure 10. Two-stage cumene process

Acknowledgment

The authors wish to thank all of the technicians who were
associated with this project.   Special thanks are due to
Mr. Robert S. Clerk who was with the project the longest and who
prepared all the figures for this paper.

Glossary of Terms

LWHSV = Liquid weight hourly space velocity = g total
feed/g total catalyst-hour

τ = Space time = 1/LWHSV

Superficial Velocity = Weight rate of total feed/cross-
sectional area

= g total feed/cm^2 hr.

Literature Cited

(1) Wise, John J., U.S. Patent 3,251,897.

24

Some Aspects of the Direct Alkylation of Pyridine and Methyl Pyridines

C. V. DIGIOVANNA, P. J. CISLAK, and G. N. CISLAK

Reilly Tar & Chemical Corp., Indianapolis, IN 46241

A significant volume of literature exists regarding the alkylation of pyridine rings. For the purposes of this report, only those processes will be considered which involve intermolecular substitution directly on the pyridine ring with other ring substituents participating in merely a directive or steric role. This is a substantial limitation since there are many processes which, after a sequence of reactions, result in an alkyl group becoming a substituent on a pyridine ring.

Quaternization reactions have been disregarded, except when the quaternary salt represents an intermediate, since they are addition rather than substitution reactions. Further, reactions involving preliminary formation of the N-oxide of the pyridine ring have been eliminated from consideration since in these materials the fundamental aromatic character of the ring is altered.

Within the limits imposed there exists a number of available procedures for introducing alkyl substituents onto the pyridine ring. All of these involve either nucleophilic or homolytic substitution since alkylation via electrophilic substitution, e.g. Friedel Crafts alkylation, is not possible with the π-deficient pyridine nucleus.

Nucleophilic Substitutions

Rearrangement of Quaternary Salts. Ladenburg (1-4) reported rearrangement of methylpyridinium iodide to a mixture of 2-methylpyridine and similar rearrangements of other quaternary salts have been reported by others (5-19). The product mix obtained from thermolysis of 1-methylpyridinium iodide is given in Table I.

397

TABLE I (12)

THERMOLYSIS OF 1-METHYLPYRIDINIUM IODIDE

Product	Yield(%)
2-Methyl	30.1
3-Methyl	5.8
4-Methyl	34.6
2-Ethyl	3.6
4-Ethyl	1.4
2,4-Dimethyl	12.8
2,5-Dimethyl	1.9
2,6-Dimethyl	9.1

The formation of 3-methylpyridine and the ethylpyridines has been suggested as an indication of a free radical mechanism for this process (11, 12).

Use of Lithium Derivatives. There are only a few reports of alkylation procedures which provide a means for preparing predominately 3-substituted pyridines. Some (16-19) involve reaction of the pyridine compound with a lithium derivative, e.g. phenyllithium or lithium aluminum hydride (LAH), followed by reaction with an alkyl or arylhalide. The reaction steps are as follows:

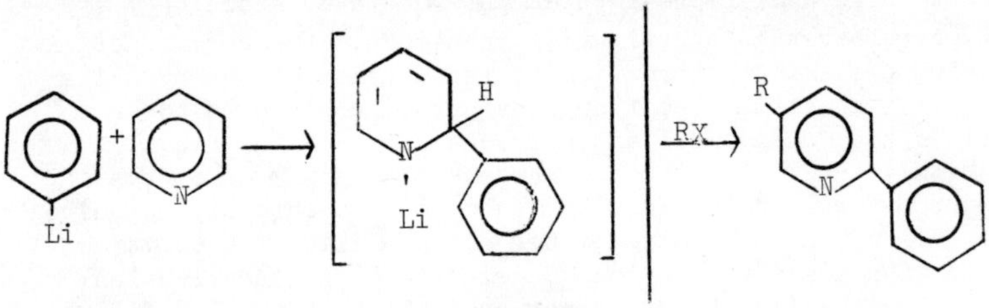

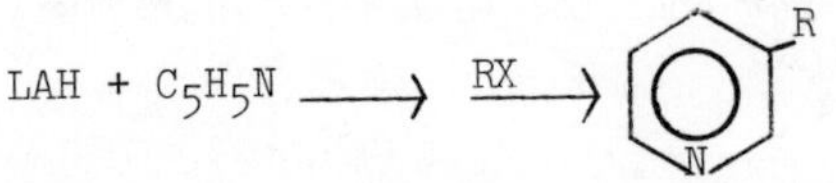

The phenyllithio intermediate has been isolated (20) and treated with styrene and ethylene oxides, and N-benzylidene aniline to provide various 2-phenyl-5-substituted pyridines. Addition of perhalomethanes affords bis(6-phenyl-3-pyridyl)methane and 6,6'-diphenyl-3,3'-dipyridyl (18). The yields of 3-substituted pyridines range, in the LAH reaction, from 89% for 3-picoline to 63% for 3-benzylpyridine. Reaction of the LAH intermediate with bromine formed 3-bromopyridine in 41% yield (17).

Treatment of pyridine with excess aryl and alkyllithium reagents is reported to give good yields of 2-,2,6-, and, with t-butyllithium, 2,4,6-substituted products (21-33).

The 1-lithio-2-t-butyl-1,2-dihydropyridine intermediate (similar to the 1-lithio-2-phenyl-1,2-dihydropyridine intermediate shown previously)' has been isolated (21), but though this material should lend itself to reaction with alkylhalides to form 2-t-butyl-5- substituted pyridines, this reaction has not been reported. An anomolous reaction is reported to occur with benzyllithium which forms the 4-benzylpyridine product exclusively (31).

<u>Use of Alkali and Alkaline Metal Compounds</u>. The reaction of pyridines with Grignard reagents has been used to alkylate (and arylate) pyridines in the 2- and 4- position (34-43). An anomalous alkylation of the pyridine system under the influence of sodium in anhydrous ammonia has been reported (44) as shown:

I: Ar = Ø, R = Et, X = H.
II: Ar = CH$_3$O-⟨O⟩-, R = Me, X = Br.

Homolytic Alkylations

Free Radical Processes. Homolytic alkylation and arylation of pyridines has been studied extensively and reviewed (45). The products are almost invariably mixtures of several isomers depending on the nature of the pyridine substrate, the free radical (and its method of generation) and the type of medium in which the reaction is carried out. Though early reports suggested 2-, and 4-substitution exclusively, more sensitive analytical techniques have shown the earlier claims to be erroneous and studies into the various factors affecting product formation have been reported (46-62).

TABLE II (51)

ISOMER RATIOS FOR HOMOLYTIC ALKYLATION OF PYRIDINE

Ref.	Radical Source	Medium	ISOMER RATIO			Yield
			2-	3-	4-	
58	t-Bu$_2$O$_2$	Non-Acid	58.0	23.0	19.0	-
51	"	"	62.0	22.9	15.1	13.5
55	Ac$_2$O$_2$	"	62.7	20.3	17.0	-
55	Pb(OAc)$_4$	"	62.1	20.5	17.4	-
51	"	"	62.7	21.7	15.6	5.0
51	"	HOAc	76.4	21.9	20.7	12.9
51	t-Bu$_2$O$_2$	HOAc	77.9	2.7	19.4	28.3
51	"	HOAc+HCl	93.2	0	6.8	11.8

From a commercial point of view, the methods described above do not lend themselves to the production of alkyl pyridines on a sufficiently large scale or in an economically attractive fashion. This is principally because the reagents involved are expensive and the mixtures formed present an overwhelming problem of separation.

Catalytic Processes. Commercially, the most attractive preparations involve catalytic processes and there are three such processes reported. The first involves the reaction of alkyl alcohols with pyridine in the presence of "boron phosphate" catalyst at 200-250 atm. at 300-500°C. (63-64). It should be noted that reaction with 2-, or 4-methylpyridine led to side chain alkylation but that reaction with 3-picoline produced 3,5-lutidine.

The second catalytic method involves reaction of pyridine with alkyl alcohols at both atmospheric (65) and elevated pressures (66-68) over alumina or alumino-silicate catalysts. The atmospheric pressure reaction afforded a mixture of all three picoline isomers in approximately equal amounts with multi-substitution also reported. At elevated pressures, the substitution is claimed to be primarily in the 3- position with low conversions.

Considerable research has been reported on the third process which involves reaction of pyridine compounds with alcohols or potential alcohols in the presence of various nickel catalysts. The reaction was first reported in 1964, (69), as a reaction of 3-hydroxypropylpyridines in the presence of Raney nickel (RaNi). (Figure 1)

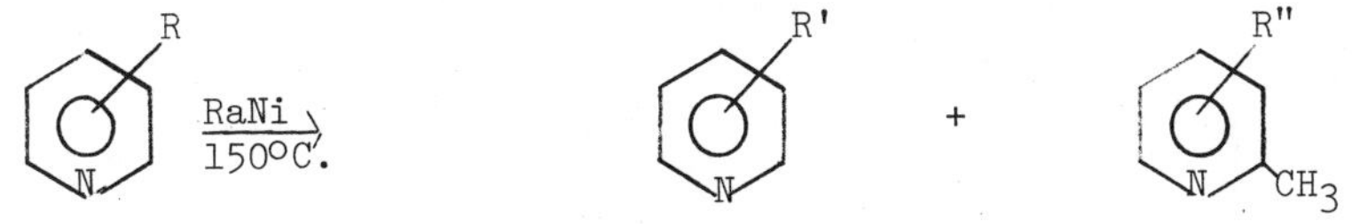

R	R' (% conv.)[a]	R" (% conv.)[a]
a 2-$(CH_2)_3OH$	2-C_2H_5 (27)[b]	6-C_2H_5 (24)[b]
a 2-$(CH_2)_3OH$	2-C_2H_5 (50)	6-C_2H_5 (0)
b 3-$(CH_2)_3OH$	3-C_2H_5 (28)	5-C_2H_5 (12)
c 4-$(CH_2)_3OH$	4-C_2H_5 (35)	4-C_2H_5 (15)

a) after 120 hrs.; b) 210° @ 3000 PSIG hydrogen

Journal of the American Chemical Society

Figure 1. Reaction of pyridyl alcohols with Raney nickel (69)

The mechanism proposed was as follows:

$R\text{-}CH_2CH$ ———— RCHO + 2H*

RCHO ———— RH + CO*

CO* + H* ———— "CH_3"*

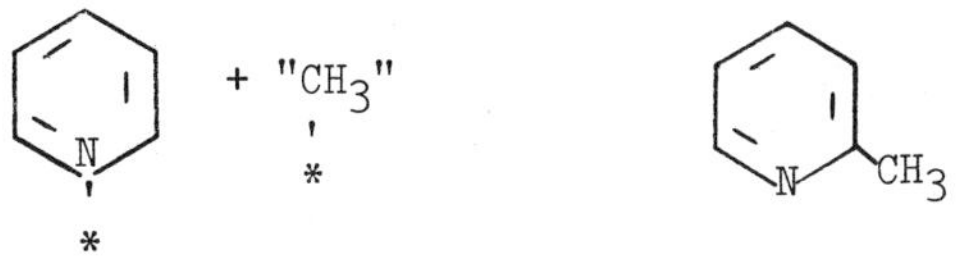

* = catalyst surface

The investigation of Reinke and Kray (13) was followed by that of Myerly and Weinberg (70) who revealed the scope of the reaction with regard to both methyl sources and pyridine substrates (Tables III and IV). The reactions were performed in the vapor phase using supported nickel.

TABLE III (70)

ALKYLATION OF PYRIDINE COMPOUNDS WITH METHANOL[a]

Pyridine Compound	Temp °C.	Mole Ratio Methanol- Pyridine Cmpd.	Contact Time (sec.)	Product (Yield %)
2-Picoline	260-5	3.6:1	12.2	2,6-Lutidine (11)
3-Picoline	265-9	3.7:1	12.2	2,5-Lutidine (54)
4-Picoline	260	3.6:1	11.5	2,4-Lutidine (21)
Quinoline	260	4.0:1	13.4	2-Methylquinoline (65)
2,4-Lutidine	260	4.0:1	12.2	s-Collidine (5)

a Harshaw catalyst N-0104T was used in all experiments reported
   in this table

TABLE IV (70)

METHYLATION OF PYRIDINE[a]

Methylating Agent	Temp. °C.	Mole Ratio Methylating Agent Pyridine	Contact Time (Sec.)	2-Picoline %Yield
Methanol	295	15:1	11.5	57[b]
Ethanol	263	3:1	11.6	61
1-Propanol	300	2.5:1	11.4	52
2-Propanol	295	2.5:1	12.2	38
1-Butanol	298	2.1:1	14.0	48
2-Methyl-1-Propanol	298	2.1:1	14.0	25
Acetone	325	2.5:1	12.2	38
Methylal	258	2.3:1	14.0	55
Ethyl Orthoformate[c]	256	1.3:1	20.6	57
Ethyl Acetate	298	1.6:1	16.2	58
Methane	260	3.8:1	9.8	10
CO-Hydrogen	270	2:1:1[d]	11.0	30

a Harshaw catalyst Ni-0104T was used in all experiments reported
   in this table. b 2,6-Lutidine formed in 24% yield as coproduct.
c By-products from this reaction were diethylether, carbon
   monoxide, carbon dioxide, and methane. d Mole ratio H_2-CO-
   pyridine.

   The methylation reaction has been reported to be catalyzed by
nickel-nickel oxide (71) and to be reversible (72). A further
variety of suitable pyridine substrates has also been revealed
(73). Concurrently, work on the mechanism of the reaction was
reported (74,75) and a study of the effects of various catalytic
substrates has been reported (76).

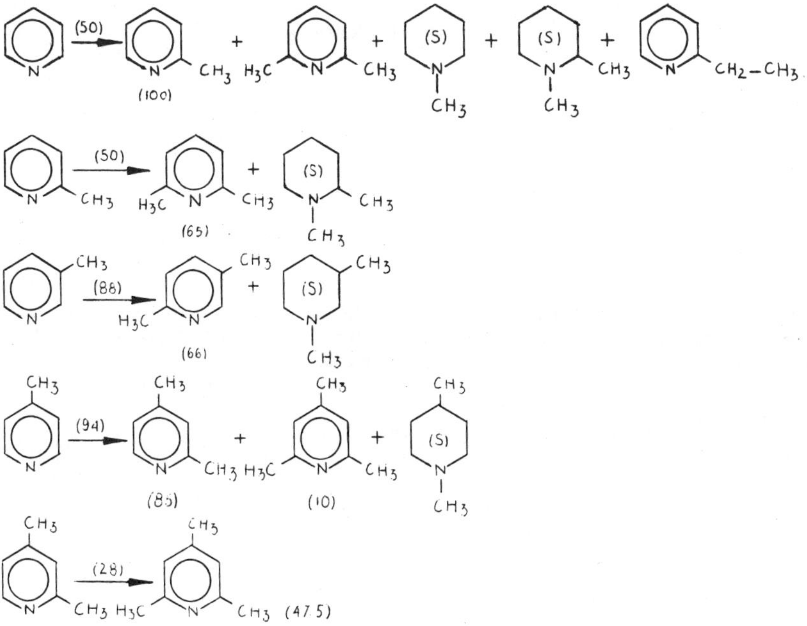

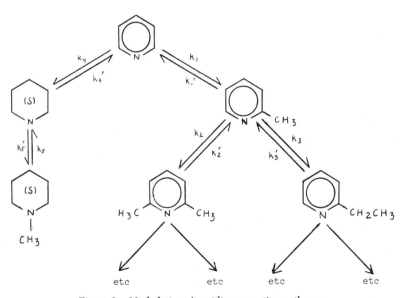

Figure 2. *Methylation of pyridines (Numbers on arrows denote conversion percents; others denote yield percents)*

Figure 3. *Methylation of pyridine—reaction pathways*

In our laboratories all possible picolines and lutidines plus pyridine, itself, have been investigated as precursors for the methylation reaction catalyzed by nickel. The reactions are summarized in Figure 2.

In addition, 2-ethylpyridine was methylated with a 39.8% conversion of starting material affording a 36% yield of 2-methyl-6-ethylpyridine, a 2.6% yield of 2,6-lutidine and a 3.14% yield of 2-picoline.

The major reaction paths in the methylation of pyridine may be depicted as in Figure 3.

With the alkylation steps, at least, reversible, as shown in the methylation of 2-ethylpyridine, the rate equation for the combination of parallel and series reactions leading to production of 2-picoline is very complicated. Further, the presence of piperidones in the final product mix complicates the rate equation. Indeed, rigorous elucidation of the kinetics of the methylation reaction presents an overwhelming task. This task is further complicated by the fact that preliminary studies in our laboratories involving material resulting from large scale methylations have indicated the formation of 2,4-, 2,3- and, possibly, 2,5-lutidine as a result of the methylation of 2-picoline. This evidence suggests that another mechanism may be operating at the same time as the usually accepted reactions.

It is rather surprising, therefore, that the reaction proceeds to give good yields of, for example, 2-picoline from pyridine. Evidently k_1 is larger than any of the other rate constants so that it is only when the concentration of the product, 2-picoline, becomes significant that formation of by-products, other than hydrogenation products, begins to become a factor.

Since methyl pyridines are readily available from ring formation reactions, the methylation reaction is useful, primarily, as a means of preparing polymethyl pyridines especially 2,4-, 2,5- and 2,6-lutidine and 2,4,6-collidine.

Acknowledgement

The authors wish to express their appreciation to Drs. W. H. Rieger, F. A. Karnatz and Mr. R. A. Kattau.

Literature Cited

1. A. Ladenburg, Ber., 16, 1408 (1883)·
2. Ibid., 2059 (1883)·
3. A. Ladenburg, Ann., 247, 1 (1888)·
4. Ibid., Ber., 18, 2961 (1885)·
5. A. E. Chichibabin and P. F. Ryumshin, J. Russ. Phys. Chem. Soc., 47, 1297 (1915); Chem. Abstr., 9, 3057 (1915)·
6. J. von Braun and W. Pinkernelle, Ber., 64, 1871 (1931)·
7. J. Overhoff and J. P. Wibaut Rec. trav. Chim., 50, 957 (1931)·
8. P. C. Teague, J. Am. Chem. Soc., 69, 714 (1947)·

9. K. E. Crook and S. M. McElvain, J. Am. Chem. Soc., 52, 4006 (1930).
10. K. E. Crook, J. Am. Chem. Soc., 70, 416 (1948).
11. J. Kuthan, et al., Coll. Czech. Chem. Commun., 35, 2787 (1970); Chem. Abstr., 73, 109091r (1970).
12. Yu. I. Chumakov and V. F. Novikova, Khim. Prom. Ukr., 2, 47 (1968); Chem. Abstr., 69, 43743m (1968).
13. N. S. Prostakov, et al., Khim. Geterotsikl. Soedin., 6, 779 (1970); Chem. Abstr., 73, 109635u (1970).
14. C. Mercier and J. P. Dubosc, Bull. Soc. Chim. Fr., 12, 4425 (1969); Chem. Abstr., 72, 90221m (1970).
15. Yu. I. Chermakov and B. D. Kabulov, Ukr. Khim. Zh., 41 (1), 99 (1975); Chem. Abstr., 83, 58622j (1975).
16. C. S. Giam and J. L. Stout, Chem. Comm., 1970, 478.
17. C. S. Giam and S. D. Abbott, J. Am. Chem. Soc., 93 (5), 1294 (1971).
18. C. S. Giam, et al., Can. J. Chem., 53, 2305 (1975).
19. R. Grashey and R. Huisgen, Ber., 92, 2641 (1959).
20. C. S. Giam and J. L. Stout, Chem. Comm., 1969, 142.
21. R. F. Francis, et al., J. Org. Chem., 39 (1), 59 (1974).
22. R. F. Francis, et al., Chem. Comm., 1971, 1420.
23. F. V. Scalzi and N. F. Golob, J. Org. Chem., 36 (17), 2541 (1971).
24. H. C. Brown and B. Kanner, J. Am. Chem. Soc., 88 (5), 986 (1966).
25. R. A. Abramovitch and C. S. Giam, Can. J. Chem., 42, (7), 1627 (1964); Chem. Abstr., 61, 5477a (1964).
26. Ibid., 41 (12), 3127 (1963); Chem. Abstr., 60, 3961f (1964).
27. Ibid., 40, 213 (1960); Chem. Abstr., 57, 12425d (1962).
28. K. Ziegler and H. Zeiser, Ann., 485, 174 (1931).
29. Ibid., Ber., 63B, 1847 (1930).
30. T. Kauffman, et al., Ber., 109 (12), 3864 (1976).
31. H. Gilman and H. A. Mc Ninch, J. Org. Chem., 27, 1889 (1962).
32. H. Gilman and G. C. Gainer, J. Am. Chem. Soc., 71, 2327 (1949).
33. T. Taguchi, et al., Chem. Lett., 12, 1307 (1973); Chem. Abstr., 80, 59904c (1974).
34. F. W. Bergstrom and S. H. Mc Allister, J. Am. Chem. Soc., 52, 2845 (1930).
35. W. von E. Doering and V. Z. Pasternak, J. Am. Chem. Soc., 72, 143 (1950).
36. E. Bergmann and W. Rosenthal, J. Prakt. Chem., 135 (2), 267 (1932); Chem. Abstr., 27, 982 (1933).
37. W. L. C. Veer and S. Goldschmidt, Rec. trav. Chim., 65, 793 (1946); Chem. Abstr., 41, 3100g (1948).
38. R. A. Benkeser and D. S. Holton, J. Am. Chem. Soc., 73, 5861 (1951).
39. D. Bryce - Smith, et al., Chem. Ind., 495 (1964).
40. Ibid., J. Chem. Soc., Perkin I, 1976, 1977.

41. R. A. Abramovitch and G. A. Poulton, J. Chem. Soc. (B), 1969, 901·

42. N. Goetz-Luthy, J. Am. Chem. Soc., 71, 2254 (1949)·

43. M. Gilman, et al., J. Am. Chem. Soc., 81, 4000 (1959).

44. F. J. Villani, et al., J. Org. Chem., 36 (12), 1709 (1971)·

45. R. A. Abramovitch and J. G. Saha in "Advances in Heterocyclic Chemistry," A. R. Katritzky and A. J. Boulton, Eds., Vol. 6, Academic Press. New York, 1966, p.229.

46. A. Clerice, et al., Tetrahedron, 29, 2775 (1973)·

47. F. Minisci, et al., Tetrahedron, 28, 2403 (1972)·

48. G. P. Gardini, et al., Chim. Ind. (Milan), 53, 263 (1971)·

49. R. Galli, et al., Tetrahedron, 26, 4083 (1970)·

50. K. C. Bass and P. Nababsing, J. Chem. Soc. (c), 1969, 388·

51. Ibid., 1970, 2169·

52. R. Noyovi et al., Tetrahedron, 25, 1125 (1969)·

53. F. Minisci, et al., Tetrahedron Letters, 54, 5609 (1968)·

54. H. Nozaki, et al., Tetrahedron Letters, 43, 4259 (1967)·

55. R. A. Abramovitch and K. Kenaschuk, Can. J. Chem., 45, 509 (1967)·

56. R. A. Abramovitch and M. Saha, J. Chem. Soc. (B), 1966, 733·

57. R. A. Abramovitch and J. G. Saha, J. Chem. Soc., 1964, 2175·

58. K. Schwetlick and R. Lungwitz, Z. Chem., 4(12), 458 (1964); Chem. Abstr., 62, 9101a (1965)·

59. J. H. P. Utley and R. J. Holman, Electrochim. Acta. 21 (11), 987 (1976)·

60. F. Minsci, et al., Ital. 906,418 (1972); Chem. Abstr., 83, 58667c (1975)·

61. F. Minsci, et al., Gazz. Chim. Ital., 105 (9-10), 1083 (1975); Chem. Abstr., 84, 168736j (1976)·

62. T. Caronna, et al., Tetrahedron Letters, 1976 (20), 1731.

63. H. Moll and H. J. Uebel, German (East) Patent 54,006 (1967); Chem. Abstr., 68, 59436d (1968)·

64. Ibid., Chem. Tech. (Berlin), 18 (10), 629 (1966); Chem. Abstr., 66, 55348n (1967)·

65. N. M. Cullenane, et al., J. Soc. Chem. Ind., 67, 148 (1948)·

66. A. K. Dariev, et al., Tr. Buryat. Kompleks. Nauch Issled. Inst. Akad. Nauk. SSSR. Sib. Otd., 20, 3 (1966); Chem. Abstr., 68, 78091d (1968)·

67. A. D. Dariev, et al., Izv. Sib. Otd. Akad. Nauk. SSSR. Ser. Khim. Nauk. SSSR. Ser. Khim. Nauk., 1966 (3), 105; Chem. Abstr., 67, 64193f (1967)·

68. V. N. Gudz, et al., Tr. Buryat. Kompleks Nauch. - Issled. Inst. Akad. Nauk. SSSR, Sib. Otd., 20, 3 (1966); Chem. Abstr., 68, 104924z (1968)·

69. M. G. Reinke and L. R. Kray, J. Am. Chem. Soc., 86, 5355, (1964)·

70. R. C. Myerly and K. G. Weinberg, J. Org. Chem., 31, 2008 (1966)·

71. Ibid., U.S. Patent 3,354,165 (1967)·

72. Ibid., U.S. Patent 3,334,101 (1967)·

73. Ibid., U.S. Patent 3,428,641 (1969)·
74. M. G. Reincke, et. al., Ann. N.Y. Acad. Sci., 145 (1), 116 (1967)·
75. G. Grins "Catalytic α-Methylation of Pyridines", Univ. Microfilms, Ann Arbor, Mich., (1970)·

25

Friedel–Crafts Alkylation of Bituminous Coals

W. HODEK, F. MEYER, and G. KOLLING

Bergbau–Forschung GmbH, Frillendorfer Str. 351, 4300 Essen 13, West Germany

Coal and mineral oil are not only used as fuel, but they are moreover important raw materials for the chemical industry, especially as carbonaceous materials for the production of organic chemicals. For this use, mineral oils show two important advantages as compared with coal: (a) they are liquids and are therefor easier to handle and process; (b) their compositions are complicated but easier to analyze.

Bituminous coals are relatively insoluble solids, and their chemical compositions, which are based on aromatic structures, are still partially unknown. It is expected that a better knowledge of the chemical structure of coal may open new ways to develop new processes for an economic use of coal. The elucidation of the chemical structure of coal is hampered by the fact that only a small part of the coal is soluble in suitable solvents. Three reasons are thought to cause this low extractability:

a) Coals consist of some crosslinked macromolecules;
b) Hydrogen bonding - especially at phenolic groups - hold the coal molecules together;
c) Physical attractive forces, caused by the high aromaticity of the bituminous coals, have an additional bonding effect.

In order to obtain insight into the chemical structure of coal, methods are needed to convert coal into substances which are soluble in suitable solvents. This conversion should be conducted at mild conditions so that the coal molecules remain unchanged as much as possible. So far, the most usual method applied to make bituminous coals more soluble in simple liquids is hydrogenation which, however, requires temperatures exceeding 350° to 400°C; consequently cracking reactions also occur.

There are, however, chemical reactions that make it possible to improve the extractability of coal at lower temperatures, in particular those involving Friedel-Crafts catalysts. Heredy and Neuworth (1) for example, used boron trifluoride and phenol for this purpose; and Ouchi, Imuta and Yamashita (2) employed

p-toluene sulphonic acid and phenol. Kröger (3) studied the
reactions between bituminous coals and propyl chloride in carbon
disulphide at 45°C, using aluminum chloride as catalyst. Quite
a different method of achieving this aim was used by Sternberg
et al (4), namely reductive alkylation of bituminous coal with
sodium and alkyl halides.

Heredy and his co-workers (1,5) have attempted to interpret
the reaction with phenol and boron trifluoride as a depolymerization process followed by saturation of the fragments. In fact,
they found that considerable quantities of phenol reacted with
the coal. In low-rank coals, the tendency towards this reaction
was greater than with coking coals which did not show any notable
improvement of extractability.

There has also been considerable additional work by many investigators to investigate depolymerization and alkylation of
coal. In a recently published paper by Larsen (6), the most
interesting results in this subject are assembled and discussed.
The investigations to be described in our paper also started from
the well-known fact that Lewis acids, such as aluminum chloride,
can exert not only a condensing but also a cracking effect on
aliphatic-aromatic molecules. We hoped, however, that it would
be possible to prevent most of these side reactions by using mild
conditions.

Our results for the Friedel-Crafts acylation of coal (7)
indicate that bituminous coal becomes extractable not only by
depolymerization but also by the introduction of long chain
substituents into the coal molecules. By the substitution with
long chain aliphatic groups, the separation between the aromatic
structures of adjacent molecules can be increased so that the
physical attractive forces are reduced. We tried alkylation in
order to make bituminous coal more extractable. Possibly in the
future extracts can be obtained economically by reaction of coal
with mineral oil fractions under relatively mild conditions.

Alkylation with Alkyl Chlorides and $AlCl_3$

Initially we used alkyl chlorides in the presence of aluminum
chloride as catalyst for alkylation. 10 grams of a medium volatile coal with 24.6% volatile matter were finely ground and suspended in 50 cc of carbon bisulfide. 10 grams of powdered aluminum chloride were then added. The mixture was then treated at
45°C with 0.25 moles of alkyl chloride. The reaction time was
3 hours. Alkyl chlorides of different chain length - 3 up to 18
carbon atoms in the chain - were used. The chlorides with 16 and
18 carbon atoms required a reaction time of 24 hours. In all
reactions, alkyl groups were inserted into the coal which could
be easily determined by the weight increase of all samples. From
the increase in weight, the number of alkyl groups per 100 carbon
atoms which have reacted with the coal were calculated (Figure 1).

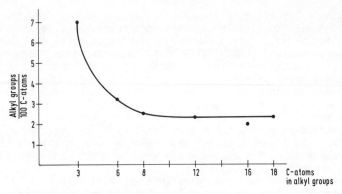

Figure 1. Alkylation of medium volatile coal with alkyl chlorides of different chain length

Two to three alkyl groups reacted as a rule per 100 carbon atoms in the coal. A remarkable exception was propyl chloride. In this case, more than 7 propyl groups per 100 carbon atoms reacted. Obviously, multiple substitution of the aromatic coal constituents was favoured by the relatively small propyl group. With increasing size of the alkyl groups, the possibilities of substitution were limited with the result that the degree of alkylation decreased rapidly until it remained at a constant value.

Alkylation conducted in this way had only a small influence on the extractability of the coal. 27.2% of our untreated coal could be extracted with pyridine. After alkylation, a maximum extractability of 35% was reached. The extractability in chloroform increased from 4 to 16%. This increase in extractability is not only caused mainly by the aluminum chloride as we found in a blank test.

The course of reaction during the alkylation of coal can be postulated as follows: In the primary step, the coal is depolymerized by the aluminum chloride; other investigators have also shown this phenomenon in the presence of Lewis acids. Now, two reactions come into competition with each other:

1)   The recombination of the fragments in at least our medium volatile coal resulted in materials with low extractability being formed again.

2)   The capture of the fragments by reaction with the alkyl chlorides thus preventing recombination reactions.

When reaction 2 predominates, the coal has split to form products with high extractability. In our case, reaction 2 took place at low rates. Thus, reaction 1 predominated with the result that the extractability of the medium volatile coal was only slightly increased by alkylation. The system alkyl chloride/aluminum chloride was not reactive enough in order to make the coal extractable by alkylation.

Alkylation with Olefins and HF/BF₃

We used for the further tests the system hydrogen fluoride/ boron trifluoride as catalyst and olefins as means of alkylation. A 5% solution of boron trifluoride in hydrogen fluoride seemed to be, for several reasons, an advantageous catalyst for the alkylation of coal.

Liquid hydrogen fluoride is an excellent solvent for many organic compounds such as aromatics, alcohols, acids, ethers, etc.; these latter substances behave in the presence of hydrogen fluoride as weak bases and add one proton. Thus, hydrogen fluoride can act as a reaction medium and as a catalyst. Boron trifluoride combines with hydrogen fluoride to produce Fluoboric acid which is a very strong acid with a catalytic activity much higher than that of hydrogen fluoride. Besides that, the low viscosity and the low surface tension of the hydrogen fluoride permit good mixing in heterogeneous reactions. A disadvantage is the strong corrosive effect of HF-BF₃ mixtures. A "monel" autoclave proved to be sufficiently resistant even at higher temperatures.

Four different types of coal were used for the alkylation tests (Figure 2): a high volatile coal with 34.4% (volatile matter), a medium volatile coal with 24.2%, a low volatile coal with 19.4% and a semi-anthracite with 14.4%.

The tests were made in a 2 liter-shaking autoclave. Initially the influence of hydrogen fluoride and HF-BF₃ was investigated on the extractability of coals with pyridine. 20 grams of finely ground coal were treated with about 1000 grams of hydrogen fluoride or a 5% solution of BF₃ in hydrogen fluoride for four hours at temperatures of 20°C and 80°C.

The extractability of coals with pyridine after this treatment is shown in Figure 3. The left column indicates the extractability of the untreated coal. The two columns on the right-hand side show the influence of the hydrogen fluoride. As a comparison, the influence of aluminum chloride has been plotted, too. It can be seen that the extractability of high volatile coal decreased, probably due to the condensing effects of the catalysts on the coal molecules.

It seems that the medium volatile coal remains unchanged. Cleaving and condensing effects probably balance each other. The extractability of the low volatile coal is however increased, especially by hydrogen fluoride at 80°C. The semi-anthracite is essentially unaffected by such treatements because of its highly condensed aromatic structure.

Aluminum chloride and hydrogen fluoride have comparable effects on the coals. The addition of boron trifluoride to the hydrogen fluoride has no significant influence on the extractability of the coals. After we ascertained the influence of the catalyst system on the extractability of the coals, alkylation

	Ash (dry basis)	VM (daf)	C (daf)	H (daf)	N (daf)	S (daf)	Cl (dry basis)	O (daf)
High volatile coal	3,4	34,4	82,7	5,04	1,59	0,95	0,25	9,4
Medium volatile coal	2,1	24,2	87,6	4,94	1,82	0,86	0,12	4,6
Low volatile coal	2,5	19,4	89,7	4,54	1,59	1,20	0,23	2,7
Semi-anthracite	10,7	14,4	91,0	3,88	1,74	1,03	0,07	2,2

Figure 2. Analytical data of the coals

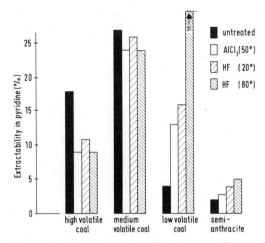

Figure 3. *Extractability of bituminous coals after treatment with Lewis acids*

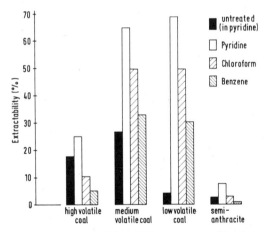

Figure 4. *Extractability of bituminous coals after alkylation with 1-hexadecene (HF/BF₃, 80°C)*

tests were begun using olefins. The reaction conditions were the same as described before. 20 grams of coal and 20 grams of olefins were added to the autoclave. Preliminary tests with short chain olefins such as ethylene, propylene and hexene indicated only minor reactions with coal. Some oligomers of the olefins were produced, but they did not react with the coal. Under the conditions investigated, however, long chain olefins reacted readily

with the coal. The alkylation was conducted with olefins with
12 up to 20 carbon atoms in the chain at 80°C. We used a 5%
solution of boron trifluoride in the hydrogen fluoride as catalyst
which, as expected, yielded better results than pure hydrogen
fluoride.

The coals were reacted with hexadecene and the extractabilities of the reaction products were determined. The results are
shown in Figure 4. The extractabilities of the low and medium
volatile coal were increased significantly by this reaction. 68%
of the low volatile coal and 65% of the medium volatile coal were
extracted with pyridine. The extractability in chloroform and
benzene, too, increased to 50 and 30% respectively. The high
volatile coal and the semi-anthracite showed no essential increase
of extractability.

As the next step, the extractabilities of the alkylated coals
were determined as a function of the chain length of the olefins
used (Figure 5). The extractability increased with the length of
chain and reached a maximum value of about 70% for 18 carbon atoms
in the case of both coals.

The number of alkyl groups inserted into the coals during
alkylation was determined by the weight increase of the samples
and is related, as described above, to 100 carbon atoms in the
coal (Figure 6). In case of the low and medium volatile coals,
two to three alkyl groups per 100 carbon atoms were inserted
regardless of the length of chain. For the high volatile coal,
this value amounted only to 0.6 to 0.8, and with the semi-anthracite, it was below 0.5. These results show that low and medium
volatile coals have a similar chemical structure and that these
coals have a maximum of aromatic C-H groups, compared with
other types of coal. The existence of these groups is a condition for the application of a Friedel-Crafts-alkylation as
conducted in this work.

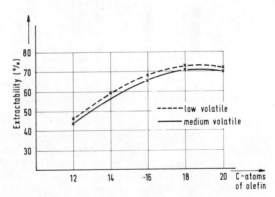

*Figure 5.   Extractability with increasing chain length of
olefin*

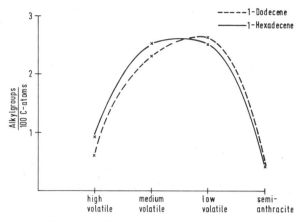

Figure 6. Alkylation of coals of different ranks

In case of younger coals, the aromatic units are more sub-stituted with aliphatic and naphthenic groups. Therefore, a lower aromaticity is given. With increasing rank of the coals, the percentage of aromatic carbon not bound to hydrogen increases. There are still more highly condensed systems which cannot be alkylated sufficiently. The aforementioned tests showed that bituminous coals, especially low and medium volatile coals, can be alkylated by the very reactive system of HF/BF_3/olefins and that the extractability of such coals can be greatly increased.

What is the reason for the high extractability of the alky-lated coals? We indicated above that Friedel-Crafts catalysts caused cleaving and condensing reactions in the coal molecules. According to what we presently know, bituminous coal is essential-ly a mixture of high molecular materials, the molecules of which consist of aromatic units linked by aliphatic or by ether bridges (Figure 7). Aliphatic bridges in particular are easily cleft by Friedel-Crafts catalysts as can be confirmed by model tests. Diphenyl-methane structures, which had been found in coal, are cleft by aluminum chloride or $HF-BF_3$. In order to demonstrate such cleavages in the coal, our test coals were first extracted with pyridine in order to remove all low-molecular constituents. Then, they were alkylated with $HF-BF_3$ and hexadecene, and the product was extracted with acetone. By liquid-solid chromato-graphy, a fraction was obtained from the acetone extract, the infra-red spectrum of which was very similar to that of the alkylated coal and which consisted obviously of alkylated aro-matics. Related to the quantity of coal extracted the quantity of this fraction was 0.35% in the low and medium volatile coal and 0.2% in the high volatile coal. Semi-anthracites did not

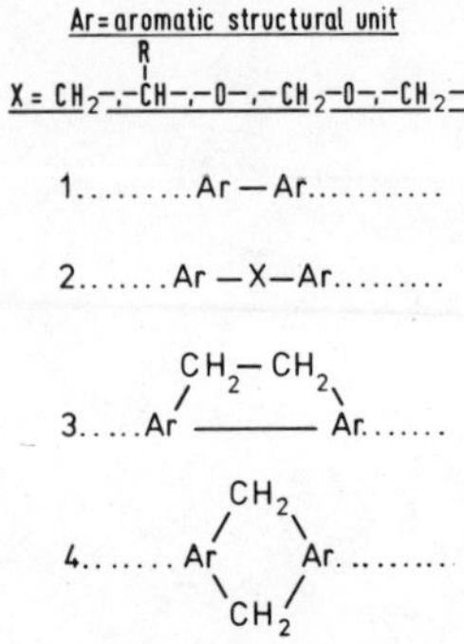

Ar=aromatic structural unit

$$X = CH_2-,-CH-,-O-,-CH_2-O-,-CH_2-CH_2-$$
(with R above the CH)

1........Ar—Ar..........

2.......Ar—X—Ar.........

```
          CH₂— CH₂
        /           \
3.....Ar ———————— Ar......
```

```
         CH₂
        /    \
4.......Ar    Ar.........
        \    /
         CH₂
```

Figure 7. Connections between aromatic units in coal

show this fraction. This fraction can be formed only by splitting from the high-polymeric coal molecules. The quantitative proportions obtained from the individual coals correspond to the observations made before; both low and medium volatile coals are very well suitable for the alkylating-cracking sequence.

Summary:

Alkylation with Friedel-Crafts catalysts can promote production of soluble products from coal. This method is currently not economical but the information obtained has clarified several points. The extracts obtained from the alkylation of coal as described above, are black, shiny, solid substances which are not distillable and apparently still have very high molecular weights. They are very similar in their properties to the extracts which we have obtained previously by acylation of the same coals with carboxylic acid chlorides and aluminum chloride.

References:

1. L. A. Heredy and M. B. Neuworth, Fuel, Lond. 41, 221, (1962)
2. K. Ouchi, K. Imuta and Y. Yamashita, Fuel, Lond. 44, 29, (1965)
3. C. Kröger, Forschungsber. Nordrhein-Westfalen No. 1488 (1965)
4. H. W. Sternberg, C. L. Delle Donne, P. Pantages, E. C. Moroni and R. E. Markby, Fuel, Lond. 50, 432 (1971)
5. L. A. Heredy, A. E. Kostyo and M. B. Neuworth, Adv. Chem. Ser. 55, 493 (1966); Fuel, Lond. 43, 414 (1964); Fuel, Lond. 44, 125 (1965)
6. J. W. Larsen and E. W. Kuemmerle, Fuel, Lond. 55, 162 (1976)
7. W. Hodek and G. Kölling, Fuel, Lond. 52, 220 (1973)

Coal Liquefaction by Aromatic Interchange with Phenol and Catalytic Hydrogenolysis

R. H. MOORE, E. C. MARTIN, J. L. COX, and D. C. ELLIOTT

Pacific Northwest Laboratories, Division of Battelle Memorial Institute, Richland, WA 99352

Not only can coals be alkylated readily, they also can be used to alkylate aromatic compounds. L. A. Heredy, et al. (1-3) realized that coal could be depolymerized by using it as an alkylating agent. The mechanism of the reaction was first demonstrated using model compounds to show that methylene groups joining two aromatic groups could be cleaved at the ring and the methylene group would then alkylate another aromatic substrate. The first step involves protonation of an aromatic ring adjacent to the methylene group. This is followed by nucleophilic bimolecular substitution of the aryl group by phenol and regeneration of the proton.

$$(1)$$

$$(2)$$

$$(3)$$

L. A. Heredy, et al. employed BF_3 as a catalyst. BF_3 forms a stable adduct with phenol which yields the proton catalyst due to the dissociation equilibrium:

$$(4)$$

In addition to depolymerization Heredy, et al. showed that alkyl
side chains were removed from the coal (dealkylation) and added
to the phenol (alkylation). They were able to demonstrate the
existence of isopropyl groups in coal by this means.
    K. Ouchi, et al. (4-8) later investigated the reaction using
p-toluenesulfonic acid as catalyst near the reflux temperature of
phenol (185°C). Based on subsequent pyridine extraction, Ouchi
reported depolymerizations of Yubari coal as high as 90 percent
with a molecular weight of 312 for the depolymerized coal monomer.
For coals of increasing rank (70-93 percent maf C) the molecular
weight of the depolymerized products increased from about 300 to
>1000. The yield of material soluble in pyridine approached 100
percent for coals of low to intermediate rank and then fell off
sharply for the high ranking coals.
    Darlage, et al. (9) employed a variation on Heredy's tech-
niques. They oxidized coal with 2M HNO$_3$ at 60-70°C for 24 hours,
depolymerized using BF$_3$ as catalyst, and acetylated using acetic
anhydride to block the phenolic hydroxyl groups and prevent intermolecular hydrogen bonding. Following this treatment, some coals
exhibited 95-96 percent solubility in chloroform while another was
only 7 percent soluble (all Hvab rank). A Lvb coal from West
Virginia was only 14 percent soluble. None of the coals exhibited
more than a few percent solubility in chloroform prior to treatment. Throughout the referenced work, this variability in the
behavior of coals from different sources is noticeable, though
perhaps not so dramatically as in this case.
    The solutions of depolymerized coals are complex mixtures.
In a few cases, pure compounds have been isolated and identified
though they comprise <1 percent of depolymerized coal substance.
    In all cases cited the depolymerized coal product contained
phenol. Ouchi showed that as depolymerization progressed the
intensity of infrared absorption attributed to phenolic hydroxyl
increased regularly.
    Ouchi also showed that the aliphatic content of benzene-
alcohol extracts increased during successive cycles of depolymerization and the residues showed an increasingly dense network
of aromatic structures. Increased aliphatic content implies
alkylation has occurred. On-going work at Battelle Northwest
is in general agreement with the results of Ouchi (10).
    Acid catalyzed depolymerization of coal with phenol affords
a means for dissolution of coal under relatively mild conditions
(185°C, ambient pressure). Once dissolved, separation of ash
constituents and unreacted char is accomplished by filtration or
centrifugation (also under mild conditions). Depolymerized coal
recovered as a low ash product from excess phenol could be dissolved in a coal derived solvent and hydrogenated to stable
liquids. It might be anticipated that access to hydrogen and
contact with the catalyst would be more efficient in the case of
the solubilized coal substance than for coal particle slurries.
Hydrogenation might proceed more efficiently and with less

fouling of catalysts due, in part, to the absence of heavy metals.
    This study was conducted to assess the behavior of various
coals with respect to solubilization by acid catalyzed depolymerization with phenol. Ash removal, phenol uptake, and reactions
of other solvent systems were considered and the hydrogenolysis
of a number of coals and of their depolymerized products were
compared.

Coal Preparation

    The coals used in this work were first crushed to -1/4 inch
using a gyratory crusher. They were then pulverized and sieved.
All coals were dried under N_2 at 80°C for 24 hours prior to
pulverizing. The pulverizer produced fractions of -7+20 and
-60+115 mesh. Most samples employed in this work were of -7+20
mesh. Only a few tests were made with coal of finer grind.
It was found that phenol not only depolymerizes coal to colloidal
and molecular size, but it also disintegrates coarse coal producing a smaller screen size distribution. This last effect is
believed to be similar to that reported by workers at Syracuse
University Research Corporation. (11) As a result, the extent
and rate of depolymerization were nearly independent of initial
particle size.
    The coals used in these tests were characterized by proximate
and ultimate analyses (Table I). Samples of four coals were
furnished us through the courtesy of the College of Mineral
Sciences, Pennsylvania State University and where received too
late for analysis (except for moisture and ash); however, these
coals have been very completely characterized by Given, et al.
(12,13) In addition, a few other coals received late in the
program were not analyzed, but typical analyses provided by the
mines are included in Table I.
    Procedures were used in this work which could be converted
to industrial process steps. Coal was refluxed with a phenol-
catalyst mixture at atmospheric pressure.* The hot solution was
filtered through a tared filter which had been precoated with
filter aid. Contents of the filter were washed, first with hot
phenol-methanol mixture (20/1, v/v) to remove adhering depolymerized coal, and finally, with methanol to remove phenol and
catalyst. The filter cake was dried at 10 mm Hg pressure and
140-160°C and weighed.
    The residue on the filter contains ash constituents from the
coal, unreacted coal, and possibly phenol combined with partially
depolymerized coal of molecular size too large to be soluble.
From the weight of the residue the extent of depolymerization is
given by:

---

*   Unless otherwise stated the catalyst was p-toluenesulfonic acid.

TABLE I. Characterization of Coals

Coal Identification	Proximate Analysis				Ultimate Analysis					Supplemental Analysis
	Moisture	Volatiles	Fixed Carbon	Ash	C	H	N	S	O	Pyritic Sulfur
Consol. Pittsburgh #8, Hvb	1.10	33.9	50.5	14.48	69.50	4.82	1.17	4.21	6.21	2.41
Georgetown #12 Pittsburgh #8 Hvb	2.68	--	--	13.70	69.8	5.26	1.54	2.64	7.64	1.82
Montour #4 Pittsburgh Seam Hvb	1.84	--	--	10.01	75.34	5.43	1.44	1.90	8.6	1.05
Kaiser Steel Raton, NM, Hvb	2.07	28.7	50.1	19.1	69.27	4.55	1.2	0.37	6.54	0.10
Midland, Trivoli, IL HVC	8.35	29.9	46.0	15.7	60.14	4.50	1.36	2.62	10.0	0.68
Knife River Beulah, ND Lignite	5.47	39.7	45.2	9.67	59.42	3.72	1.21	0.67	20.23	0.22
Dave Johnston Glenrock, WY Sub-Bitum.	24.8	29.0	29.5	16.7	63.7	4.7	0.7	0.6	21.2	--
Big Horn Sheridan, WY, Sub-Bitum.	22.1	28.9	42.9	6.10	--	--	--	--	--	--
[a]Queen, WA hv Ab	1.89	--	--	21.5	--	--	--	--	--	--
[a]Big Seam Strip Roslyn, WA hv Bb	3.50	--	--	25.2	--	--	--	--	--	--
[a]N. Fork Virgin R. Iron Co. Utah hvCb	9.34	--	--	22.4	--	--	--	--	--	--
[a]Big Mine Crested Butte, CO hv B to Cb	4.04	--	--	4.19	--	--	--	--	--	--
Savage Lignite Montana	34.5	25.4	35.1	4.98	--	--	--	--	--	--

a   Samples furnished through courtesy of Pennsylvania State University. [12][13]

$$\% \text{ depolymerization} = \frac{(g \text{ maf coal}) - (g \text{ maf residue})}{(g \text{ maf coal})} \times 100 \quad (5)$$

where (g maf coal) is the mass of coal charged corrected for its moisture and ash content.

The filtrate contains a dispersion of particles too small to be trapped by the filter and solubilized coal. Microscopic measurements indicate the particles are slightly less than 0.5 micron average diameter.

The determination of depolymerized coal in this filtrate was accomplished by diluting a weighed aliquot with 20-30 times its weight of water. Phenol and catalyst dissolve whereas the depolymerized coal product precipitates.* The precipitate was filtered, water washed, and dried in a vacuum at 110°C. The weight of this precipitated product was designated as maf (moisture and ash free) depolymerized product.

The amount of phenol which has become chemically combined with the coal is given by:

$$\% \text{ phenol} = \frac{(\text{maf residue}) + (\text{maf depoly. prod.}) - (\text{maf coal})}{(\text{maf coal})}$$

$$\times 100. \quad (6)$$

Equation (6) shows that the combined phenol is derived from the total increase in the weight of coal. No distinction is made between phenol trapped in the residue or combined with soluble depolymerized coal. The weight gain is a measure of the extent of reaction. Ouchi, et al. also utilized this total weight increase as a measure of the extent of reaction.

Coal Screening Tests. Effect of Coal Rank

Screening tests were conducted over a considerable period of time as the various coals became available for study so test conditions exhibit slight variations. These variations are not regarded as sufficient to modify conclusions which may be drawn from the data in Table II.

The extent of depolymerization of these coals (Table II) ranges from 25.3 to 88.7 percent. This defines the amount of maf coal solubilized. With this particular array of coals, the coal solubilized bears almost an inverse relationship to the combined phenol. This appears to be dictated by chance; nevertheless, the coals to the right side of Table II take up more phenol in relation to the amount solubilized than those to the

---

* With certain coals slight solubility of the depolymerized coal product is evidenced by color imparted to the aqueous phase.

TABLE II. Comparison of the Reactivity of Various Coals to the Depolymerization Reaction

	Big Horn	Dave Johnston	Consol	Montour	George-town	Queen	Savage	Knife River	Kaiser Steel	Roslyn	Virgin River	Midland	Colorado
Run Number	193	209	134	108	89	178	211	64	78	180	181	153	179
Catalyst	H_2SO_4	H_2SO_4	H_2SO_4	PTSA*	PTSA*	H_2SO_4	H_2SO_4	PTSA*	PTSA*	H_2SO_4	H_2SO_4	H_2SO_4	H_2SO_4
Temperature, °C	155	164	169	168	164	164	164	163	162	163	161	171	164
Reaction Time, hr	1.0	1.0	1.0	1.0	1.0	1.0	1.0	1.33	1.0	1.0	1.0	1.0	1.0
Phenol: Catalyst: Coal	90/.92/1	30/.72/1	30/.92/1	24/1/1	48/2/1	30/.92/1	48/2/1	48/2/1	30/.92/1	30/.92/1	30/.92/1	30/.92/1	30/.92/1
Coal Depoly. %	83.3	88.7	77.3	72.5	67.0	64.7	61.7	58.0	49.7	46.7	37.9	36.8	25.3
Combined Phenol g/g coal charged	.174	.08	.201	.137	.179	.263	.16	.113	.100	.298	.147	.319	0.235
Combined Phenol g/g coal depoly.	.173	.083	.207	.158	.211	.290	.22	.163	.168	.390	.278	.464	.482
Figure of Merit (Coal Depoly. / Combined Phenol / g/g Coal charged)	(4.79)	(11.09)	(3.85)	(5.29)	(3.74)	(2.44)	(3.52)	(5.13)	(4.97)	(1.57)	(2.58)	(1.15)	(1.08)
Coal Rank	sub bituminous	sub bituminous	hv bituminous	hv A to B bituminous	hv A to B bituminous	hv A bituminous	lignite	lignite	Hv A to B bituminous	hv B bituminous	hv C bituminous	hv C bituminous	hv B to C bituminous

left side of the table. This can be expressed numerically as a
"Figure of Merit" by dividing the percent depolymerized by the
percent combined phenol. The numerical value of this "Figure
of Merit" ranges from 1.08 to 11.09. It is obviously desirable
to achieve maximum solubilization with minimum phenol consumption.

About half of the coals tested are suitable for a process
which entails solubilization of coal as an initial step. These
dissolve readily to yield solutions which can be filtered to
remove ash (including inorganic sulfur compounds). Eighty to 90
percent of the maf coal goes into solution to become available
for subsequent hydrogenation to oils.

Some coals take up large amounts of phenol (>25 percent)
but do not dissolve. Phenol has good solvating power for relatively polar solutes of up to about 1,000 in molecular weight.
Perhaps these coals do not depolymerize to coal fragments small
enough to enter solution in phenol. A co-solvent might help
dissolve these moieties. If so, it may be premature to conclude
that such coals would be unsuitable for processing.

The data of Table II were taken at 1.0 hour reaction time.
The reaction continues at a rapidly diminishing rate as shown
by data in Figure 1. Here the relative rates of depolymerization
for a few coals are compared over a 6-hour period. These tests
were conducted at reflux temperature (160-170°C) and at comparable
phenol:catalyst:coal ratios in the range 50:1:1. No line is
drawn for the Knife River lignite coal. Its solutions filter with
difficulty and accurate data are difficult to acquire. Data for
this coal (solid squares) scatter about the line for the Kaiser
Steel coal.

Reaction Kinetics

The reaction of phenol and the ensuing dissolution of coal
in the phenol catalyst mixture appears to follow no simple rate
mechanism. Ouchi, *et al*. were able to fit their data to a
simple first order rate equation which upon integration is of
the form:

$$\ln (1-\alpha) = kt \qquad\qquad (7)$$

where

$$\alpha = \frac{w_t}{w_\infty} . \qquad\qquad (8)$$

w_t is the weight increase due to reaction with phenol at time,
t and w_∞ is the maximum weight increase. When the data taken on
Montour coal are plotted in accordance with this simple rate
mechanism, the result shown in Figure 2 is obtained. The open
circles represent the rate of dissolution, the shaded circles

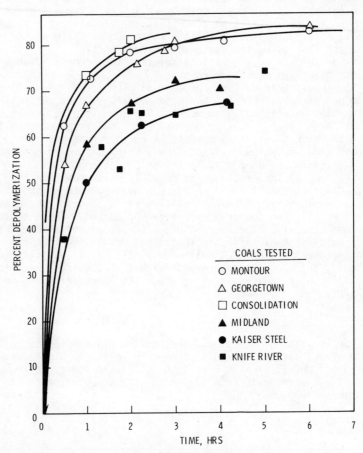

Figure 1. Relative rates of depolymerization (160°–170°C) with p-toluene-sulfonic acid catalyst

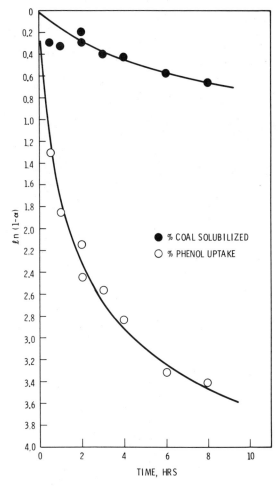

*Figure 2.   Depolymerization of Montour coal in accord-
ance with first-order rate mechanism*

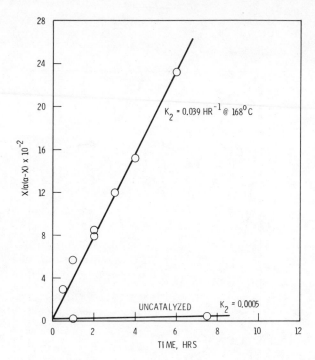

Figure 3. Dissolution of Montour coal in accordance with a second-order rate mechanism

represent the rate of phenol combination.

Clearly, the rate process is not first order and this is true of all the coals tested. Wiser (14) has compared the kinetics of coal pyrolysis and solvent extraction. In solvent extraction, the yield curves of fraction extracted versus time closely resemble the curves shown in Figure 1. Wiser showed such data exhibit compliance with second order kinetics in the early period of reaction and revert to compliance with first order kinetics in later stages. Hill, et al. (15) also showed that activated extraction of coal with hydrogen donor solvents behaved in this manner.

Data for dissolution of the Montour coal can be fitted to a second order equation of the form:

$$\frac{dx}{dt} = k_2 (a-x)^2 \tag{9}$$

where k_2 = reaction constant, a = maximum percent depolymerization attainable with this coal (arbitrarily estimated to be 87 percent) and x = the percent depolymerization at time, t. The integrated equation with a lower limit of x = o when t = o has the form:

$$\frac{x}{a(a-x)} = k_2 t.\tag{10}$$

Figure 3 shows how well the data fit second order kinetics during the first six hours. Beyond six hours, the data do not lie along this line, but instead follow simple first order kinetics as shown in Figure 4. The equation which describes this portion of the reaction is:

$$\frac{dx}{dt} = k_1 (a-x).\tag{11}$$

Integrating, with limits x = o when t = o, leads to:

$$\ln \frac{a}{(a-x)} = k_1 t.\tag{12}$$

Thus, a plot of $\ln \frac{a}{a-x}$ versus t yields the result shown in Figure 4, where "a" has the same definition as above. Similar results are obtained upon plotting the data for Consolidation, Georgetown, Midland, Kaiser Steel and Knife River coals.

In the case of the Consolidation coal, a few data were obtained at different temperatures. Figure 5 shows the result of plotting the data as a second order rate process at temperatures

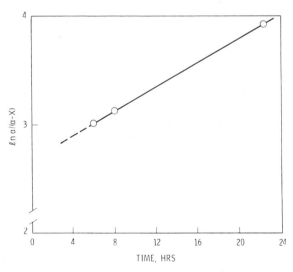

Figure 4.   Compliance of Montour dissolution rate at long time intervals with a first-order rate mechanism

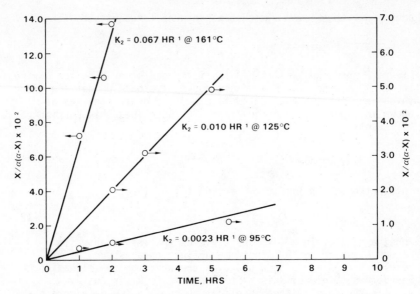

Figure 5. Temperature dependence of second-order dissolution of consolidation coal, PTSA catalyst

of 95°C, 125°C, and 161°C. The slopes of these lines give the second order reaction rate constants.

Applying absolute reaction rate theory as demonstrated by Hill, *et al.* the rate constant can be expressed as:

$$k_2 = \left(\frac{KT}{h}\right)\left(e^{-\Delta H/RT}\right)\left(e^{\Delta S/R}\right) \tag{13}$$

where k = Boltzmann's constant
      h = Plank's constant and
      R = Universal Gas constant.

A plot of ln (k_2/T) versus $1/T$ (Figure 6) allows evaluation of the enthalpy of activation, ΔH, from the slope, and the entropy of activation, ΔS, from the intercept.

The value of ΔH, 15.2 Kcal/mole, is about half that which Wiser (14) obtained (28.8 Kcal/mole) for dissolution of coal by hydrogen donor solvents in the 350-450°C temperature regime and by the second order rate process. On the other hand, it is about twice the value obtained by Hill, *et al.* (15) for low temperature dissolution of coal in a hydrogen solvent under the driving force of ultrasonic energy, i.e., 8.7 Kcal/mole.

Extrapolation of the line drawn in Figure 6 to 250°C ($1/T$ = 1.912 x 10^{-3}) leads to a value of ln k_2/T = -582, from which k_2 = 1.55 hr^{-1}. Since the data are in good compliance with second order rate theory at short time periods, extrapolation should predict the time required for solubilization of the coal at 250°C.

Equation (10) shows that complete dissolution, where x = a requires infinite time. However, at finite conversions to soluble products (e.g., 80 percent of maximum and 90 percent of maximum [maximum = 87 percent]) 1.8 and 4.0 min., respectively, are required at 250°C.

Experiments at 250-300°C confirm that nearly complete solubilization of coal in such short time periods can be achieved.

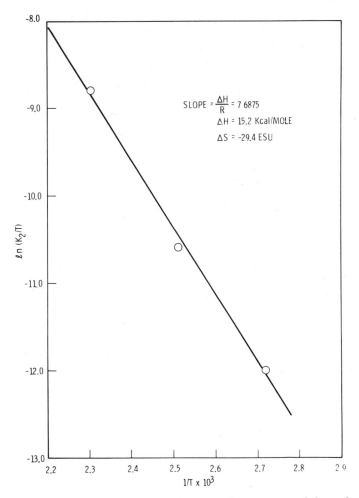

Figure 6. Ln K_2/T vs. $1/T$ for evaluation of the activation enthalpy and entropy

The data (Table III) also indicate that repolymerization may occur if the system is maintained at such high temperatures for too long a time interval.

The experiments at 250°-300°C were performed in a 300 ml stirred autoclave. Reaction times refer to the "time at temperature." Some 30 min. was required to bring the autoclave and its contents to temperature and about 15 min. was required to cool it again to ambient temperatures.

TABLE III. Depolymerization of Consolidation Coal at High Temperatures and Pressures of 100-200 psi Using PTSA Catalyst

Temp., °C	Time, Min	% Depolymerization	% Phenol Uptake
250	5	87.6	29.9
300	7	77.8	37.7
300	20	57.6	24.2
300	5	78.9	29.7
300	120	55.3	7.8

The extrapolation from data at lower temperatures, together with data in Table III, indicate nearly complete solubilization can be achieved in time periods of practical interest for flow through reactors.

Behavior of Other Solvents

Solvents other than phenol have been tested for effectiveness in coal depolymerization. The results are summarized in Table IV. Both percent depolymerization (solubilization) and weight increases (%) can be used as a measure of the extent of reaction. Data using phenol solvent are included for comparison.

Data for the weight increase are lacking in cases where the product was not isolated. This is not easy to accomplish when the solvent is insoluble in water. This lack is unfortunate in the case of cresote oil and benzaldehyde. Here the large negative value for percent depolymerization means the residue weighed more than the maf coal charged. This is evidence of solvent-coal reaction.

It is apparent that many of the solvents undergo reaction with coal, but the reaction may not depolymerize the coal to soluble fragments. Alternatively, the solvent may be a poor one in which to dissolve the depolymerized coal product.

The xylenes, e.g., yield a larger percent weight increase than does phenol and yet no significant solubilization of the coal results. In order to compare the products from depolymerization with m-xylene to those using phenol, products from these reactions were extracted with pyridine for 24 hours using a soxhlet extractor. The phenol depolymerization product proved

TABLE IV. Comparison of Different Solvents in the Depolymerization of Consolidation Coal

Run No.	Rx. Temp., °C	Rx. Time, hr	Catalyst, (g)	Coal, g	Solvent, (g)	% Coal Depoly.	% Wt. Increase*
134	169	1.0	H_2SO_4 (3.68)	4.0	Phenol (120)	77	125
154	185	2.0	PTSA·H_2O (1.0)	2.0	Naphthalene (30) Phenol (30)	65	--
162	176	1.0	H_2SO_4 (3.68)	4.0	Cresylic Acid (120)	61	--
161	158	2.0	H_2SO_4 (3.68)	4.0	Creosote Oil (100) Phenol (25)	9	--
160	170	3.0	H_2SO_4 (3.68)	4.0	Creosote Oil (120)	-22	--
159	170	1.0	H_2SO_4 (3.68)	4.0	Creosote Oil (120)	-28	--
164	160	1.0	H_2SO_4 (3.68)	4.0	Benzaldehyde (135)	-43	--
37	165	2.0	PTSA·H_2O (5.0)	10.0	Aniline (120)	18	--
41	211	2.0	PTSA·H_2O (2.5)	5.0	3,5-Xylenol (78)	56	149
46	190	2.0	PTSA·H_2O (5.0)	10.0	2-Naphthol (184)	41	--
47	212	2.0	PTSA·H_2O (5.0)	10.0	Naphthalene (117.5) Phenol (2.5)	(a)	--
48	200	2.0	PTSA·H_2O (5.0)	10.0	Naphthalene (108) Phenol (13.4)	10.5	--
105	108	2.0	H_2SO_4 (2.6)	5.0	Toluene (93)	(a)	128(b)
189	127	22.25	H_2SO_4 (3.68)	4.0	Xylenes (135)	(a)	134
190	127	22.75	H_2SO_4 (3.68)	4.0	m-Xylene (136)	(a)	150

*% Wt. Increase = $\dfrac{\text{maf depoly. product} + \text{maf char} - \text{maf coal}}{\text{maf coal}} \times 100$

(a) Weights of maf char exceeded maf coal charge indicating chemical reaction between coal and solvent.

(b) Assuming 0.15g of 3.56% of the maf coal tracks with the filtrate during filtration.

to be 61 percent soluble in pyridine whereas the xylene depolymerization product was only 30 percent soluble.

Thus, there appears to be something unique about the weakly
acidic phenolic materials. In addition to phenol itself, 3,5-
xylenol, 2-naphthol, phenol-naphthalene (1:1), and cresylic
acid all show ability to solubilize the coal. Recently, Darlage
and Bailey have studied the phenol catalyzed depolymerization of
a Kentucky coal (Pond Creek Seam, Pike County). (16) This coal
does not depolymerize efficiently and would compare with the
poorest coals tested with data shown in Table II. These authors
also show that phenolic solvents in general cause solvation of
coal whereas non-phenolic aromatics add to coal but do not solubilize the coal.

If the depolymerization reaction proceeds by the mechanism
of equations (1), (2) and (3) it should be possible to predict
a solvent reactivity sequence. The relative reactivity is thus
expected to be 3,5-xylenol > naphthol $\sim$ phenol > xylene > toluene
> naphthalene > aniline $\sim$ benzaldehyde. This order, as measured
by the indices for reaction used here, is not followed.

Hydrogenolysis Experiments

Hydrogenolysis was done using a one liter rocking autoclave
to which 30g coal, 15g catalyst, and 200 ml of mixed decalins
were charged. This was pressurized with H_2 heated to temperature
within a 4-5 hour period, rocked for a predetermined reaction
period (usually 4 hours) and cooled to ambient temperature (within
1.5 - 2.0 hours). Gases, liquids and solids were recovered
separately and analyzed. The yield of oil was determined from
the weight of unreacted coal adjusted for ash and moisture, e.g.

$$\% \text{ yield} = \frac{(\text{maf coal charged}) - (\text{coal residue})}{(\text{maf coal charged})} \times 100. \qquad (14)$$

The decision to employ decalin as a solvent was made to
circumvent difficulties in handling small quantities of coal and
catalyst in a large autoclave and to facilitate separation of
solvent and coal reaction products. In some respects this was
an unfortunate choice. Some reaction with decalin occurred, so
accurate measurement of H_2 consumption by reaction with coal was
not possible. Further, both coal and depolymerized coal were
virtually insoluble in this solvent so the hoped for advantage
of solvent solubility anticipated for depolymerized coal may
have been lost.

Commercially available catalysts were used. With but few
exceptions these were Harshaw-0402, Co-Mo, 1/8-inch tablets and
Harshaw-Ni-4301, Ni-W, 1/16 in. extrusions. At constant coal:
catalyst:solvent ratio and with H_2 in large excess the principal
variables which effect hydrogenolysis yield are temperature and
catalyst type. As shown by data in Figure 7, the conversion

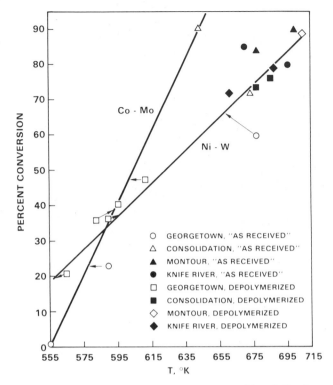

Figure 7.   Effect of temperature on conversion yield with Harshaw Ni–W and Co–Mo catalysts

yield is primarily a function of temperature for a given catalyst irrespective of coal type or depolymerized coal. There appears to be no significant difference in the conversion yield for depolymerized coal versus the coal from which it was derived with the possible exception of the Georgetown coal where the depolymerized coals performed better.

Details of the hydrogenolysis runs are collected in Table V. In all cases hydrogenolysis was conducted under mild conditions. Attempts were made to determine the H_2 consumed during the course of reaction, but interference by the solvent obscures the results in most cases. This interference is most pronounced at temperatures exceeding 400°C.

The process of hydrogenolysis is designed to fragment the coal structure into smaller molecular weight units: light oil, heavy oil, and asphaltenes. A small amount of gaseous products also result.

Table V
Hydrogenolysis of Coal and Depolymerized Coal

Run No./ Coal	Sample Description	wt Sample maf Basis g	Temp. °C	Time, hr	Initial P, psig (obs)	Final P, psig (calc.)	Catalyst Type	Yield %	Moles H₂/ g maf Coal Charged x 10⁻⁴	Moles H₂/ g maf Coal Convert. x 10⁻⁴
20 Georgetown	As Received (no solvent)	25.06	371	6.2	400	864	None	30.4	**	**
22 Georgetown	Depoly. Ash Not Sep. (no solvent)	26.43	420	36.0	400	930	None	53.0	**	**
15 Georgetown	As Received	25.06	404	4.0	400	908	Ni-W	59.8	130	220
12 Georgetown	Depolymerized Ash Free	30.0	292	4.2	1390	2170	Ni-W	20.7	3	16
13 Georgetown	Depolymerized Ash Free	23.5	316	4.0	1310	2590	Ni-W	36.5	94	26
11 Georgetown	As Received	25.06	281	4.2	1420	2640	Co-Mo	0.9	16	1700
10 Georgetown	As Received	25.06	316	4.3	1500	2965	Co-Mo	22.9	68	300
8 Georgetown	Depolymerized Ash Free	30.0	309	10.5	1700	3320	Co-Mo	35.8	**	**
9 Georgetown	Depolymerized Ash Free	24.2	322	4.0	1800	3595	Co-Mo	40.5	**	**
3 Georgetown	Depolymerized Ash Free	30.0	338	4.0	1800	3690	Co-Mo	47.4	**	**
5B Montour	As Received	26.45	404	1.0	1000	2272	Ni-W	84	350	420
7B Montour	As Received	26.45	426	4.2	1000	2346	Ni-W	90	410	460
6B Montour	Depolymerized + Ash	27.0	435	4.4	1000	2375	Ni-W	89	440	490
2B Knife River	As Received	25.44	397	7.6	1000	2248	Ni-W	85	800	950
4B Knife River	As Received	25.44	422	4.5	1000	2332	Ni-W	80	810	1010
3B Knife River	Depolymerized + Ash	27.09	>388[4]	4.8	1000	∿2225	Ni-W	72	440	610
8B Knife River	Depolymerized + Ash	27.09	414	3.0	1000	2305	Ni-W	79	410	530
21 Consolidation	Depolymerized (No Solvent)	30	397	5.6	400	900	None	33	156	192
10B Consolidation	As Received	25.3	400	4.25	300	680	Ni-W	72	180	249
14 Consolidation	Depolymerized	30	404	4.2	400	910	Ni-W	73.8	110	150
16 Consolidation	Depolymerized	30	412	4.0	300	690	Ni-W	76.3	110	150
1B Consolidation	As Received	25.3	370	4.8	1000	2158	Co-Mo	90	380	420
17 Consolidation	Depolymerized	30	410	4.0	300	690	G-3A[1]	62	90	240
18 Consolidation	Depolymerized	30	421	4.0	400	930	T-T1571[2]	72	110	150
19 Consolidation	Depolymerized (Tetralin Solv.)	30	412	4.2	400	920	G-68[3]	98	93	95

(1) Girdler G-3A, Chromium promoted iron oxide, 3/8 x 3/16 in. Tablets
(2) Girdler G-T1571, Silica Carrier, 3/16 in. Spheres.
(3) Girdler G-68, Promoted Pd on Gamma Alumina, 3/16 x 1/8 in. Tablets.
(4) Recorder Malfunctioned.
** Data Not Measured.

TABLE VI. Principal Features of IR and NMR Spectra

Brief Summary of Major IR Absorption Bands

Heavy Oil	Asphaltenes	Probable Band Assignment
3546 cm^{-1}	3401 cm^{-1}	OH Band (small)
3012	3012	Aromatic C-H Stretch
2924	2933	Aliphatic C-H Stretch
2857	2865	Aliphatic C-H Stretch
1905	1905	Overtones and Combination for
1754	1776	Multi Sub Benzenes
1605	1605	Aromatic Ring Unsaturation
1449	1451	Asym Methyl Band and Aromatic Ring Vibration
1374	1377	Sym Methyl Band
1252	1250	
1221	--	
1032	--	
877	840	Penta-Hexa-Sub Benzene
698		Di Sub Benzene
675		Benzenoid (?)

Brief Summary of NMR Spectrum

Heavy Oil	Asphaltenes	Description
0.9 ppm	0.75, 0.84, 0.93, 1.09	Methyl Protons
1.23	1.20	
1.50	1.27	
1.71	1.47	
2.22		Methylene Protons
2.39		
2.60	3.58	
3.76	(Broad Bands)	
6.92		Aromatic Protons
7.05		
7.20	7.21	
7.39	(Broad Bands)	
7.42		
7.75		

The light oil consisting mainly of mononuclear aromatic compounds, was separated during steam distillation to remove decalin. The actual amount was quite small and codistilled with decalin in which it appears as an impurity. An effort was made to distill this material through a Podbielniak distillation apparatus, but the amount of each component was so small, (i.e., in the range of tenths of milliliters), that clean separations were not obtained. Probably a spinning band column would have more accurately isolated these materials. Gas chromatographic analysis indicated the presence of roughly 60 different compounds, more than half of which had shorter retention times than decalin. Among those tentatively identified are benzene, cyclohexane, cresols, dimethyl phenol, phenol and diphenylether.

The steam distillation residue contains heavy oils and asphaltenes. These were separated by solubility differentiation. Both of these materials were sufficiently soluble in CCl_4 to permit examination by both IR and NMR. Principal features of the IR and NMR spectra are shown in Table VI. The possibility of colloidal dispersion of the asphaltenes instead of true solubility may have caused some loss of fine structure for the aromatic absorption regions.

The ratio of aliphatic protons to aromatic protons for the heavy oil was 4.01/1 and of methylene to methyl 1/1.75. For the asphaltenes, the ratio of aliphatic protons to aromatic protons was 3.49/1 and of methylene to methyl 1/1.1. The asphaltenes were observed to melt at 146°C. Molecular weights obtained by vapor phase osmometry (o-xylene solvent) were 407 for the heavy oil and 638 for the asphaltenes. There appears to be little difference in the relative yield of heavy oils and asphaltenes from depolymerized coal as compared with the "as received" coals.

Acknowledgements

The authors wish to express their appreciation to Dr. Ernest E. Donath for advice and encouragement throughout this work and to the Battelle Energy Program for support and permission to publish. The authors are also indebted to Messrs. W. A. Wilcox and G. L. Roberts who performed many of the measurements involved in this work.

Literature Cited

1. Heredy, L. S., and M. B. Neuworth, Fuel (1962) 41, 221.
2. Heredy, L. A., et al., Fuel (1964) 43, 414.
3. Heredy, L. A., et al., Fuel (1965) 44, 125.
4. Ouchi, K., et al., Fuel (1965) 44, 29.
5. Ouchi, K., et al., Fuel (1965) 44, 205.
6. Ouchi, K., Fuel (1967) 46, 319.
7. Ouchi, K., et al., Fuel (1967) 46, 367.
8. Ouchi, K., et al., Fuel (1973) 52, 156.

9. Darlage, L. J., J. P. Wiedner, and S. S. Block, Fuel, (1974) 53, 54.
10. Franz, J. A., Private Communication.
11. Annonymous, C. And E. News, (September 2, 1974) 16-17.
12. Given, P. H., et al., Fuel (1975) 54, 34.
13. Given, P. H., et al., Fuel (1975) 54, 40.
14. Wiser, W. H., Fuel (1968) 47.
15. Hill, G. R., et al., Fuel (1974) 53.
16. Darlage, L. J. and M. E. Bailey, Fuel (1976) 55.
17. Donath, E. E., "Coal Hydrogenation Vapor Phase Catalysts," Advances in Catalysis, Volume III, Academic Press, 1956, p. 258.

27

Liquefaction of Western Subbituminous Coals with Synthesis Gas

CHARANJIT RAI
University of Wyoming, Laramie, WY 82071

HERBERT R. APPELL and E. G. ILLIG
Pittsburgh Energy Research Center (ERDA), 4800 Forbes Ave., Pittsburgh, PA 15213

The recent energy crisis on the east coast and continued concern regarding environmental effects have once again emphasized the importance of producing low-sulfur fuels by liquefaction. Most liquefaction processes involve some form of hydrogen transfer, where coal is heated and liquefied at 400-500°C in a solvent. Falkum and Glen (1) suggested that the initial reaction of coal with molecular hydrogen involved dealkylation to reactive unsaturated materials which are either stabilized by hydrogenation or repolymerize. Van Krevelen (2) reported that coal pyrolysis involved depolymerization and disproportionation in which free radicals are formed. Hill et al. (3) discussed the liquefaction of a high volatile bituminous coal in tetralin in terms of dissolution and stated that "the hydrogen transfer reaction at 350-450°C is a second order process, in which the activation energy increases as the reaction proceeds." Curran et al. (4) described the transfer of hydrogen from tetralin to coal as a free radical reaction involving thermal cleavage of the coal molecules.

Heredy and Neuworth (5) were apparently the first investigators to report the depolymerization of coal using Friedel-Crafts catalysts. They observed that the solubility of coal could be considerably increased by treatment with boron trifluoride in phenol at 100°C. Subsequent studies (6) indicated that para-toluenesulfonic acid in phenol was the most efficient catalyst for the depolymerization reaction at 185°C. Lower rank coals were found to be more reactive resulting in higher solubility (7).

More recently Sternberg (8) has studied the dissolution of coal by reductive alkylation, which involves a variety of reactions, including free radical and elimination reactions. The reductive alkylation proceeds by the addition of electrons to the aromatic nuclei to form the corresponding anions followed by C-alkylation. Cleavage of ether bonds results in phenolate anions and subsequent O-alkylation of the anions solubilizes coal. Coal has also been solublized by a grafting technique which can be carried out at mild temperatures (140-150°C) and atmospheric pressure (9). Pulverized bituminous or anthracite coal is mixed with a chemical graft initiator and an appropriate monomer, and the reaction product is liquefied either by heating or by addition of solvents derived from coal or petroleum. Non-polar alkyl monomer substituents enhance the solubility of coal in aliphatic hydrocarbons; whereas aromatic substituents make coal soluble in aromatic solvents. The polar groups make the grafted coals more soluble in polar solvents such as ketones and alcohols.

Vigorous efforts are being made to develop catalytic hydrogenation processes for producing liquid fuels from coal. The role of hydrogen in liquefaction of coal is to saturate the free radicals formed by bond cleavage of coal at reaction temperature. This can occur directly or through hydrogenation of smaller solvent molecules which then donate their hydrogen to reactive coal species. Desulfurization and saturation of double bond and aromatic ring structures also occurs in the presence of hydrogen. Enormous amounts of hydrogen are needed for hydrogenation processes and without a technological breakthrough for production of hydrogen at low costs, an economical process for liquefaction will not be achieved. An alternative approach has been investigated by Appell et al. (10) consisting of liquefying lignite and bituminous coal with low cost synthesis gas. Fu and Illig (11) have attempted liquefaction and desulfurization of high sulfur bituminous coals with synthesis gas at temperatures of 400-450°C, and operating pressures of 3000-4000 psi in the presence of cobalt molybdate and sodium carbonate catalysts, steam and a recycle oil.

The effect of mineral matter present in western coals was investigated in an attempt to develop an economical non-catalytic liquefaction process using low cost synthesis gas. The minerals present in the subbituminous coals offer an adequate, inexpensive and naturally occuring source of catalytic materials for liquefaction and hydrodesulfurization reactions involved in coal conversion. Mitchell and Gluskoter (12) have identified bassanite, quartz, kaolinite, calcite and pyrite in the Rocky Mountain subbituminous coals by low temperature ashing technique. Morooka and Hamrin (13) observed that the mineral matter from two Kentucky coals exhibited catalytic activity for the dehydrosulfurization of thiophene and butene isomerization.

In this study subbituminous coals were liquefied and desulfurized effectively by hydrotreating with synthesis gas at moderate operating conditions in the presence of steam and anthracene oil used as a solvent. The data were compared when the same coals were liquefied with pure hydrogen under similar experimental conditions. Pyritic and organic sulfur present in these coals appear to be responsible for the formation of catalytic active form of iron sulfide that functions as an effective hydrotreating catalyst (14). The presence of

aluminosilicates and carbonates in these coals promote the conversion of complex organic structures to soluble liquid products.
The water-gas shift reaction and the reduction of carbonyl
groups in the coal structures are catalyzed by alkali metal
carbonates. The reduction of carbonyl groups proceeds via
donation of the aldehydic hydrogen in the formate ion to the
carbonyl group. The formate ion is continually regenerated
by the reaction of carbon monoxide with alkaline materials
present in coal ash.

$$M^+ + OH^- + CO \rightarrow HC\overset{O}{\underset{O\ominus\oplus}{\big\langle}}$$

$$\underset{O\ M}{}$$

$$\rangle C = O + H-\overset{O\ominus\oplus}{\underset{}{C}}-O\ M \rightarrow -\overset{H\ominus\oplus}{\underset{|}{C}}-O\ M + CO_2$$

$$-\overset{H\ominus\oplus}{\underset{|}{C}}-O\ M + H_2O \rightarrow H-\overset{}{\underset{|}{C}}-OH + OH^\ominus + M^\oplus$$

Experimental

  Equipment. The liquefaction of coal was studied in a 1
liter magnetically stirred stainless steel autoclave. The autoclave was equipped with a thermowell; a gas inlet port connected
to a synthesis gas (H_2:CO ratio of 1:1), or to a hydrogen cylinder; and to a pressure gauge; a port for connection to a
rupture disk assembly; and a port for gas discharge. The vertical magentic rod with dashers stirred the coal, solvent and
ash or catalyst particles. The autoclave temperature was
measured by a thermocouple placed into the thermowell.

  Materials. Twelve subbituminous coals from four major
coal-bearing regions of Wyoming were studied in this program.
Chemical and physical analyses of these coals are presented in
Table I. The coal samples were ball-milled to minus 100 mesh
under nitrogen and stored in a refrigerator before use.
  A Co-Mo-Al$_2$O$_3$ catalyst (Harshaw CoMo 0402T, 3% CoO - 15%
MoO$_3$) was used in some experiments either with or without potassium carbonate. Pyrite isolated from coal was also used as
a catalyst with potassium carbonate in some experiments. Anthracene oil obtained from Crowley Tar Products Company was
used as the start-up solvent. In the recycle runs with Sheridan
Field Coal (W-74-45), 80% of the anthracene oil was gradually
replaced by coal-derived oil after nine recycles. The benzene
and pentane used for separation of oil, and asphaltene were
Fisher solvent grade.

  Procedure. In a typical experiment, the autoclave was
charged with subbituminous coal (80-110g), anthracene oil (140g),
water as needed to provide 25 wt% of the coal charge with synthesis gas or hydrogen as the reducing agent, and with or without

TABLE I. ANALYSES OF SUBBITUMINOUS COALS FROM MAJOR WYOMING COAL FIELDS

| SAMPLE NO. | COAL NAME | MOISTURE | PROXIMATE ANALYSIS (PERCENT) | | | ULTIMATE ANALYSIS (PERCENT) | | | | | BTU/ LB |
			VOLATILE MATTER	FIXED CARBON	ASH	C	H	N	O	S	
HAMS FORK COAL REGION											
WYO-74-3	ADAVILLE #10	Ar[1] 22.1	31.1	37.9	8.9	50.8	6.1	1.4	31.0	1.8	8980
		Maf[2]	45.1	54.9		73.6	5.3	2.1	16.4	2.6	13010
WYO-74-4	ADAVILLE #6	Ar 23.8	33.1	37.8	5.3	50.7	6.1	1.4	35.9	0.6	8560
		Maf	46.7	53.3		71.5	4.9	2.0	20.7	0.9	12070
WYO-74-12	ADAVILLE #4	Ar 20.9	34.0	41.2	3.9	55.9	6.1	1.5	32.2	0.4	9750
		Maf	45.1	54.9		74.3	5.1	1.9	18.2	0.5	12960
WYO-74-14	ADAVILLE #3	Ar 20.3	37.1	37.8	4.8	56.3	6.1	1.4	31.1	0.3	9780
		Maf	49.5	50.5		75.1	5.2	1.9	17.4	0.4	13050
WYO-74-19	ADAVILLE #1 (Elkol)	Ar 16.7	36.5	42.8	4.0	60.1	6.2	0.9	27.5	1.3	10530
		Maf	46.0	54.0		75.8	5.5	1.2	15.9	1.6	13270
GREEN RIVER REGION											
WYO-74-20	DEADMAN	Ar 19.5	32.6	42.0	5.9	55.6	5.6	1.1	31.3	0.5	9270
		Maf	43.8	56.2		74.6	4.6	1.5	18.7	0.6	12440
HANNA COAL FIELD											
WYO-74-24	BED #80	Ar 12.4	39.2	39.6	8.8	59.7	5.7	1.3	23.3	1.2	10450
		Maf	49.7	50.3		75.7	5.5	1.7	15.6	1.5	13260
POWDER RIVER COAL BASIN											
WYO-74-40	WYODAK (Anderson-Canyon)	Ar 21.1	35.0	38.4	5.5	54.9	5.9	0.9	32.3	0.5	9480
		Maf	47.7	52.3		74.8	4.9	1.2	18.4	0.7	12920
WYO-74-45	UNNAMED DIETZ (Sheridan Field)	Ar 19.5	32.9	32.8	14.8	48.0	5.9	1.1	28.0	2.2	8560
		Maf	50.0	50.0		73.1	5.7	1.6	16.2	3.4	13040
WYO-74-50	MONARCH (Upper Bench)	Ar 18.0	36.8	37.4	7.8	53.6	6.3	1.3	29.8	1.2	9450
		Maf	49.6	50.4		72.3	5.8	1.7	18.6	1.6	12750
WYO-74-52	MONARCH (Lower Middle Bench)	Ar 18.3	37.5	36.1	8.1	53.7	6.1	1.2	28.9	2.0	9600
		Maf	51.0	49.0		72.9	5.5	1.6	17.2	2.8	13030
WYO-74-53	MONARCH (Lower Bench)	Ar 19.1	28.3	30.7	21.9	42.1	5.4	1.0	27.7	1.9	7380
		Maf	48.0	52.0		71.4	5.5	1.7	18.3	3.1	12520

catalysts. The autoclave with the reactants was flushed with
nitrogen followed by synthesis gas or hydrogen and charged with
the appropriate gas with an initial pressure of 1600 to 2000 psi.
The autoclave temperature was then increased to the desired
operating temperature (400° to 450°C) over a period of 50 to 60
minutes, maintained at the temperature from 15 to 75 min. After
the experiment was completed, rapid internal cooling of the autoclave to the ambient temperature was achieved by cold water.
The final pressure in the autoclave was recorded the next day,
the difference in pressure indicating the volume of gas consumed in the experiment. The products were filtered at ambient
or higher temperatures to obtain liquid oils. Filter cakes
containing unreacted carbon, ash and water were extracted with
benzene or acetone. The water was removed by distillation and
the remaining oil was recovered by removing the solvent in a
rotary vacuum evaporator. Gaseous products were analyzed by
mass spectrometry. Data on conversion, oil yield and gas consumption are given as weight percent based on moisture-free and
ash-free (maf) coal.

Results and Discussion

Liquefaction with Synthesis Gas ($1H_2:1CO$) and Hydrogen.
Subbituminous coals from four major coal-bearing regions of
Wyoming: Hams Fork, Green River, Hanna and Powder River Coal
Basin having ash from 4 to 22 wt%, moisture from 12 to 25 wt%
and sulfur from 0.6 to 3.4 wt% (maf) were liquefied and desulfurized with synthesis gas in anthracene oil and in absence
of added catalysts. The major elements present in the mineral
matter of these coals are shown in Table II (15). Some of these
metals act as catalysts for depolymerizing and liquefying coals.
When Monarch (Lower Bench) coal (W-74-53) was hydrotreated
with synthesis gas in anthracene oil and 25 wt% moisture at
425 ± 5°C, an operating pressure of 3900-4000 psi, and a reaction
time of one hour, a 99% conversion and 72% selectivity to oil on
a moisture- and ash-free basis was observed. The viscosity of
the coal oil in anthracene solvent was remarkably low, 53 centipoises at 60°C. At lower residence time from 15 min. to
30 min., the product oil resulted in a high viscosity, even-
though the conversions were not greatly affected. These data
are shown in Table III and Figure 1.
Essentially all the subbituminous coals when hydrogreated
with synthesis gas in anthracene oil and adequate amounts of
moisture resulted in high conversions varying from 90 to 99%
on moisture- and ash-free basis at a temperature of 425 ± 5°C,
an operating pressure of 3500 to 4500 psig, and a residence time
of an hour. The product oil invariably had a low viscosity.
These are optimum conditions for autoclave experiments. Higher
temperatures resulted in increased gas consumption, lower conversions, and a product oil of low viscosity. Likewise,

TABLE II. ELEMENTAL COMPOSITION OF WESTERN SUBBITUMINOUS COALS

SAMPLE NO.	COAL NAME COAL FIELD	MAJOR ELEMENTS AS PERCENT OF WHOLE COAL							
		Si	Al	Ca	Mg	Na	K	Fe	Ti
WYO-74-3	ADAVILLE #10 Kemmerer	2.60	0.44	0.59	0.17	0.01	0.034	1.50	0.026
WYO-74-4	ADAVILLE # 6 Kemmerer	1.83	0.17	0.21	0.17	0.005	0.020	1.40	0.011
WYO-74-12	ADAVILLE # 4 Kemmerer	1.33	0.20	0.23	0.12	0.004	0.016	0.32	0.012
WYO-74-14	ADAVILLE # 3 Kemmerer	1.49	0.42	0.23	0.12	0.008	0.040	0.30	0.019
WYO-74-19	ADAVILLE # 1 Kemmerer	1.20	0.52	0.31	0.08	0.004	0.006	0.17	0.017
WYO-74-20	DEADMAN Rock Springs	1.45	0.55	0.53	0.14	0.014	0.006	0.21	0.032
WYO-74-24	BED #80 Hanna	1.70	0.92	1.10	0.196	0.014	0.083	0.86	<0.001
WYO-74-40	WYODAK Powder River	0.86	0.53	1.00	0.150	0.072	0.017	0.23	0.055
WYO-74-45	UNNAMED DIETZ Sheridan	3.11	1.80	0.80	0.23	0.150	0.20	2.10	0.068
WYO-74-50	MONARCH (Upper Bench) Sheridan	1.43	0.85	0.62	0.27	0.12	0.063	0.77	0.036
WYO-74-52	MONARCH (Lower Middle Bench) Sheridan	0.88	0.69	0.59	0.230	0.075	0.041	1.20	0.206
WYO-74-53	MONARCH (Lower Bench) Sheridan	5.50	2.50	0.49	0.340	0.086	0.370	1.50	0.11

TABLE III. NON-CATALYTIC LIQUEFACTION OF WYOMING SUBBITUMINOUS COALS

Monarch (Lower Bench, 74-53)

Effect of Reaction Time on Conversion and Oil Yield with Synthesis Gas

RUN NO.	CONTACT TIME, MINUTES	TEMPERATURE, °C	PRESSURE, psig	CONVERSION, PERCENT	OIL YIELD, PERCENT	VISCOSITY AT 60°C cp
5	60	420-25	3900	99.4	72.0	136
26	30	425-26	4100	95.9	72.0	355
27	15	424-25	3900	91.1	78.6	595
28*	30	424-26	4150	97.8	73.4	97.83

*FeS_2, K_2CO_3 was used as a catalyst.

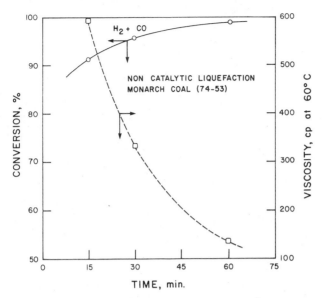

Figure 1.   *Effect of reaction time on conversion and viscosity*

EFFECT OF PRESSURE ON COAL LIQUEFACTION

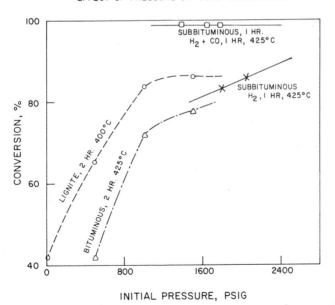

Figure 2.   *Effect of pressure on coal liquefaction*

TABLE IV. NON-CATALYTIC LIQUEFACTION OF WYOMING SUBBITUMINOUS COALS WITH SYNTHESIS GAS AND HYDROGEN

(SOLVENT: ANTHRACENE OIL; CONTACT TIME: 1 HOUR)

COAL NAME RUN NO.	GAS	TEMPERATURE °C	PRESSURE, psig MAXIMUM	ΔP	CONVERSION, PERCENT	OIL YIELD, PERCENT	VISCOSITY AT 60°C cp
ADAVILLE #10 (W-74-3, No. 17)	HYDROGEN	450	4200	350	75.3	47.2	124
ADAVILLE #10 (W-74-3, No. 6)	SYNGAS	425	4250	125	89.1	55.4	372
ADAVILLE #6 (W-74-4, No. 19)	HYDROGEN	425-26	4130	220	82.4	34.4	517
ADAVILLE #6 (W-74-4, No. 18)	SYNGAS	425-25	4400	250	95.2	63.3	242
ADAVILLE #4 (W-74-12, No. 25)	HYDROGEN	425-30	3900	325	85.1	59.4	215
ADAVILLE #4 (W-74-12, No. 24)	SYNGAS	426-28	4100	375	89.0	71.6	191
ADAVILLE #3 (W-74-14, No. 23)	HYDROGEN	420-55	4350	300	45.9	21.6	185
ADAVILLE #3 (W-74-14, No. 22)	SYNGAS	425	4650	150	72.8	38.2	392
ADAVILLE #1 (W-74-14, No. 14)	HYDROGEN	420-25	3510	200	77.4	44.4	379
ADAVILLE #1 (W-74-19, No. 7)	SYNGAS	425-28	3975	205	93.2	54.9	128
DEADMAN (W-74-20, No. 15)	HYDROGEN	422-34	3650	210	83.2	55.4	726
DEADMAN (W-74-20, No. 10)	SYNGAS	425-30	3800	420	89.5	54.3	208.7

TABLE IV. CONTINUED

COAL NAME RUN NO.	GAS	TEMPERATURE °C	PRESSURE, psig MAXIMUM	PRESSURE, psig ΔP	CONVERSION, PERCENT	OIL YIELD, PERCENT	VISCOSITY AT 60°C cp
HANNA BED #80 (W-74-24, No. 2)	HYDROGEN	420-30	4510	320	88.9	60.7	252
HANNA BED #80 (W-74-24, No. 1)	SYNGAS	425-28	4400	240	93.6	62.0	172
WYODAK (W-74-40, No. 11)	HYDROGEN	425-35	4000	290	80.6	34.4	155.6
WYODAK (W-74-40, No. 8)	SYNGAS	425-27	4160	275	90.4	71.0	97.2
MONARCH (Upper Bench) (W-74-50, No. 21)	HYDROGEN	425-30	3820	290	88.7	53.3	243
MONARCH (Upper Bench) (W-74-50, No. 20)	SYNGAS	425-26	4450	350	98.2	47.4	56.9
MONARCH (Lower Middle Bench) (W-74-52, No. 36)	HYDROGEN	425	3950	380	94.0	68.0	79
MONARCH (Lower Middle Bench) (W-74-52, No. 32)	SYNGAS	426-28	4250	260	96.2	76.7	45
MONARCH (Lower Bench) (W-74-53, No. 3)	HYDROGEN	420-27	4175	250	86.2	50.0	348
MONARCH (Lower Bench) (W-74-53, No. 12)	SYNGAS	420-25	4030	300	98.7	72.2	53

TABLE IV. CONTINUED

COAL NAME RUN NO.	GAS	TEMPERATURE °C	PRESSURE, psig MAXIMUM	ΔP	CONVERSION, PERCENT	OIL YIELD, PERCENT	VISCOSITY AT 60°C cp
UNNAMED DIETZ (W-74-45, No. 38)	SYNGAS	425	4140	270	96.0	70.0	46
UNNAMED DIETZ (W-74-45, No. 39)	HYDROGEN	425	3930	570	95.0	71.0	49

higher pressures resulted in higher conversions and the best
results were obtained at an initial pressure of 1800-2000 psig,
with a corresponding operating pressure of about 3600-4200 psig.
The effect of pressure is shown in Figure 2 and compared with
the published data (10).

Liquefaction of these coals with pure hydrogen under similar
experimental conditions invariably gave lower conversions,
lower selectivity to oil and a product with higher viscosity in
the absence of added catalysts. The sulfur content and the
viscosity of the product oil both decrease with the amount of
hydrogen consumed in each case, however, less total hydrogen is
required for the same oil quality with synthesis gas than with
pure hydrogen. These data are shown in Table IV and compared
with the results obtained by using synthesis gas.

Catalytic Liquefaction with Co-Mo-Al$_2$O$_3$ Catalyst and
Constituents of Coal Mineral Matter. Some low ash coals such as
Adaville #3 (W-74-14), and Adaville #10 (W-74-3) from Kemmerer
field exhibited poor conversion and selectivity to oil under
typical hydrotreating conditions with synthesis gas. By conducting the liquefaction in the presence of potassium carbonate,
pyrite or cobalt molybdate (Co-Mo-Al$_2$O$_3$), a remarkable increase
in overall conversion and selectivity to oil was achieved while
the viscosity of product oil in anthracene solvent was also
considerably lowered. With Adaville #3 (W-74-14), using synthesis gas at 425°C, and an operating pressure of from 4150 to
4225 psig, and a residence time of an hour, the conversion
increased from 73% to 94% with K$_2$CO$_3$ and or pyrite, whereas the
selectivity to product oil increased from 38% to 68%. The
viscosity of the product oil was also considerably lowered from
392 to 56-80 centipoises at 60°C. With Co-Mo-Al$_2$O$_3$ catalyst
and K$_2$CO$_3$ a similar trend was observed, the conversion being 97%,
selectivity to oil 55% and the product oil viscosity being 35
centipoises at 60°C. Monarch (Lower Bench) coal from the
Sheridan Field also exhibited improved conversion, higher product
oil yield and lower viscosity of the product oil. These data
are shown in Table V.

Effect of Recycle on Liquefaction and Product Distribution.
It is important that the product oil from a coal liquefaction
process should have a low viscosity so that it could be used
for preparing coal-oil slurries for recycling in a continuous
liquefaction process. The product oil should also exhibit
reactivity or solvency for the coal so that the viscosities do
not deteriorate with prolonged recycling operation.

The recycle characteristics of the liquefied coal-oil were
determined by conducting nine consecutive recycle runs. In this
set of experiments a freshly pulverized subbituminous coal
(W-74-45) with 3.4 wt% sulfur (maf) from the Sheridan field of
the Powder River Coal Basin was used. In each of the experiments

TABLE V.　CATALYTIC LIQUEFACTION OF WYOMING SUBBITUMINOUS COALS WITH SYNTHESIS GAS

(Solvent: Anthracene Oil)

COAL MINE	CATALYST	CONTACT TIME, MIN.	TEMPERATURE °C	PRESSURE, psig MAXIMUM	ΔP	CONVERSION PERCENT	OIL YIELD PERCENT	VISCOSITY AT 60°C cP
ADAVILLE NO. 3 (W-74-14, Run 22)	None	60	425	4650	150	72.8	38.2	392
ADAVILLE NO. 3 (W-74-14, Run 33)	FeS_2	60	425	4250	275	91.9	68.14	56.2
ADAVILLE NO. 3 (W-74-14, Run 31)	K_2CO_3	60	420-25	4150	175	92.3	56.7	80.5
ADAVILLE NO. 3 (W-74-14, Run 29)	K_2CO_3 FeS_2	60	425-26	4150	220	94.4	56.1	83.5
ADAVILLE NO. 3 (W-74-14, Run 30)	K_2CO_3 CoMo	60	425-26	4225	380	96.9	54.6	35
MONARCH (Lower Bench) (74-53, Run 26)	None	30	425-26	4100	220	95.9	71.9	355
MONARCH (Lower Bench) (74-53, Run 28)	K_2CO_3 FeS_2	30	424-26	4150	220	97.8	73.5	136

TABLE VI. RECYCLE RUNS USING SUBBITUMINOUS COAL (74-45) WITH SYNTHESIS GAS AT 425°C

RECYCLE #	PERCENT CONVERSION	PERCENT OIL YIELD	OPERATING PRESSURE, psig	ΔP	VISCOSITY CPA AT 60°C	PERCENT COAL DERIVED
1	94	55.3	3840	720	22	18.9
2	94	53.7	3960	250	26	33.9
3	92	47.3	4010	290	29	45.1
4	87	48.3	3900	305	40	54.4
5*	90	41.3	3710	440*	52	61.3
6*	88	38	3500	550*	92	66.7
7	85	33.3	4020	230	80	70.5
8	88	40	4070	250	62	74.8
9	87	37.3	4020	250	62	78.3

* Pressure Loss

the product oil derived form coal was separated by filtration
and then used in the next experiment, thus gradually becoming
richer in the coal-derived oil until about 80% of the anthracene
oil used as a start-up solvent, was replaced by subbituminous
coal-derived oil. The recycle run data are shown in Table VI.
The viscosity of the product oil after nine recycle runs was
quite low, 62 centipoises at 60°C, and the sulfur level was
reduced to 0.22 wt% compared to 3.4 wt% (maf) in the starting
subbituminous coal as shown in Table VII. High coal conversions
from 85 to 95% on maf basis were also observed when the product
oil was recycled.

TABLE VII. LIQUEFACTION OF SUBBITUMINOUS COAL USING RECYCLE OIL

(1hr, 425°C; 1:1 synthesis gas, 3500-4000 psig)

SHERIDAN FIELD, BIG HORN MINE - COAL (74-45)

	1ST RUN	9TH RUN
CONVERSION, %	94	87
OIL YIELD, %	55.3	37.3
PRODUCT:		
C		88.24
H		7.04
O		3.03
S	0.22	0.22
N		1.63
CENTIPOISES AT 60°C	22	62

The product oil derived by recycle has the following
components: oil, 61.3%; asphaltenes, 26.5%; and benzene
insolubles, 12.2%. The oil fraction was analyzed by mass
spectroscopy. The organic compounds identified in the oil
fraction are shown in Table VIII.

TABLE VIII. IDENTIFIED ORGANIC COMPOUNDS PRESENT IN COAL-
             DERIVED OIL FRACTION (74-45) BY MASS SPECTROSCOPY

COMPOUND	PERCENT
INDANOL	1.0
PHENYL NAPHTHALENES	6.7
PHENOLS	9.1
ANTHRACENES/PHENANTHRENES	25.4
ACENAPHTHYLENES/FLUORENES	5.1
ACENAPHTHENES	8.5
BENZOFURANE	0.1
BENZOPYRENE	0.5
NAPHTHALENES	21.4
INDENES	0.1
CHRYSENE	2.0
INDANES AND TETRALINS	2.1
PYRENES	18.0

Conclusions

Western subbituminous coals can be readily liquefied and desulfurized by non-catalytic liquefaction with synthesis gas and steam at 400-450°C and an operating pressure of 3800-4400 psi. The mineral matter present in these coals functions as effective catalysts for promoting the water gas shift reaction and the reduction of carbonyl groups to oil soluble products. The sulfur content and the viscosity of the product oil decrease with the amount of hydrogen consumed with both synthesis gas and hydrogen, however, much less total hydrogen is required for the same quality oil product with synthesis gas than with pure hydrogen.

The recycle characteristics of coal-derived oil are very important for the commercialization of a liquefaction process. One subbituminous coal from the Sheridan field of the Powder River Coal Basin was studied for its recycle characteristics and found to be quite acceptable. After nine recycles the viscosity of the product oil remained low and about 95% reduction in sulfur level to 0.22 wt% was achieved. Such a liquefaction process holds promise for those Western coals that have high sulfur and high ash and cannot be economically transported to distant markets.

Acknowledgement

The authors are grateful to Gary B. Glass of the Geological Survey of Wyoming for providing the coal samples. C. Rai gratefully acknowledges the financial support from the Oak Ridge Associated Universities and the Amax Foundation under which this study was carried out. The authors are also indebted to the Pittsburgh Energy Research Center (ERDA) for the use of the facilities during the summer of 1976. The analytical support was provided by Forrest E. Walker of the Bureau of Mines and Jane V. Thomas of the University of Wyoming.

Abstract

The effect of mineral matter present in subbituminous coals was investigated by carrying out non-catalytic liquefaction with hydrogen and synthesis gas. Most of the twelve coals studied could be readily liquefied to a low viscosity and low sulfur oil in the absence of added catalysts with synthesis gas at temperatures of 400-450°C and operating pressures of 3800-4400 psi. Comparison with coal liquefaction using pure hydrogen at optimum liquefaction temperature of 425°C and pressure of 3800-4000 psi resulted in lower conversions, lower selectivity to oil and a product with higher viscosity.

Some low ash coals exhibited poor conversion and selectivity to oil under typical liquefaction conditions with synthesis gas.

By conducting the liquefaction in the presence of potassium
carbonate, pyrite or cobalt molybdate a remarkable increase in
the overall conversion and selectivity to oil was achieved while
the viscosity of product oil was considerably lowered. The recycle characteristics of liquefied coal after nine recycles,
when 80% of the start-up solvent was replaced with coal-derived
oil, were quite acceptable. The viscosity of the recycled
product oil remained remarkably low, and the sulfur level was
reduced from 3.4 maf% in the starting coal to 0.22 wt%.

Literature Cited

1. Falkum, F. and Glen, R. A., Fuel 1952, 31, 133.
2. Van Krevelen, D. W., Coal, Elsevier, Amsterdam, 1961, 339.
3. Hill, G. R., Hairiri, H., Reed, R. I., and Anderson, L. L.,
    Am. Chem. Soc., Advances in Chem. Ser. 1966, 55, 427-447.
4. Curran, G. P., Struck, R. T., and Gorin, E., Ind. Eng.
    Chem., Process Design Dev. 1967, 6 (2), 166.
5. Heredy, L. A. and Neuworth, M. B., Fuel 1962, 41, 221.
6. Ouchi, Imuta K., and Yamashita, Y., Fuel 1973, 52, 156.
7. Heredy, L. A., Kostyo, A. E., and Neuworth, M. B., Fuel
    1965, 44, 125.
8. Sternberg, H. W. and Delle Donne, C. L., Fuel 1974, 53,
    172.
9. Dichter, M., Mehta, M., and Nowacki, T., Chem. & Eng. News,
    Jan 24, 1977, 26-27.
10. Appell, H. R., Moroni, E. C., and Miller, R. D., Preprints,
    Div. Fuel Chem., Am. Chem. Soc. 1975, 20(1), 58-65.
11. Fu, Y. C. and Illig, E. G., Ind. Eng. Chem. Process
    Design Dev., 1976, 15 (3), 392-96.
12. Mitchell, R. S., and Gluskoter, H. J., Fuel (London),
    1976, 55, 90.
13. Morooka, S. and Hamrin, C. E., Jr., Chem. Eng. Sci., 32,
    125, 1977.
14. Mukherjee, D. K. and Chowdbury, P. B., Fuel (London),
    1976, 55, 4.
15. Glass, G. B., Geological Survey of Wyoming, Report of
    Investigation No. 11, (1975), 32.

INDEX

INDEX